AF333804

DRUG DETERMINATION IN THERAPEUTIC AND FORENSIC CONTEXTS

METHODOLOGICAL SURVEYS IN BIOCHEMISTRY AND ANALYSIS

Series Editor: Eric Reid

Guildford Academic Associates
72 The Chase
Guildford GU2 5UL, United Kingdom

The series is divided into Subseries A: Analysis, and B: Biochemistry
Enquiries concerning Volumes 1–11 should be sent to the above address.

Volumes 1–10 edited by Eric Reid

Volume 1   (B): Separations with Zonal Rotors

Volume 2   (B): Preparative Techniques

Volume 3   (B): Advances with Zonal Rotors

Volume 4   (B): Subcellular Studies

Volume 5   (A): Assay of Drugs and Other Trace Compounds in Biological Fluids

Volume 6   (B): Membranous Elements and Movement of Molecules

Volume 7   (A): Blood Drugs and Other Analytical Challenges

Volume 8   (B): Cell Populations

Volume 9   (B): Plant Organelles

Volume 10 (A): Trace-Organic Sample Handling

Volume 11 (B): Cancer-Cell Organelles
                 edited by Eric Reid, G. M. W. Cook, and D. J. Morré

Volume 12 (A): Drug Metabolite Isolation and Determination
                 edited by Eric Reid and J. P. Leppard
                 (includes a cumulative compound-type index)

Volume 13 (B): Investigation of Membrane-Located Receptors
                 Edited by Eric Reid, G. M. W. Cook, and D. J. Morré

Volume 14 (A): Drug Determination in Therapeutic and Forensic Contexts
                 Edited by Eric Reid and Ian D. Wilson

DRUG DETERMINATION IN THERAPEUTIC AND FORENSIC CONTEXTS

Edited by

Eric Reid

Guildford Academic Associates
Guildford, Surrey, United Kingdom

and

Ian D. Wilson

Hoechst Pharmaceutical Research Laboratories
Milton Keynes, United Kingdom

PLENUM PRESS • NEW YORK AND LONDON

Library of Congress Cataloging in Publication Data

Main entry under title:

Drug determination in therapeutic and forensic contexts.

(Methodological surveys in biochemistry and analysis; v. 14. Subseries A, Analysis)
Based on a Bioanalytical Forum held at the University of Surrey in Sept. 1983.
Includes bibliographies and index.
1. Drugs—Analysis—Congresses. 2. Body fluids—Analysis—Congresses. 3. High performance liquid chromatography—Congresses. 4. Chemistry, Clinical—Technique—Congresses.   I. Reid, Eric, 1922–   II. Wilson, Ian D. III. Bioanalytical Forum (1983: University of Surrey) IV. Series: Methodological surveys in biochemistry and analysis; v. 14. V. Series: Methodological surveys in biochemistry and analysis. Subseries A, Analysis. (DNLM: 1. Chemistry, Clinical—methods—congresses. 2. Chromatography —congresses. 3. Drugs—analysis—congresses. W1 ME9612NT v. 14 / QV 25 D7935 1983)
RB56.D78 1984   616.07′56   84-15151
ISBN 0-306-41809-6

©1984 Plenum Press, New York
A Division of Plenum Publishing Corporation
233 Spring Street, New York, N.Y. 10013

Printed in the United States of America

Senior Editor's Preface

In common with its four predecessors, which are still available
as a 'package', this volume is based on a Bioanalytical Forum held
at the University of Surrey (in September 1983) whilst not being a
mere patchy 'Proceedings' of ephemeral character. The book, like the
Forum, is in effect a pool of know-how and lore. Analysis of biological samples for drugs or other trace-organics still calls for
experience and for awareness of pitfalls, even though sample preparation for HPLC is simpler than that traditionally needed for GC.

Coupled with gratitude for the publication texts, belated as
well as prompt ones, long-felt misgivings are now aired on behalf of
the editorial fraternity, whose efficiency and goodwill may nowadays
be taken for granted. High-calibre work deserves good presentation.
Editors can take in their stride and rectify irrelevance, repetiti-
veness, occasional errors, and infelicities in English, punctuation
or paragraphing. Dark thoughts do, however, arise about an investigator's 'life-style' when, amongst a growing minority, one encounters
inconsistencies, omissions or other 'rough spots' hardly attributable to work-pressures and manifest in various publication fields.
It would help if journal referees were asked to comment critically
on textual 'hiccups' as well as on scientific content. Thorough
editorial effort was devoted to both aspects in the present exercise
and earned some tributes from authors concerned.

Forensic and cancer-chemotherapy investigators are amongst those
catered for in the present volume. The usual analyte-oriented articles are preceded by articles on techniques that are or could be
applicable to biological samples. The 'Analyte Index' (hitherto
designated 'Compound Index'; that in Vol. 12 was cumulative) is not
a mere A-to-Z listing but entails chemical categorization which
brings together, with inevitable shortcomings, kindred molecular
features relevant, for example, to pH-dependent solvent extractabi-
lity and to GC detection. Interest in possible responsiveness to a
GC 'nitrogen detector' may have waned, but the presence of a primary
amino group is relevant to derivatizability for HPLC as well as GC,
and to ion-pairing potentialities. Patient scrutiny of the appropriate Index category should enable anyone faced with a novel analyte
to track down precedents helpful to method-development, particularly
where the mention of a sample type may imply pre-isolation.

Acknowledgements.- Valuable support for the Forum came from the Cancer Research Campaign, from Johnson Matthey & Co., and from U.K. pharmaceutical companies - Beechams, Glaxo, ICI and Smith, Kline & French. Moreover, some speakers came without full financial coverage. The choice of presentations was guided by Honorary Advisers including Drs. S.H. Curry (Chairman), J.A.F. de Silva, L.E. Martin, J. Chamberlain and G.G. Skellern. Drs. Jim Leppard and Joan Reid are thanked for Index drafting. As mentioned in the text, some Figs. have already appeared in journals, whose publishers (e.g. Elsevier, Dekker, Preston) are thanked: sources include *Journal of Chromatography, Journal of Liquid Chromatography* and *Journal of Chromatographic Science,* also (art. #E-5) a Wiley book edited by M. Trimble.

Abbreviations.- In connection with HPLC ('LC' is a pet aversion) this Editor has often deplored the upstart use of 'ECD'- a term hallowed by its GC usage as in art. #F-2 later in the book. To connote 'electrochemical' the term 'EC' is now used, but 'ECD' is reserved for the electron-capture detector. Other abbreviations which, although well known, are generally defined in each article concerned include NP, normal-phase [HPLC]; RP, reverse(d)-phase; i.s., internal standard; MS, mass spectrometry (EI, electron-impact; CI, chemical-ionization); RIA, radioimmunoassay; UV, ultraviolet (usually absorbance).

'Derivatization'.- This is an apt term for introducing a 'TMS' or EC-responsive group into a molecule, but hardly for altering it photolytically or oxidizing it (e.g. alcohol → aldehyde). The Editors kept this in mind, and now float a terminological idea: *transformation?*

Guildford Academic Associates   ERIC REID
72 The Chase, Guildford GU2 5UL
Surrey, U.K.   10 April 1984

Contents

The NOTES & COMMENTS ('NC' items) at the end of each section comprise some comments made at the Forum on which the book is based, together with some supplementary material.

[†] *Sub-listed on this and subsequent 'NC' title pages: comments on particular articles in the section.*

List of Authors

Primary author	*Co-authors, with relevant name to be consulted in left column*

<table>
<tr><td>

H.K. Adam – pp. 211-217
ICI Pharm., Alderley Edge, U.K.

G.W. Aherne – (i) pp. 247-248,
(ii) pp. 249-250, (iii) pp. 313-314
Univ. of Surrey, Guildford, U.K.

U.A.Th. Brinkman – (i) pp. 99-100,
(ii) pp. 183-184; *see also* de Jong
Free Univ., Amsterdam,
The Netherlands

J. Caldwell – pp. 47-52
St. Mary's Hosp. Med. Sch., London

J. Chamberlain – pp. 399-403
Hoechst Pharm. Res. Labs.,
Milton Keynes, U.K.

G.S. Clarke – pp. 59-60
Squibb Int^1. Dev. Labs., Moreton, U.K.

G.B. Cox – pp. 71-80
DuPont (UK), Stevenage, U.K.

J. Cummings – pp. 245-246
Univ. of Glasgow, U.K.

S.H. Curry – (i) pp. 275-285,
(ii) pp. 367-368
Univ. of Florida, Gainesville, U.S.A.

H. de Bree – (i) pp. 65-70,
(ii) pp. 305-310, (iii) pp. 395-397
Duphar Res. Labs., Weesp,
The Netherlands

G.J. de Jong – pp. 101-106
as for H. de Bree

J.A.F. de Silva – pp. 385-392
Hoffmann–La Roche, Nutley, NJ, U.S.A.

K. Ensing – pp. 287-296
State Univ., Groningen,
The Netherlands

V. Goldberg – pp. 297-302
Nat. Hosps. for Nervous
Diseases, London

</td><td>

A.R. Aitkenhead – Aherne (iii)
G. Algozzini – Curry (ii)
W.H.R. Auld – Skellern

R.R. Bain – Jordan
G.B. Baker – I.L. Martin
B. Bidlingmeyer – Krull (ii)
K. Bratin – Krull (iii)
R.R. Brown – Jordan
N.K. Burton – Aherne (iii)

K.C. Calman – Cummings
S. Colgan – Krull (ii)
I.A. Cotgreave – Caldwell
R.T. Coutts – I.L. Martin

C.P. Daniel – Leith
H.J. Dengler – von Unruh
G. de Vries – Brinkman (i)
R.A. de Zeeuw – Ensing
L.W. de Zoeten – de Bree (ii)
X-D. Ding – Krull (i) & (iii)
D. Dixon – L.E. Martin

M. Eichelbaum – von Unruh
G. Forcier – Krull (iii)
R.W. Frei – Brinkman (ii)

S.J. Gaskell – Leith

</td></tr>
</table>

Primary author

Co-authors, with relevant name to be consulted in left column

V. Craig Jordan - pp. 219-225
Univ. of Wisconsin, Madison, U.S.A.

I.S. Krull - (i) pp. 139-144,
(ii) pp. 173-179, (iii) pp. 365-366
Northeastern Univ., Boston, U.S.A.

J. Landon - p. 311
St. Bartholomew's Hosp., London

E.P.Lankmayr - pp. 81-89
Technical Univ., Graz, Austria

Heather M. Leith - pp. 235-240
Tenovus Inst., Cardiff, U.K.

R.P. Maickel - pp. 3-16
Purdue Univ., Lafayette, IN, U.S.A.

C.A. Marsden - pp. 319-330
Medical Sch., Nottingham, U.K.

I.L. Martin - pp. 331-336
MRC Centre, Cambridge, U.K.

L.E. Martin -pp. 191-194
Glaxo Res., Ware, U.K.

M.R. Montgomery - pp. 269-274
VA Hosp. & Univ. of S. Florida, U.S.A.

D.R. Newell - pp. 145-153
Inst. of Cancer Res., Sutton, U.K.

J.S. Oliver - pp. 17-26
Univ. of Glasgow, U.K.

J.W. Paxton - (i) pp. 201-209, (ii) 251-254.
Univ. of Auckland Med. Sch., N. Zealand

D. Perrett - pp. 111-121
St. Bartholomew's Hosp., London

J.D. Ramsey - pp. 357-362
St. George's Hosp., London

G.G. Skellern - pp. 337-342
Univ. of Strathclyde, Glasgow, U.K.

R.N. Smith - pp. 261-268
Met. Police Forensic Science Lab.,
London

L.A. Sternson - (i) 161-172,
(ii) 227-234
Univ. of Kansas, Lawrence, U.S.A.

P. Hajdu - Chamberlain
K.R. Harrap - Newell
F. Hochberg - Krull (i)
M.R. Howlett - Skellern
H. Keuker - de Bree (iii)
R. King - Krull (ii)
N. Lammers - de Jong
P.T. Lascelles - Goldberg
D.J. Lawson - Aherne (ii)
P.A. Lewis - Woodward
L.K. Liu - Clarke
S.D. Lyman - Jordan

J. Maddock - Woodward
V. Marks - Aherne (i) & (ii)
C.S. McArdle - Cummings

U. Neue - Krull (ii)
A. Newhard - Krull (ii)
R.I. Nicholson - Leith
Janet Oxford - L.E. Martin

A. Peters - de Bree (iii)
E. Piall - Aherne (i) & (iii)
M. Quinton - Aherne (ii)

N. Ratnaraj - Goldberg
M.L. Robinson - Clarke
H.M. Ruijten - de Bree (i)

C. Santanasia - Krull (ii)
R. Schuster - L.E. Martin
C. Selavka - Krull (i) & (iii)
Z.H. Siddik - Newell
J.F.B. Stuart - Cummings

Primary author

*Co-authors, with relevant name
to be consulted in left column*

D. Stevenson – pp. 243–244
Univ. of Surrey, Guildford, U.K.

P.C. Uden – pp. 123–138
Univ. of Massachusetts, Amherst,
U.S.A.

M.Th.M. Tulp – de Bree (ii)
P.H. van Amsterdam – de Bree
                    (i) & (ii)
D.J.K. van der Stel – de Bree (ii)
W.R. Vincent – de Bree (ii)

C.E. von Unruh – pp. 27–37
Med. Universitätsklinik Bonn,
Bonn, W. Germany

Roberta J. Ward – pp. 155–160
MRC Clin. Res. Centre, Harrow, U.K.

R. Whelpton – (i) pp. 39–44,
(ii) pp. 189–190
London Hosp. Med. Coll., London

W. Wegscheider – Lankmayr
D. Witts – Wilson (ii)

J. Williams – pp. 377–384
Welsh Nat. Sch. of Med., Cardiff,
U.K.

Reed C. Williams – pp. 53–58
E.I. du Pont de Nemours,
Wilmington, NJ, U.S.A.

R.L. Williams – pp. 343–356
Met. Police Forensic Science Lab.,
London

I.D. Wilson – (i) pp. 91–96,
(ii) pp. 185–188
Hoechst Res. Labs., Milton Keynes,
U.K.

K-H. Xie – Krull (ii)
Wing Yu – Curry (ii)

A.J. Woodward – pp. 369–370
Simbec Res., Merthyr Tydfil, U.K.

J. Zeeman – de Jong

Section #A

SAMPLE HANDLING AND USEFULNESS OF ISOTOPES

#A-1

SEPARATION SCIENCE APPLIED TO ANALYSES ON BIOLOGICAL SAMPLES

Roger P. Maickel

Department of Pharmacology &
Toxicology, School of Pharmacy
& Pharmacal Sciences
Purdue University
W. Lafayette, IN 47907, U.S.A.

BetaMED Pharmaceuticals
Inc.
6925 Guion Road
Indianapolis
IN 46248, U.S.A.

The historical development, present 'state-of-the-art' and future projections of separation science in the analytical context are presented, using specific examples to illustrate various aspects. Despite notable advances in the separation technology that must precede the application of a specific measurement procedure and/or detection device, many problems and difficulties remain. The difficulties are compounded by the development of instrumentation often presenting an apparently high degree of automation but possibly with unsuspected jeopardy to the reliability of analytical results.

Human beings tend to think in terms of the development of virtually anything as a temporal sequence of events. Man progresses in life from infancy, through youth, to adulthood, then on to old age; books are often written with a prologue at the beginning and an epilogue at the end; scientific papers have a progression from introduction to discussion. Indeed, in preparing to write a scientific manuscript or to make a presentation at a scientific meeting, one is often torn between reporting the results of a series of experiments in the temporal sequence in which they were performed, or re-ordering them to make more 'sense' to the audience.

This temporal frame of development has also existed for analytical chemistry in general and for separation science in particular (Fig. 1). Less than 50 years ago, the analytical chemist was virtually restricted to two technologies for separation of substances from the complexity that characterizes most biological materials. Distillation procedures date back to the days of alchemy: if the

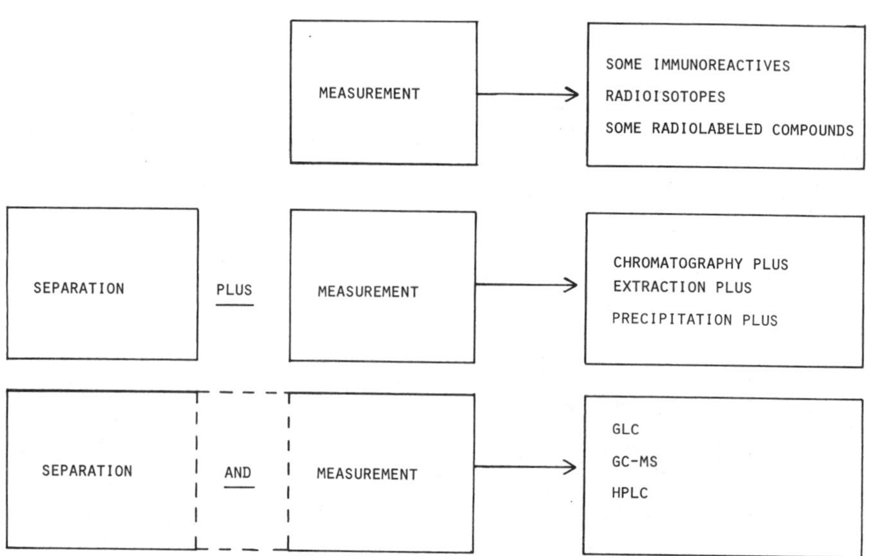

Fig. 1. Separation and measurement combinations.

desired substance was sufficiently volatile, it could be isolated by heating the entire mass, then cooling the vapours to produce a condensate. Sophisticated versions of such procedures are still in use today: micro-distillation apparatus, cold-finger procedures, Kjeldahl procedures remain in use for specialized purposes.

Along with distillation procedures that took advantage of the differing vapour pressures or boiling points of chemicals, the early separation scientists made use of various precipitation procedures to separate and isolate substances on the basis of differing solubilities. Such techniques were versatile: they served well in both research and teaching. I can still remember my first chemistry laboratory experiment as an undergraduate. We were given an unknown mixture of NaCl, naphthalene and sand. The only equipment issued consisted of an analytical balance, beakers, a Bunsen burner, a porcelain crucible, a filter funnel with papers, and a supply of benzene and water. The recommended initial procedure was to weigh the unknown, then add a modest amount of benzene and stir the entire mass for 15 min. After pouring this through a filter paper, the filtrate was collected in a beaker and allowed to evaporate in a fume hood overnight; the naphthalene crystals were then weighed. The residue remaining in the first filter paper cone was then dissolved in water. The process of filtration and evaporation of that filtrate overnight led to crystals of NaCl that could be weighed. Finally, the material remaining in the second filter paper cone was transferred to a crucible and heated with the burner; this procedure left dry sand to be weighed.

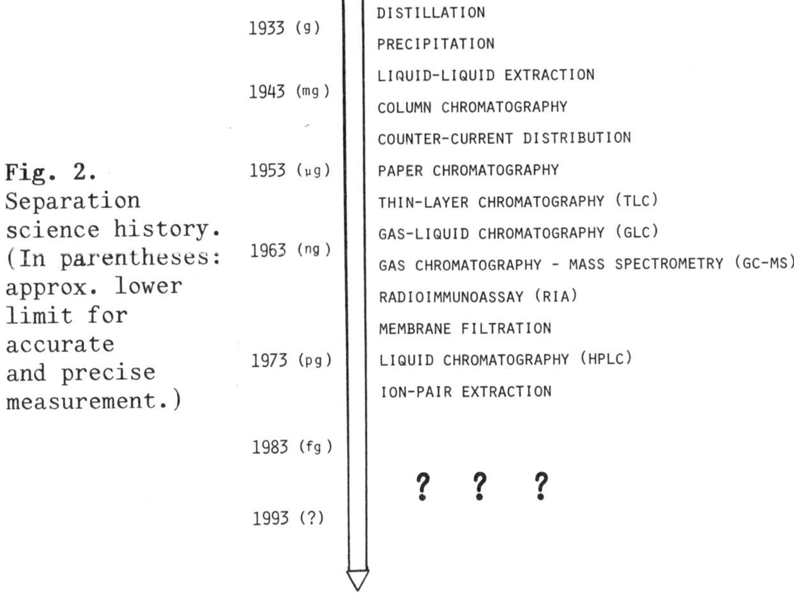

Fig. 2.
Separation
science history.
(In parentheses:
approx. lower
limit for
accurate
and precise
measurement.)

Crude though that experiment may seem by modern standards of
analytical technology, the basic principles are those common to
virtually all analytical procedures, viz. *separation, identification*
and *measurement*. This article aims to discuss 'separation science',
entailing some allusion to the other two basic principles on occasion
since all three are intimately related (Fig. 1). This is especially
true with biological samples, in which the analyte may be one of
hundreds (or even thousands) of chemicals present over a wide range
of concentrations in the sample.

Let us first examine a few aspects of separation science that
are critical for procedures in general and for analyses of biological
samples in particular. Fig. 2 gives a temporal view of the development of various separation procedures; the parentheses indicate the
approximate working limits of satisfactory measurement. As can be
seen, science has increased these 'limits of sensitivity' by $\sim 10^3$
with each decade. While such gains in sensitivity are obviously of
great advantage - making possible analyses that only two decades
earlier would have been impractical, if not impossible - they also
are responsible for new problems. Thus, suppose that the analyst of
1943 wished to determine the concentration of substance 'A', usually
present at 1.0 mg/ml in blood serum. Another substance, 'M', is also
present at 1.0 mg/ml in serum; it would interfere in the method for
quantitative determination of 'A' (by spectrophotometry) unless

removed by an appropriate separation procedure. If such a procedure were successful in removing 99% of 'M', the residual amount would represent only 0.01 mg/ml, a 1% error in the determination of 'A'.

Consider, however, the analyst of 1963 who wished to determine substance 'B', normally present at 1.0 ng/ml in serum, and structurally similar to 'A'; with 'M' still present at 1.0 mg/ml, obviously it will again interfere. The separation procedure that removes 99% of 'M' would now be completely useless, since 0.01 mg/ml would remain, an amount 10^4 times that of 'B'! If the analyst of 1983 wishes to determine the level of substance 'C' in serum, normally present at 1.0 fg/ml and still with 'M' present at 1.0 mg/ml, the 99% successful method for getting rid of 'M' would leave 10^{10} times more 'M' than 'C'!

It may be argued that use of a more suitable and/or more specific method for measuring 'B' or 'C' would be the answer. In some situations this may be the case, but even highly specific technologies such as sophisticated mass fragmentography or unique immunoassay procedures may be subject to failure when the concentration ratio of interfering to desired substances exceeds 10^3; when the ratio exceeds 10^6, such procedures are virtually doomed to failure. The only answer is better methods and technologies for separation of desired from undesired substances.

Another difficulty that has developed concurrently with greater sensitivity and hence determination of smaller amounts relates to sample loss during separation procedures. Adsorption onto surfaces (e.g. the inner walls of glass containers) has been known for many years. It is not generally great: usually below pg/cm^2 in order of magnitude. When the analyst is determining concentrations of 1 mg/ml, such adsorption may represent less than 10^{-8} of the total present – an insignificant fraction. However, for 1.0 pg/ml, the potential adsorption loss may represent *over 100%* of the sample, thereby rendering the analytical procedure valueless.   (See Index entry for 'Adsorption' in Vols. 5, 7, 10 & 12, this series.-*Ed.*)

FEATURES OF THE ANALYTE AND THE SAMPLE

When one considers the basic characteristics of any analytical method, those portions of the technology concerned with separation science generally depend on some combination of the physicochemical properties of the substance to be determined and those of the components of the separation system. Thus, dielectric constant, ionizability, mol. size/wt., vapour pressure and solubility, just to name a few properties, may be involved, and to varying extents. However, such properties are common to all analytical methodologies, virtually regardless of the sample source. With biological samples their very nature introduces an additional dimension. They can run the gamut

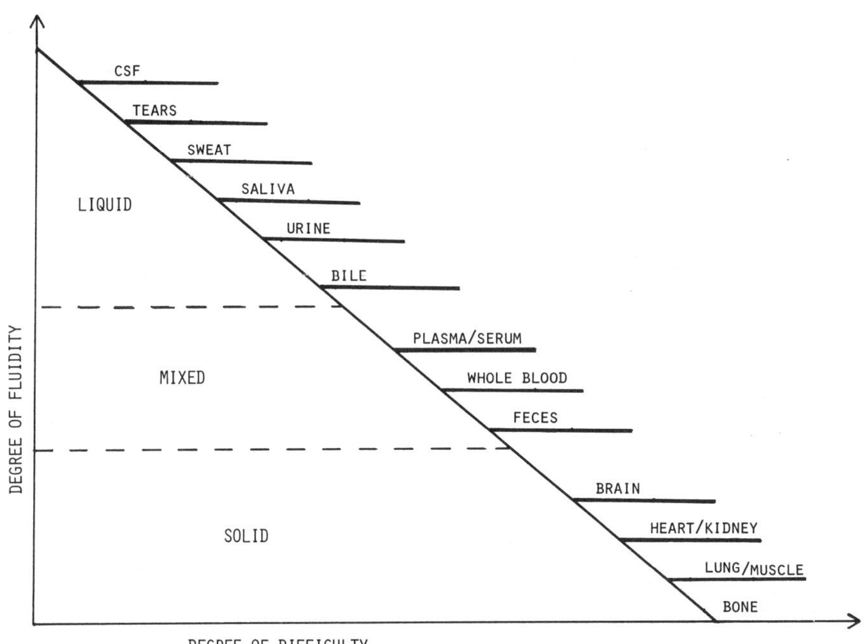

Fig. 3. Biological sample characteristics.

from relatively simple aqueous solutions such as CSF, through more
complex heterogeneous suspensions such as whole blood, to highly
organized solid materials such as bone (Fig. 3).

Such a variety of samples leads to a concomitant variety of
problems and difficulties. Thus, whereas sampling from CSF, tears
or sweat poses no great problems, sampling from a heterogeneous suspension such as plasma or blood may be fraught with difficulties.
Faeces may represent anything from a solid, through a semi-solid amor-
phous material, to a suspension. Tissues represent the same range
of consistencies except that none can be considered as 'liquid'.
While the designation of the axes in Fig. 3 may be considered somewhat non-scientific, and the positioning of samples with regard to
these axes may be considered arbitrary, they represent a compendium
of input and experiences accumulated over almost 30 years of the
author's experience! Such a listing cannot be found in any textbook
presently marketed.

Since many biological samples represent solids or semi-solids,
it should be obvious that some procedures must be applied to render
them amenable to sampling and consequent introduction into an appropriate analytical procedure. The usual generalized term for such
procedures is destructive disruption; the end result is generally a
heterogeneous suspension, often called a homogenate, that is amenable

Table 1. Destructive disruption procedures.

CHARACTER-ISTICS	MORTAR-PESTLE HOMOGENIZERS		BLADE-CONTAINER HOMOGENIZERS		OTHER PROCEDURES	
	POTTER-ELVEJEHEM	TEFLON-GLASS	WARING BLENDER	VIRTIS/SORVALL	SONICATION	CHEMICAL DISSOLUTION
MECHANISM	Shear stress	Shear stress	Cut + shear	Cut + shear	Vibration	Hydrolysis, saponification, and others
COOLING POSSIBLE	Easily	Easily	No	Easily	Perhaps	Usually at elevated T^o
SPEED VARIATION	Continuous	Continuous	Stepped, or continuous	Continuous	N/A	N/A
CONTAMINATION WITH METALS	None	Minimal	Severe	Severe	Minimal	None
FOAMING PROBABILITY	Slight	Slight	Moderate	Moderate	Moderate	Severe
CONSISTENCY FACTOR	Poor	Poor	Good	Good	Good	Good
LIMITATIONS	Hard tissues are difficult	Hard tissues are difficult	Container size	Minimal	Type of sample	Harsh conditions
SAFETY RATING	Poor	Fair	Good	Excellent	Good	Fair

to ready sampling. Table 1 summarizes the major 'types' of destructive disruption procedures and some of their important characteristics.   As can be readily concluded, no one procedure is perfect, although some are less perfect than others. However, from a consideration of the biological sample involved, the substance to be analyzed, the procedures to be used after disruption, and the features of the various disruption technologies, one may generally select the most appropriate (I have often used the term "least worst") for any given situation.

Since, implicitly, some liquid will be used as the medium in which the semi-solid or solid sample will be physically disrupted, the choice of liquid needs proper consideration. Table 2 aims to summarize the characteristics of some commonly used disruption liquids or, as they are often called in laboratory jargon, homogenization solvents.   As with the choice of disruption/homogenization

Table 2. General characteristics of homogenization media.

SOLVENT	ADVANTAGES	DISADVANTAGES
DISTILLED WATER	Relatively good solvent Does not destroy tissue constituents Final pH near 7.0	Degree of ionization may vary Does not denature enzymes consistently Final pH may vary with tissue
DILUTE ACID (<0.5N)	Relatively good solvent Denatures many enzymes Final pH <7.0 Foaming is minimal	Considerable protein denaturation Compound(s) may be acid sensitive
DILUTE ALKALI (<0.5N)	Relatively good solvent Denatures many enzymes Final pH >7.0	Considerable protein denaturation Compound(s) may be alkali sensitive May cause foaming (soaps)
STRONG ACID (>0.5N)	Good solvent Denatures all enzymes Precipitates proteins Final pH <4.0	Clumping and aggregation may occur Compound(s) may be acid sensitive Tissue constituents may break down
STRONG ALKALI (>0.5N)	Relatively good solvent Denatures all enzymes Precipitates proteins Final pH >10.0	Clumping and aggregation may occur Compound(s) may be alkali sensitive Tissue constituents may break down Foaming generally serious (soaps)
ORGANIC SOLVENTS	Relatively chemically inert Denatures most enzymes Precipitates proteins	Clumping and aggregation generally severe May be poor solvents

procedures, selection of an appropriate solvent for a given situation involves a variety of factors. What is the analyte's chemical stability? Will denaturation of proteins be advantageous or detrimental? Might there be clumping, or soap formation ?

With an appropriate combination of disruption process and homogenization solvent settled, and a suitable sampling procedure at hand, the next step for virtually all tissue samples (and even for plasma or serum in some cases) is to find a way to get rid of extraneous material such as proteins, cellular debris and subcellular particles. Any need to measure a protein in the sample would of course be a contra-indication; but even for smaller, soluble polypeptides the presence of cellular debris or larger polypeptides such as albumins often precludes assay. With the advent of membrane filters, vacuum or pressure filtration may be applicable; previously available procedures such as ordinary filter funnels or even Buchner funnels were virtually useless. Yet the commonest ways of removing large polypeptides or particles from biological samples involved the technology termed protein precipitation. This basically entails either denaturation or co-precipitation, as indicated in Table 3.

Table 3. Common protein-precipitation procedures. **D** denotes denaturation, and **C** co-precipitation.

AGENT	TYPE	CONDITIONS	CHARACTERISTICS/COMMENTS
HEAT	D	90-100 C° for 5-15 minutes	Efficiency is poor Precipitate often amorphous Compound(s) may be heat sensitive
COLD	D	Complete freezing/thawing Cycle may be repeated	Efficiency is poor Precipitation generally incomplete Time consuming
AMMONIUM SULFATE	D	Added to saturation point Final pH about 7.0	Efficiency is moderate High salt concentration in supernate
ZINC HYDROXIDE	CP	Zinc sulfate + sodium hydroxide Final pH about 7.0 Better at low temperature	Efficiency is excellent Fine (microcrystalline) precipitate
BARIUM SULFATE ZINC HYDROXIDE	CP	Zinc sulfate + barium hydroxide Final pH about 7.0 Better at low temperature	Efficiency is excellent Fine (microcrystalline) precipitate
META-PHOSPHORIC ACID	D	Final concentration 5-10% Final pH <3.0 Low temperatures required	Efficiency is excellent Reagent must be kept cold Compound(s) may be acid sensitive
PERCHLORIC ACID	D	Final concentration 0.2-0.4 M Final pH <3.0 Low temperature required	Efficiency is excellent Reagent must be kept cold Reagent must be removed (explosion hazard) Compound(s) may be acid sensitive
TRICHLORACETIC ACID	D	Final concentration 5-10% Final pH <3.0 Better at low temperature	Efficiency is good Reagent should be kept cold Reagent difficult to remove (organic) Compound(s) may be acid sensitive

A denaturing agent or process hinges on insolubilization at extremes of temperature, osmotic or ionic strength, or pH. With a co-precipitation agent, a quite flocculent precipitate is formed, entrapping any molecules not in true solution and also coprecipitating proteins by modest denaturation. Again the choice of procedure depends on the particular situation as indicated above.

NON-CHROMATOGRAPHIC SEPARATION APPROACHES

With success in getting a modest volume of an aqueous solution containing the substance desired along with dozens or hundreds of other solutes, both organic and inorganic, the analyst comes to the core (perhaps even the hard core) of separation science. The undesired substances have to be separated from the desired substance so that the latter can be measured accurately, precisely and specifically. This may call for partitioning between two immiscible solvent phases, based on the Nernst Distribution Law. Such partitioning underlies liquid-liquid extraction and chromatographic procedures (including paper and thin-layer) with exceptions such as ion-exchange and affinity chromatography which hinge on interactive bonding.

In liquid- liquid extraction, with the sample in aqueous solution, a choice has to be made amongst solvents such as those listed as follows in order of 'relative' polarity (< water): n-pentane; n-heptane, cyclohexane, benzene, dichloroethane, dichloromethane, chloroform, dibutyl ether, diethyl ether, acetone, ethyl acetate, cyclohexanol, isoamyl alcohol, n-butanol, acetonitrile, pyridine (then water, salt solutions). Relative 'polarity' cannot be derived from any single physicochemical constant but rather reflects a melange of properties. One must develop a mode of thinking and a set of semantic concepts - *lipophilicity = hydrophobicity = non-polarity;* cf. *lipophobicity = hydrophilicity = polarity*, with the overriding tenets that "like dissolves like" and "like partitions into like". Thus, benzene would be apt for naphthalene (lipohilic, non-dissociable), or for benzoic acid if undissociated (say pH 1 in the aqueous phase), but if dissociated (as at pH 10) it would remain in the aqueous phase. This differential partitioning is of inestimable value analytically.

A second influence on partitioning is 'salting out', well known to synthetic organic chemists. Thus, at its pKa (hence 50% dissociated) benzoic acid will be preferentially in the benzene phase if NaCl is added, but in the aqueous phase if naphthalene is added. The analytical literature is replete with examples. Thirdly, there is the newer concept of ion-pair extraction, whereby a quaternary nitrogen or protonated amino compound partitions into a non-polar phase through combination with an appropriate anion to give a neutral and relatively non-polar 'ion-pair'. The converse applies to a strongly anionic analyte. The chosen ligand commonly is a large lipophilic molecule, thereby enhancing the lipid partition: e.g. methyl orange, bromthymol blue, or dipicrylamine as anionic ligands.

There are variants on the theme of liquid-liquid extraction.
(i) It may be used to concentrate a sample by preferential extraction
into a smaller volume, e.g. (x 10; ~100% recovery) benzoic acid from
benzene into one-tenth vol. of N NaOH. (ii) It can remove unwanted
constituents selectively: a chloroform extract of plasma containing
both oestrogens and corticosteroids can be freed from the former for
assay of the latter by 'washing' with 0.1 N NaOH. (iii) Sequential
separations can be done, e.g. where chlorpromazine and its sulphoxide
are each to be assayed, one washed with buffer of appropriate pH.

Considering the historical aspects, early procedures utilized
separatory funnels, with manual shaking by the analyst, and depended
on time and differential specific gravities to achieve phase separation. This was supplanted by stoppered glass containers, mechanical
shaking machines, and centrifuges; these developments resulted in
increased speed and reproducibility. The more sophisticated mode of
liquid-liquid extraction was developed to the point of availability
of semi-automated systems with as many as 1000 transfers. Nevertheless, all these methods are rather severely limited by the time and
labour involved; hence chromatographic technologies developed over
the past 40 years have become the current leaders for separations.

CHROMATOGRAPHIC SEPARATION APPROACHES

Chromatography is a two-phase process depending on a succession
of repetitive partitionings. The phase pairings may be gas-solid,
gas-liquid, liquid-solid, liquid-liquid, or some multiple combination
thereof. In addition, uniquely specific adsorption or related phenomena may operate, e.g. ion-exchange, complexation or chelate formation, or mol. wt. sieves. Fig. 4 indicates the applicability of
various chromatographic procedures, as related to mol. wts. It is
meant to offer general guidance, not implying that a given procedure
is guaranteed to work with a compound having a certain mol. wt. Thus
a compound of mol. wt. 250 but with two quaternary nitrogens is not
likely to be sufficiently volatile for GC, nor is methylcholanthrene
an ion-exchange candidate, being non-dissociating (mol. wt. 300).

In general, paper chromatography and classical column chromatography have lately lost favour, as has liquid-liquid extraction. The
loss in 'popularity' reflects the time and labour required for indi-
vidualized handling. The other three main procedures - GC, HPLC and
TLC - have all been supplemented with semi- or fully-automated procedures and instrumentation. Fig. 5 presents a type of sequential
decision-making process where the analyst takes account of his needs
and of features of the analyte, presuming that the sample has been
pre-treated so as to be amenable to the remaining analytical steps.

Chromatographic modes other than GC have in common that the
mobile phase is a liquid, and each individual partitioning obeys the
Nernst Distribution Law, such that the solvent ranking given above

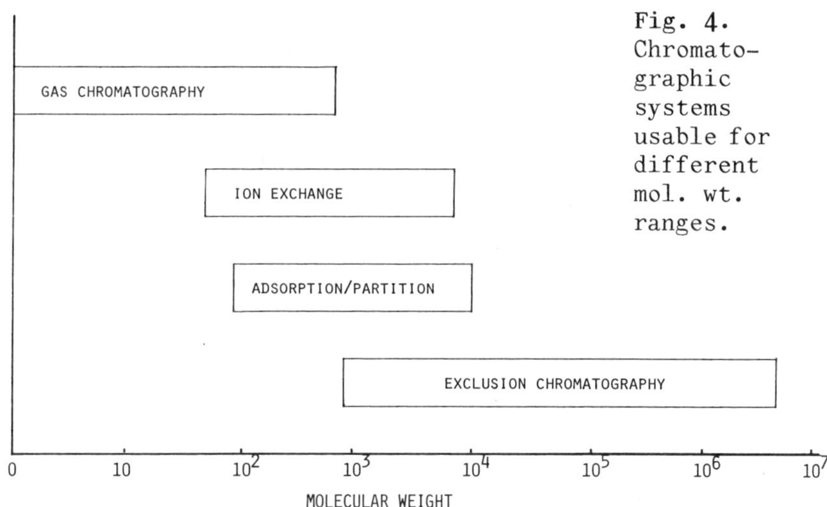

Fig. 4. Chromato-graphic systems usable for different mol. wt. ranges.

Fig. 5. Selection of chromatographic procedure.

for liquid-liquid extraction is relevant. In the following more
specific quantitative listing, the ranking is in terms of solvent
strength, ° - a measure of the polarity of the solvent for liquid-
solid adsorption chromatography, derived from the free energy of
adsorption onto a standard surface. Also listed, in parentheses ( ),
are the Hildebrand Solubility Parameter values for most of the solvents - a measure of the polarity of a solvent for liquid-liquid
partition chromatography, representing the square root of the heat
of vapourization per ml of pure substance. The list based on ° is:-
  n-pentane, 0.00 (7.1); n-hexane, 0.01; cyclohexane, 0.04 (8.2);
  carbon tetrachloride, 0.18 (8.6); xylene, 0.26 (8.9);
  benzene, 0.32 (9.2); ethyl ether, 0.38 (7.4);
  chloroform, 0.40 (9.3); dichloromethane, 0.42 (9.7);
  dichloroethane, 0.49; acetone, 0.56 (9.9); ethyl acetate, 0.58 (9.6);
  acetonitrile, 0.65 (11.7); n-propanol, 0.82 (11.5); methanol, 0.95 (14.4).
Pure solvents mixed in various proportions can furnish continuous
series of ° values in unitary steps, e.g. 0.05, 0.10, 0.15...... Thus,
new HPLC separations can be developed on a reasonably rational basis.
As with any other choice situation, the analyst must make the final
selection; no 'cookbook' is available.

GENERAL COMMENTS

Impurities in materials are a universal problem in separation
science. Problems due to abundant interfering substance(s) in a
biological sample were discussed earlier; the remedy was, and very
simply so, an appropriate separation scheme. However, every chemical
solvent, and each piece of apparatus used in a separation procedure,
may in itself contribute one or more types of 'contamination' to
the sample! Suppose, for example, that the 10 ml of an organic solvent used in a partition system contains an impurity that 'co-assays'
with the analyte: with the analyte at 1.0 µg/ml, a 1 ppm solvent
impurity would be present at a virtually equivalent concentration in
the extract [awkward involatility?-*Ed.*]. Such impurities can accumulate
from an extraction solvent, a deproteinizing reagent or a TLC or HPLC
developing solvent; they demand careful and meticulous performance
of reagent blanks in every analysis. Thus, as more sensitive methods
are developed, the problems of separation science are likely to
become more critical. This cannot be prevented; it must be accepted
and appropriate modifications or alternative procedures developed.

Finally, mention must be made of the end-point in GC and HPLC
procedures - the detector system. Even though this is not a part of
'separation science', it serves not only to *measure* the amount of
substance present, but also to add to the *confirmation of identity*
of the substance. In early GC days, relatively non-specific methods
such as the gas-density balance, thermal conductivity and flame-ioni-
zation detectors were common. Electron-capture detectors, specific
'atom' detectors (N, P, S) and, ultimately, MS improved this situation. With the advent of HPLC, absorbance and fluorescence detectors

came into use; these can be quite specific. Yet we must realize that in the combination of a specific detection system with multiple partitioning and appropriate initial disruption and deproteinization steps, it is the separation science that permits the detector system to function. Without the 'isolation' of the substance of interest, the detector would yield meaningless results.

Traditional concluding questions should be posed. Where have we been ?   Where are we now ?   Where shall we go ?   In terms of 'separation science' as applied to analyzing biological samples, I suggest a summary as follows, based on Fig. 1.

(a) We have seen, in the past half-century, tremendous advances in analytical technology. Merely in terms of the sensitivity of methods, the increase has been 12–15 orders of magnitude. Separation science, both theory and practice, has been the literal 'heart' of this development. While instrumentation, both design engineering and production, has been crucial, the instruments are really the mechanical extensions of the theoretical principles.

(b) At present we are still in the process of honing (or fine-tuning) much of analytical technology. Better methods of separation, better procedures for detection and quantitation, remain in various stages of development. The more a method, or procedure, or technology is used, the more the analyst knows about it, and the more likely is an improvement to be made. In terms of separation science, the statement holds – "Build a better method (mousetrap) and the world will beat a path to your door".

(c) The future holds nothing but promise, for more sensitive, more specific, more rapid, more economical analytical methodology. Considering that we have advanced 12–15 orders of magnitude (in sensitivity limits) over the past half-century, can we expect a lesser improvement over the next 50 years ? I think not; in fact I look forward to the day when determination of less than 100 molecules of a specific substance in a biological sample will be a routine procedure rather than an 'impossible dream'. Such an advance in analytical technology will only become a reality, however, through the continued efforts of all analysts, especially those who deal in the field of 'separation science'.

Back-up references

This listing of references is not meant to be highly specific, nor all-inclusive. It is a modest compendium of classical and recent books in various areas of separation science. The novice analyst may wish to read some or all; the experienced analyst has probably done so at one time or another.

Brodie, B.B. & Gillette, J.R., eds. (1971) *Concepts in Biochemical Pharmacology*, Part 2, *Handbook of Experimental Pharmacology*, Vol. 28/z, Springer-Verlag, Berlin, 778 pp.

Clarke, E.G.C., ed. (1969) *Isolation and Identification of Drugs*, Pharmaceutical Press, London, 870 pp.

Garrett, E.R. & Hirtz, J.L., eds. (1977) *Drug Fate and Metabolism - Methods and Techniques*, Vols. 1 & 2, Dekker, New York, 313 pp. (Vol. 1) and 382 pp. (Vol. 2).

Reid, E., ed. (1976) *Assay of Drugs and other Trace Compounds in Biological Fluids* [Vol. 5, this series], North Holland, Amsterdam, 254 pp.

Reid, E., ed. (1981) *Trace-Organic Sample Handling* [Vol. 10, this series], Horwood, Chichester, 383 pp.
(*NOTE by Senior Editor, who added this ref.:* material closely relevant to the present article includes the Editor's opening survey and articles that deal with sample problems and with contaminants in solvents. Vol. 7 is also pertinent.)

Reid, E. & Leppard, J.P., eds. (1983) *Drug Metabolite Isolation and Determination* [Vol. 12, this series], Plenum, New York, 289 pp.

Thoma, J.J., Bondo, P.B. & Sunshine, I., eds. (1977) *Guidelines for Analytical Toxicology Programs*, Vols. 1 & 2, CRC Press, Cleveland, 268 pp. (Vol. 1) and 280 pp. (Vol. 2).

#A-2

SAMPLE HANDLING IN FORENSIC TOXICOLOGY[*]

J.S. Oliver

Department of Forensic Medicine and Science
The University, Glasgow G12 8QQ, U.K.

The forensic toxicologist's case work mainly involves the detection, isolation, identification and quantitation of drugs in biological materials. The samples received for this purpose vary in quantity, variety and quality. The quantity and variety depend on the pathologist; the quality depends on the condition of the deceased and, to a lesser extent, on the pathologist. The task of the toxicologist can be assisted by good medical and circumstantial evidence as back-up. Also, a knowledge of significant post-mortem findings is helpful.

Assay for drugs is undertaken using techniques of immuno-assay, spectrometry, HPLC, GC and GC-MS. Whilst these techniques are relatively simple to use, they can only be exploited to their full potential following careful sample handling. In the past decade a variety of sample-handling techniques have been introduced which have improved the effectiveness of analysis for drugs.

When confirmatory identification and then measurement in the biological samples have been achieved, the results have to be interpreted in consultation with the pathologist in conjunction with whatever other evidence is available, and finally presented in court. When received from the pathologist, who is responsible for the collection of adequate specimens and should ensure that they are placed in clean properly labelled containers, the specimens will already have undergone deterioration to an extent depending on the circumstances of the death and thereafter. Further deterioration must be avoided by rapid transfer to the laboratory.

------

[*]*Editor's note:* This article, placed here although the sample dealt with is typically post-mortem, complements the wide-ranging forensic surveys by R.L. Williams (#F-4) and R.N. Smith (#E-1; RIA). See also concluding part of #F, 'Comments'.

SAMPLES

To ensure the continuity of evidence, the toxicologist must issue a receipt for all the specimens received. Prior to analysis, the samples can be kept refrigerated, *not frozen* unless a substantial delay is likely, whilst essential details are collected. Receipt of frozen specimens can cause unnecessary delay in the laboratory.

It is essential that adequate documentation always be received. Information required includes name, age, sex, cause of death (if known) or significant post-mortem findings and circumstances of death. Police reports generally include a list of poisons available to the deceased and details of drugs prescribed over the previous 2 years. Thereby valuable clues are obtainable to guide a search for specific poisons, although other poisons must be searched for also.

The samples received for analysis vary in quality. In parti- cular, blood specimens range from the almost perfect non-haemolyzed specimen familiar to the clinical biochemist, through grossly haemo- lyzed specimens and coagulated non-pipettable specimens to putrid dark-brown fluids. Tissue samples vary through corresponding stages. Urine samples are received only if available. Stomach contents can be revealing: occasionally recognizable fragments of capsules or tablets can be readily identified, and a distinctive odour can give a valuable clue as to which poison has been ingested (see below).

INVESTIGATION

With or without the help of the case history, the analyst has to search for a drug or a poison in a specimen that is usually far from ideal. If a drug is involved, it will probably have come on to the market in the last two decades. In general, this means that the sought-for drug will be pharmacologically active at relatively low dosages and will require the use of analytical techniques capable of detecting and measuring it at ng/ml blood levels. Active metabolites and, as a result of polypharmacy, the possibility of interaction of other drugs at therapeutic levels have also to be considered.

The toxicologist can now benefit from analytical equipment that has been developed over the past two decades with the capability of detecting small quantities of drugs in extracts. The procedures used in investigating biological samples for drugs and other poisons are described in specialized books [1-4], and should be consulted for procedural details. This article outlines illustrative procedures analyzing drugs in specific sample types.

Stomach contents

Residues of pure drug or poison can be looked for in stomach contents. Clues as to the identity of a poison may come from smell

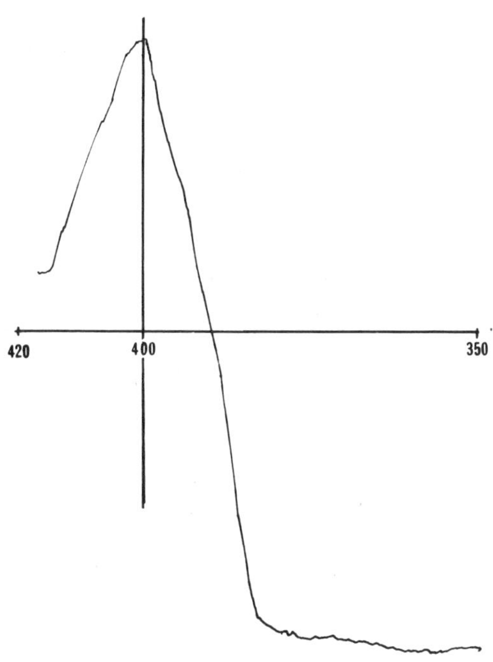

Fig. 1. Methyl-dopa determination by fluorescence spectroscopy. Emission measured at 500 nm, excitation being varied over the range 350-420 nm, as shown, evidently with 400 nm as the optimum for excitation. The trace was obtained with de-proteinized plasma from unhaemolyzed blood (see text), following oxidation to furnish the trihydroxyindole deriva-tive [4]. Cf. Fig. 2 (post-mortem samples).

(e.g. phenols), colour (e.g. the violet-blue dye used in a proprie-tary metaldehyde-based slug bait), discernible tablets etc., and spot test results for trichloro compounds, heavy metals, oxidizing agents and herbicides [1]. Otherwise poisons detection needs a methodical stepwise analytical approach [1]. Stomach contents have the advantage that the poison can often be isolated in relatively large quantities. Detection in the body proper still has to be achieved.

BLOOD AND URINE

With blood and urine techniques, forensic toxicology and clini-cal biochemistry come closest to overlapping; but the state of the specimen (see above) is one important difference, and for the foren-sic toxicologist drug identification must be absolute for evidential purposes. Thus the whole range of sensitive immunoassays available to the clinical biochemist serves merely as valuable sensitive pres-umptive tests. Positive results require confirmation with differen-tiation, by separation, between a drug, metabolites, and similar drugs.

The state of the specimen implies the possibility of analytical interference from putrefaction products [5, 6]. If the clinical bio-chemist relies on a spectrophotometric assay, the toxicologist will usually find his extract unsuitable due to pigment resulting from haemolysis or putrefaction, as exemplified by α-methyl-dopa. It can be assayed without extraction in deproteinized plasma as a fluorescent

Fig. 2. Assay of α-methyl-dopa in post-mortem blood by RP-HPLC (peak 3; 1 & 2 represent blood constituents). After adding 2 ml of 5% (w/v) trichloroacetic acid to the 1 ml blood sample, vortexing, and centrifuging after standing for 5 min, the supernatant was chromatographed.
Column: RP8 Chromosorb. Mobile phase: 16.8 g NaH_2PO_4 in 980 ml water; 20 ml methanol; 1 ml 0.1 M EDTA; 1 ml 100 nM $[(t\text{-butyl})_4N]_2SO_4$. Detection: electrochemically (0.7 V ox.; 3 nA back-off, 50 nA fsd).
Analytical result: 1.0 µg/ml was present in the authentic sample used.

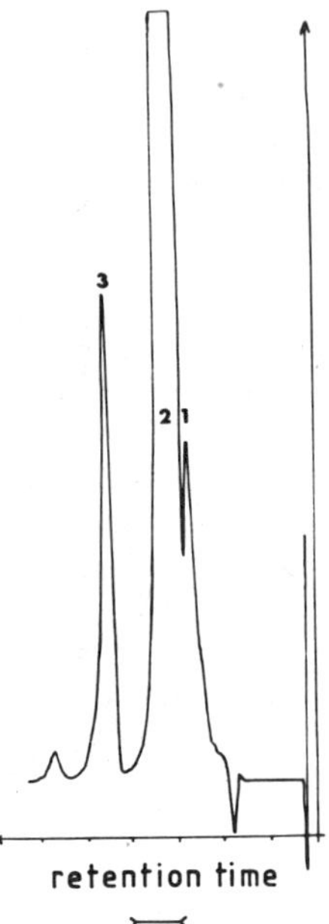

oxidation product (Fig. 1) [4]. Its emission is quenched where there is colour due to haemolysis in the blood. The toxicologist has to resort to an alternative technique such as that shown in Fig. 2 to detect and measure (by RP-HPLC with EC detection) this drug in post-mortem blood samples.

The first step in the analysis of blood specimens is to screen for the presence of groups of related drugs. The use of sensitive immunoassays enables a broad search to be undertaken confidently without expending much of the sample (cf. article by R.N. Smith, #E-1). Positive results can be confirmed by the use of appropriate analytical procedures which will differentiate between drug and metabolites. For lysergic acid diethylamide (LSD), confirmation relies on RIA. If positive, HPLC may be performed, with collection of fractions to be analyzed again by RIA; positive results should be obtained in fractions corresponding to the retention volume of LSD [8].

Acidic drugs can be identified in blood specimens using HPLC or GC. Blood or plasma samples can be extracted using short columns of 'Extrelut' (Merck, Darmstadt) for GC [9], as follows:
- to 600 µl blood add internal standard (i.s.), and mix;
- apply to Extrelut column, leave 2 min, and elute with 2 ml of dichloromethane; evaporate to dryness;
- reconstitute in 25 µl dichloromethane, and apply 3 µl to the GC column.

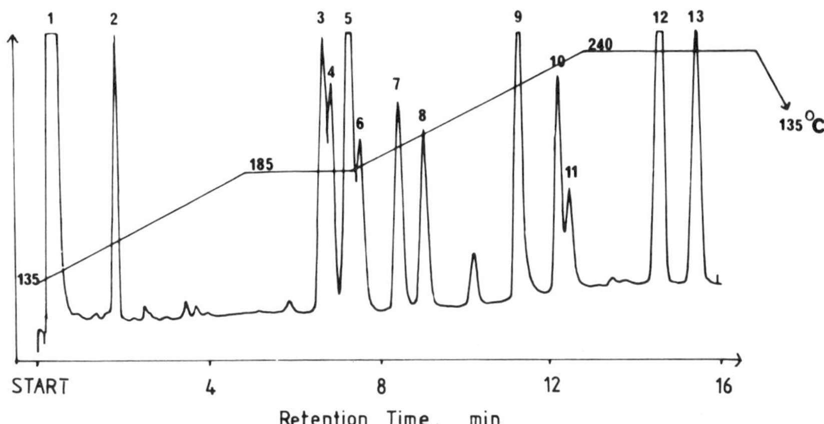

Fig. 3. GC separation of drugs, on a glass column (5 ft. x o.d.¼ in.)
packed with 2% SP-2110/1% SP-2510 DA on Supelcoport (Supelco Inc.,
Bellefonte, PA 16823), temperature-programmed as indicated.
1, solvent; 2, ethosuximide; 3, caffeine; 4, butobarbitone;
5, amylbarbitone; 6, pentobarbitone; 7, quinalbarbitone;
8, meprobamate; 9, methohexitone (i.s.); 10, phenobarbitone;
11, theophylline; 12, primidone; 13, phenytoin.-As spiked into the blood.

Fig. 3 shows a temperature-programmed GC pattern for a range of
drugs.   [Some are not regarded as 'acidic' for the purpose of the
Analyte Index.- *Ed.*] RP-HPLC gives good resolution of a similar
series of acidic and neutral drugs [10], as shown for some commonly
encountered barbiturates in Fig. 4.

Paracetamol (acetaminophen) is a widely available drug, often
found in forensic samples as a result of ingestion of purchased
analgesics or formulations such as 'Distalgesic'.   Since 1972, the
increasing numbers of deaths involving the misuse of 'Distalgesic',
a formulation of paracetamol and dextropropoxyphene, has caused concern in the forensic field and in the news media. Paracetamol can
be identified and measured in blood samples by GC (Fig. 5), after
the following sample-preparation procedure:
- to 1 ml blood add 1 ml of i.s. solution (*N*-butyroyl-*p*-aminophenol),
  and adjust to pH 7.4;
- extract into 2 ml of ethyl acetate, and evaporate to dryness;
- derivatize with 0.2 ml 'Tri-Sil Z' and apply to GC column.

Fig. 6 shows GC-ECD assay applicable to trichloroethanol in
blood or urine. With simple extraction into amyl acetate and a suitable i.s., the analyte can be rapidly detected and measured after
ingestion of dichloralphenazone. The assay can be used in reverse
to detect and measure chlorbutanol in samples following its illegal
use in sport. J. Ramsey's article (#F-5) deals with the detection
and measurement of similar chemicals in breath from solvent-abusers.

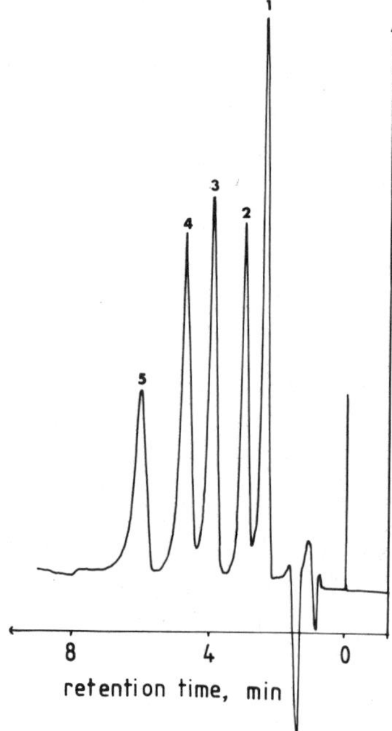

Fig. 4.   HPLC of some barbiturates. Column: ODS Hypersil. Mobile phase: methanol/water, 60:40 by vol. Detection at 220 nm. Run at room temp., not 50° as in [10] where the mobile phase was different (aceto-nitrile/phosphate buffer pH 4.4). 1, phenobarbitone; 2, butobarbitone; 3, amylobarbitone; 4, quinalbarbi-tone; 5, methohexitone (i.s.).

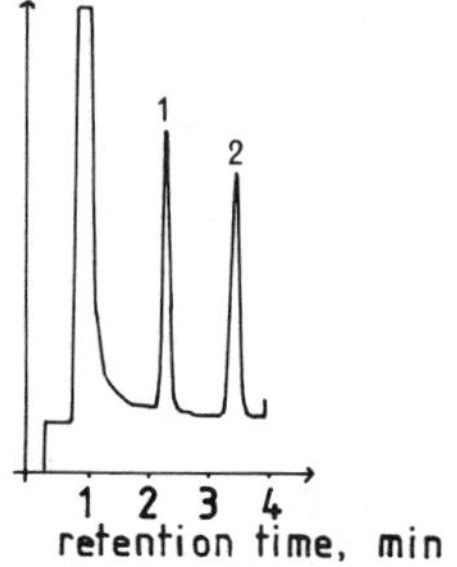

Fig. 5. Paracetamol (1; 2 =i.s.) as spiked into blood: GC-FID (SE30 column, 185-190°).

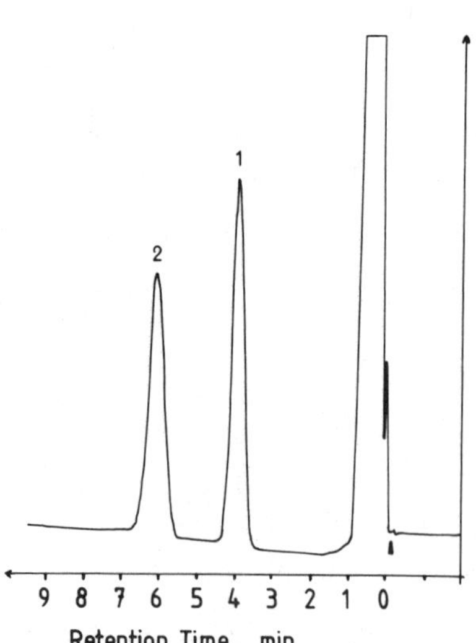

Fig. 6.   GC-ECD assay of trichloroethanol (2), also applicable to chlorbutanol (1). Column: 15% FFAP; run at 150°. Detector: 10 A current; constant-current type, run at 200°.

Analysis of blood samples for non-acidic drugs and metabolites can be carried out by GC and GC-MS following extraction into ethyl acetate after saturating the sample with powdered anhydrous ammonium carbonate [11] (a minor modification of the plasma/urine technique):
- to 1 ml blood add 2 ml water, and saturate with ammon. carbonate;
- extract with 5 ml ethyl acetate, centrifuge, and dry down;
- for clean-up, extract into dil. sulphuric acid and then re-extract;
- perform GC/GC-MS on 1-2 μl.
GC analysis with a nitrogen-specific detector and use of a comprehensive list of retention indices [12] enables most drugs encountered in forensic toxicology to be detected and identified. GC-MS permits an unknown drug or metabolite to be identified.

To determine morphine, it can be measured as its TMS derivative using MS in the multiple-ion selection mode, or HPLC can be applied with EC detection to the underivatized drug. Benzodiazepines with a 3-OH group are derivatized by treating the biological extract with *t*-butyl-dimethylsilyl-trifluoracetamide (Phase Separations Ltd.) and determined by GC-ECD or by GC-MS, monitoring the intense $M-57$ m/z fragment (multiple-ion detection mode) which can normally be discerned against a low background signal even in the biological extract.

LIVER

Amongst tissue samples received, liver is the commonest. Various analytical techniques have been developed to release drugs from tissue samples. [Cf. M.D. Osselton, Vol. 10, #C-1.-*Ed.*] A survey of the most generally used techniques revealed that HCl hydrolysis consistently gave the highest drug yields. However, the procedure was not suitable for acid-labile drugs. Although aluminium chloride produced slightly lower yields, it was preferred because it caused less hydrolysis of labile drugs. The other commonly used procedures, based on use of tungstic acid or ammonium sulphate, gave poor recoveries [13]. This study entailed spiking tissue slurries with the drugs. A more realistic comparison was made using case material and comparing use of the protease subtilisin Carlsberg with acid hydrolysis.

The enzyme digestion technique was rapid, gave high yields and was suitable for use with acid-labile drugs [14]. Experience with general case work confirmed the superiority of the enzyme digestion procedure over the tungstic acid and acid-hydrolysis techniques [15]. An interesting claim was made that the increased yields of drug from the use of subtilisin was due only to the presence of the Tris buffer at pH 10.5. With blood the authors obtained the highest yield when the sample was incubated at room temperature with the buffer for 1 h. They concluded that the increased drug yields were due to a wetting effect on the specimen which made the drug more extractable [16]. This is contrary to our own experience (see below).

Basic drugs can be readily extracted from liver as follows:
- homogenize 50 g of liver in 50 ml water;
- buffer to pH ~9 and extract with diethyl ether;
- wash the ether with dil. NaOH and with water;
- extract the drugs into dil. sulphuric acid, as below;
- determine UV absorption (as a 'screen' that costs little time in
  comparison with the benefit that can be gained);
- re-extract into solvent, and evaporate down;
- analyze by GC/GC-MS.

The UV estimate of total drug and metabolites is of value in the subsequent re-extraction and GC/GC-MS analysis.

Using surplus liver samples from analyzed case materials, the stated direct-analysis procedure was compared with the use of subtilisin Carlsberg both at pH 10.5 and at physiological pH, and with ammonium sulphate and aluminium chloride techniques as described by Stevens et al. [13]. In summary, the techniques were as follows, all starting with maceration of 20 g liver in 20 ml of (usually) water.-

Direct analysis:
- after maceration, add 2 ml 60% (w/v) KOH, then 400 ml diethyl
  ether and shake;
- decant the ether (note recovery) and wash sequentially with 10 ml
  10% (w/v) NaOH and 2 X 100 ml water;
- extract sequentially with 5 ml 0.1 M sulphuric acid and 5 ml 3.6 M
  sulphuric acid;
- examine extract by UV spectrometry.

Subtilisin + buffer:
- after maceration, with 20 ml Tris buffer (not water), 1 M, pH 10.5,
  add 5 mg subtilisin A, and incubate 1 h with shaking at 50°;
- add 5 ml 60% (w/v) KOH, then 400 ml ether and shake;
- decant the ether (note recovery) and wash sequentially with 10 ml
  10% (w/v) NaOH and 2 X 100 ml water;
- extract sequentially as in the foregoing procedure;
- similarly examine by UV.

Subtilisin without buffer:
- after maceration, with water, proceed as for *Subtilisin + buffer.*

Ammonium sulphate:
- after maceration, add 70 ml water + 10 ml 10 N HCl and excess
  solid ammonium sulphate;
- incubate at 50° for 1 h, then filter (Whatman No. 40);
- after washing with 2 X 15 ml water, cool the filtrate and add
  excess ammonium sulphate, then ammonium hydroxide to pH 9.5-10;
- add 400 ml ether, then proceed as above.

Aluminium chloride:
- after maceration, add 40 ml water and 50 ml 10% (w/v) aluminium
  chloride + 10% citric acid in 2 M HCl;
- incubate and proceed as for *Ammonium sulphate,* except for
  the treatment of the cooled filtrate: it is made alkaline with
  60% (w/v) KOH.

The yield of total drug as determined by UV was expressed relative
to that by the direct-analysis procedure as unity. It emerged that
the use of subtilisin Carlsberg as described by Osselton et al. [14;
cf. #C-1 in Vol. 10] was consistently better than the other techniques (Table 1). The direct-analysis procedure performed well in
comparison with the two protein-precipitation procedures.

CONCLUSION

The techniques outlined above are only a selection of the procedures which can be used in forensic toxicology, and the list is
far from comprehensive. It does serve, however, to illustrate the
significant differences in sample handling between the 'ideal' situation of the research pharmacologist or the clinical biochemist and
the forensic toxicologist. The requirement is for absolute identification in far from ideal samples. Rigorous approaches to the
problem must be adopted [1-3] and, where necessary, analytical techniques adapted to provide an assay procedure. An example of this is
the above-mentioned combined use of RIA and HPLC to identify LSD in
biological samples.

Table 1. Recovery factors, relative to 'Direct method' as 1.00,
for drugs in post-mortem liver samples extracted by different
methods. The drugs were 'endogenous', not spiked in.

Drug	Subtilisin + buffer	Subtilisin only	Ammonium sulphate	Aluminium chloride
Cyclizine	1.39	0.80	0.39	0.30
(2 samples)	1.12	0.85		
Diazepam	2.80	1.83	0.33	1.67
Flurazepam	1.54	0.77	0.46	0.69
Chlorpromazine	1.96	1.72	0.22	0.44
(4 samples)	2.33	2.00	0.13	0.67
	2.55	8.03	0.08	0.20
	1.45	1.38	0.01	0.15
Orphenadrine	2.96	3.18	0.15	1.08
(2 samples)	1.92	2.43	0.31	0.56
Amitriptyline	4.57	6.43	0.11	2.51
(4 samples)	2.60	3.05	0.02	0.87
	0.48	0.35	0.05	0.11
	1.35	1.15	0.03	0.16
Propoxyphene	1.66	0.94	0.69	1.16
(3 samples)	0.68	0.80	0.40	1.04
	2.84	2.22	0.47	1.51
Chlormethiazole	1.63	1.05	0.42	0.53
Thioridazine	1.47	1,40	0.20	0.53
Propranolol	0.74	0/81	0.05	0.87

References

1. Curry, A.S. (1976) *Poison Detection in Human Organs*, 3rd edn., Thomas, Springfield, IL, 356 pp.
2. Sunshine, I. (1975) *Methodology for Analytical Toxicology*, CRC Press, Cleveland, OH, 478 pp.
3. Clarke, E.G.C. (1969 & 1975) *Isolation and Identification of Drugs*, Vols. 1 & 2, Pharmaceutical Press, London, 1257 pp.
4. Baselt, R.C. (1980) *Analytical Procedures for Therapeutic Drug Monitoring and Emergency Toxicology*, Biomedical Publications, Davis, CA, 316 pp.
5. Oliver, J.S. & Smith, H. (1973) *J. Forensic Sci. Soc. 13*, 47–54.
6. Oliver, J.S., Smith, H. & Williams, D.J. (1979) *Forensic Sci. 9*, 195–203.
7. Oliver, J.S. (1982) *Human Toxicology 1*, 293–297.
8. Twitchett, P.J., Fletcher, S.M., Sullivan, A.T. & Moffat, A.C. (1978) *J. Chromatog. 150*, 73–84.
9. Werner, M., Mohrbacher, R.J. & Riendeau, C.J. (1979) *Clin. Chem. 25*, 2020–2026.
10. Kabru, P.M., Koo, H.Y. & Marton, L.J. (1978) *Clin. Chem. 24*, 657–662 [cited in ref. 4, pp. 14–15].
11. Horning, M.J., Gregory, P., Nowlin, J., Stafford, M., Lertratangkoon, K., Butler, C., Stillwell, W.G. & Hill, R.M. (1974) *Clin. Chem. 20*, 282–287.
12. Moffat, A.C., Finkle, B., Müller, R.K. & de Zeeuw, R.A. (1982) *Bull. [Special Issue] Int. Ass. Forensic Toxicologists*, 27 pp.
13. Stevens, H.M., Owen, P. & Bunker, V.W. (1977) *J. Forensic Sci. Soc. 17*, 169–176.
14. Osselton, M.D. (1977) *J. Forensic Sci. Soc. 17*, 189–194.
15. Dunnett, N. & Ashton, P. (1980) in *Forensic Toxicology* (Oliver, J.S., ed.), Croom Helm, London, pp. 272–278.
16. Kauert, G., von Meyer, L. & Drasch, G. (1980) in *Toxicological Aspects* (Kovatsis, An., ed.), University of Thessaloniki, Greece, pp. 376–382.

#A-3

STABLE ISOTOPES IN PHARMACOKINETIC STUDIES

G.E. von Unruh, M. Eichelbaum and H.J. Dengler

Medizinische Universitätsklinik Bonn
Sigmund-Freud-Str. 25
D-5300 Bonn 1, W. Germany

Drugs labelled with stable isotopes may be used in three different ways for pharmacokinetic studies.-
(1) Stable isotope dilution: 'analytical' internal standard.
(2) Administration of labelled drug alone.
(3) Co-administration of unlabelled and labelled forms of a drug - 'biological' internal standard.

The present survey includes the choice of labelled molecule and some illustrative applications.

Stable isotope dilution is frequently employed if extreme sensitivity and specificity are required for the quantification of a drug in biological material. The labelled compound - not used *in vivo* - is added to the biological sample as early as possible. It serves as internal standard (i.s.), identical to the analyte in properties (phase distribution, extraction, derivatization reactions, chromatography; minor exceptions will be mentioned later) but distinguishable by the detector [1, 2].

The emphasis in this survey is on the use of stable isotopes *in vivo*. Administration of the pure labelled drug is a seldom used way, very elegant, to quantify drug metabolites by reverse isotope dilution [3]. The i.s. can be unlabelled metabolite(s), if available, obviating the need to synthesize a labelled i.s. for every metabolite under study. Pharmacokinetic studies, clinical aspects of which have been reviewed [4], entail co-administration of unlabelled and labelled forms of a drug. This allows the comparison of different pharmaceutical or different stereochemical forms of a drug or different routes of administration under identical biological conditions. Constant biochemical conditions are especially important for drugs with a high first-pass effect or under conditions of enzyme induction. In both instances, the results obtained from classical studies may reflect mainly changes of the biological test system. With stable isotopes, the kinetics of a drug can be measured in patients at steady state without interrupting therapy by substituting a normal dose for one of the labelled drug. Pharmacokinetic parameters from single-dose studies with healthy young volunteers are essential but may not be representative of patients under long-term treatment, especially if the regimen includes enzyme-inducing drugs.

Our survey concentrates on analytical methodology, especially points that are hardly mentioned elsewhere but are critical for the outcome of a study employing stable isotopes particularly in the human context, on which the examples are focused. The methodology was developed earlier in animal studies or applied to enzyme preparations, but this is not exemplified. The co-administration approach can give dramatic insights into the fate of drugs (or other compounds) in living systems, and important guidance for the long-term treatment of patients, not obtainable by other means. Co-administration excludes biological variations with time, giving fewer but considerably more meaningful observations.

We consider in turn the choice of labelled molecule, the question of 'isotope effects', and limits of detection with 'real' samples. [Besides this vol., earlier vols. contain relevant material, indexed under 'Stable......' and starting with #E-1 by B.J. Millard in Vol. 5; only minimal attention is given to pharmacokinetic applications.-*Ed*.]

SELECTION OF THE 'BEST' LABELLED MOLECULE FOR A PHARMACOKINETIC STUDY

For a real study the ideal labelled molecule will probably be too expensive or too complicated to synthesize. But that is no reason to accept any labelled molecule rather than select with discrimination. Before the synthesis is started, the analytical chemist should raise several important points; they may mean additional work for the synthetic chemist, but may help obviate wasted effort. Five questions must be answered at the outset.

1. What is the intended role of the labelled compound ?

The compound may be used as an analytical i.s. or may be intended for human use. The former role has the fewest restrictions. The label has to be stable under the conditions of work-up and analysis. The compound has to be labelled in such a way that the detection is not disturbed under the analytical conditions used. For human use there are additional requirements, besides harmlessness and legal requirements that differ for each country. The label should be stable in the biological environment, with no loss under conditions of oxidation, reduction, transamination and possibly re-utilization or enterohepatic circulation. Also, the label should not alter properties of the molecule (isotope effects, maybe shown by deuterium).

2. What analytical instrument will be used for the detection ?

The microwave plasma detector (discussed by Uden, #C-2 later in this vol.) is a GC detector capable of simultaneously monitoring the hydrogen and deuterium content of the GC effluent. Other isotopes cannot be detected in extracts from biological samples. The anti-arrhythmic drug propafenone, labelled with one ^{14}C and 5 deuterium (D) atoms, has been used to evaluate and confirm the usefulness of this detector for samples obtained from pharmacokinetic and metabolic studies [5].

Considering only mass spectrometry (MS), the choice of labelled molecule may be constrained by the type of imstrument. If the molecular ion is the base peak under electron-impact (EI) conditions, any label may be used. If a fragment ion has to be used for quantification, this ion has to contain the label. Therefore certain parts of the molecule cannot be labelled (example in Fig. 4 below), even if labelling in such a part of the molecule is highly desirable from a synthetic point of view. During methane chemical ionization (CI), *N*-methyl groups may be exchanged. Under fast atom bombardment (FAB) conditions, reactions between the matrix and the compound under investigation may occur.

For 5 years we have used [^{2}H$_{14}$]valproic acid (VPA-D-14; Fig. 1) as the i.s. for the valproic acid determination [6]. In MS-EI the McLafferty fragment of the methyl ester is measured: this ion contains 8 D atoms (7 from the retained chain and one from the γ position) and occurs at mass 124. No severe interferences have been seen in this mass-124 trace in more than 60 kinetic investigations performed. Recently we subjected some samples, already measured under EI conditions, to re-measurement using CI with methane, isobutane or ammonia as the reactant gas. The (M + H)-ion trace (m/z 173) was severely disturbed by interferences. The 14 deuterium atoms gave the i.s. the mol. wt. of a homologue fatty acid (the work-up was for fatty acids). Thus the measurement at a higher mass and in the CI mode – both conditions generally believed to be more selective than EI –

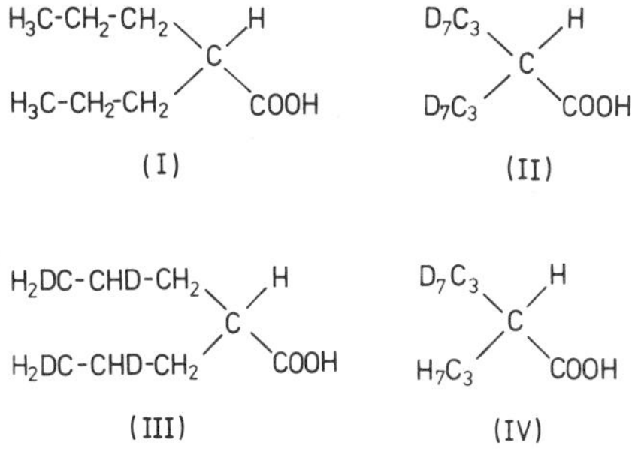

Fig. 1. Formulae of valproic acid (VPA; I) and 3 of the deuterated VPAs synthesized: [^{2}H$_{14}$]VPA (VPA-D-14; II) as i.s., [^{2}H$_{7}$]VPA (VPA-D-4; III) for pharmacokinetic studies, and [^{2}H$_{7}$]VPA (VPA-D-7; IV) as i.s. for MS-CI.

gave more interferences with this type of label. [If we were to change to a CI assay, we would use VPA-D-7 as i.s. (Fig. 1).]

3. Which isotope should serve as the label ?

The elemental formula indicates which isotopes can be used. As every organic molecule contains hydrogen and carbon, deuterium (^{2}H) and ^{13}C may serve as examples; the arguments for or against ^{15}N, ^{17}O, ^{18}O and other isotopes are similar.

Deuterium is always the first choice. It is relatively cheap and readily available (more than 1000 deuterated compounds can be selected from catalogues and considered as starting materials). Catalytic deuteration allows introduction of a label in a late step of the synthesis. Altogether 3 additional mass units are introduced by a perdeuterated methyl group (^{2}H$_5$), 5 by a perdeuterated ethyl group, and only 2 by a ^{13}C-labelled ethyl group.

The time required for the synthesis also has to be considered. If the introduction of 2 mass units via an ethyl group is required, reduction of ordinary acetic acid with lithium aluminium deuteride may be used, or else of [^{13}C$_2$]acetic acid with lithium aluminium hydride, the time required being equal. For the introduction of 2 mass units with a longer carbon chain, the deuteration of any fatty acid is as fast as that of acetic acid. But to insert 2 ^{13}C atoms in a chain requires the synthesis of the carbon chain first and is much more time-consuming. Therefore for an i.s. deuterium is always the first choice.

Deuterated drugs are the least desirable of all stable isotope labelled drugs for use in humans. Firstly, deuterium is toxic in kg amounts. However, the natural body content of 1 g of deuterium for a 70 kg person is increased by only a few percent by a pharmacokinetic study. To the best of our knowledge there is no toxicological problem, but a psychological problem remains. There are no problems for ^{13}C, ^{15}N or ^{18}O [7]. Secondly, hydrogen is lost more easily by metabolic attack than carbon atoms. The biological stability of the label is essential for most pharmacokinetic applications. Finally, if there are isotope effects, they are greatest with deuterium because of the mass increase by 100% from H to D. One consequence of isotope effects is the so-called metabolic switching, described by M.G. Horning [8], where the metabolism by one pathway is slowed down. Alternative pathways take care of the unmetabolized compound. As one consequence, the intensity pattern of metabolites changes. Metabolic switching is known to be important in oxidative N-dealkylations. Therefore one should try to avoid deuteration in positions known to be sites of such oxidative attack.

^{13}C is the best choice for psychological reasons, for reasons of biological stability and because of the essential absence of distur-

bing isotope effects. It is a reasonable candidate for the introduction of several mass units since it is available in isotopic
purity of >98%. But the synthesis of the desired compound may be
troublesome.

4. What isotopic purity is required ?

Complete isotopic purity is a highly desirable goal. ^{13}C comprises ~1.1% of any carbon in nature; therefore the 'unlabelled'
drug is always partially labelled. If the 'labelled' material contains much unlabelled material, overlap of the two forms may affect
the limit of detection (see below).

5. Which position(s) should be labelled ?

This question is discussed using valproic acid as an example.
The metabolic degradation products are known [9].   No position is
resistant to attack. Glucuronidation, introduction of hydroxyl or
oxo groups or of double bonds, and oxidative generation of carboxyl
groups occur simultaneously. The best label is undoubtedly ^{13}C. But
which positions should be selected if ^{2}H is to be used? The α posi-
ition is chemically unstable. The β position is the site of attack
by β-oxidation, the main metabolic pathway. The remaining 2 positions are the site of the less important ω-oxidation. Therefore we
introduced the ^{2}H here (Fig. 1), leaving one or two H atoms to preserve enzyme susceptibility.

These 5 questions are not independent. The success and the
elegance of an experiment with stable isotopes depends on all of
these criteria and their interaction.

ISOTOPE EFFECTS AND THEIR IMPORTANCE IN PHARMACOKINETIC STUDIES

Isotope effects may be a disturbing factor in pharmacokinetic
studies. Demonstration of their absence under the chosen experimental conditions should be the aim of the first experiment, as in
our trial of VPA-D-4 (Fig. 1) given to volunteers as a solution containing an equimolar amount of unlabelled drug. The two forms of
the drug are indeed identical in absorption, distribution and decay
[7]; hence we could use the labelled form for our pharmacokinetic
studies [10]. The fact that the overall pharmacokinetics is identical does not necessarily mean that isotope effects are not present.
It might well be that there is a kinetic deuterium isotope effect in
the synthesis of the 4-keto metabolite or during the synthesis of
the VPA carboxylic acid by ω-oxidation. But the two metabolites are
formed in small amounts only, and even if their formation was slowed
to half of the original rate, the overall pharmacokinetics of the
parent drug would not change noticeably.

There are a few interesting cases where isotope effects are actually desired. They belong to the field of metabolic or mechanistic studies [2] and are beyond the scope of this survey.

WHAT DETERMINES THE DETECTION LIMIT WITH REAL SAMPLES ?

There is the physical limit of detection imposed by the compound to be measured and by the analytical instrument. At the outset the minimum amount of pure compound to which the instrument responds has to be established in a feasibility study. The actual limit in a biological sample can be guessed as 10 times higher, if unskilled personnel are performing the work-up and analysis. Multiplication by another factor of 10 gives a safe limit of detection for routine measurements, performed by medical students on hundreds of samples.

As the problems of interferences in biological samples have been repeatedly considered in this series (notably Vol. 10, besides articles in the present vol.), we will mention only the influence of isotopic purity on the detection limit. Assuming that we measure a labelled compound 3 mass units heavier than the unlabelled one, through introduction of 3 isotopic atoms of 99% isotopic purity, $3 \times 1\%$ of the molecules are not labelled by 3 mass units. If the isotopic purity was only 95% – usual for ^{15}N until recently – 15% of the molecules are not truly labelled. If the isotopic purity is only 90% – usual for ^{13}C until recently – 30% of the molecules do not have the desired mass; the limit of detection is 30% lower.

Another important effect in certain applications is the content of natural label in the 'unlabelled' material. If labelled and unlabelled drug are present in equal amounts, the calibration curve compensates for that contribution. But measuring kinetics in the steady state after a pulse dose of labelled drug results in the measurement of very small concentrations of the labelled drug in the presence of a high concentration of the unlabelled drug. The contribution of the 'natural' label to the intensity of the peak m/z 119 (116 + 3) is 0.11% of the intensity of the McLafferty fragment peak for the 'unlabelled' VPA methyl ester [6]. This tiny contribution of 0.11% sets the detection limit for the VPA-D-4 (measured fragment ion m/z 119) to 1 part in 1000 parts of unlabelled drug.

ILLUSTRATIVE APPLICATIONS

Having considered how to design good studies with stable isotope labelled drugs, we present an example for each of the four typical types of pharmacokinetic study in man where stable isotopes offer distinctive advantages over conventional techniques. Relative and absolute bioavailability studies, the pharmacokinetics of the stereoisomers of racemic drugs, and the pharmacokinetics of drugs during long-term treatment, all rely on the co-administration of unlabelled and labelled drug.

The important step in all of these stable-isotope studies is the transformation of the mass label into a label for something else. The mass label serves thereafter as a time marker, as a label for the site or route of application, or for a galenic formulation, etc. Different pharmaceutical or stereochemical forms of a drug, or different routes of administration, can be compared simultaneously under truly identical biological conditions.

Relative bioavailability of retard preparations of VPA

If test results are largely a function of the variability of the human test system, single measurements are meaningless and very large numbers of volunteers are required to obtain statistically meaningful data. We have compared experimental and commercially available forms of sustained release formulations of VPA. (In parallel they were subjected to *in vitro* measurements by Dr. Loehr from Ciba Geigy in Wehr.) Fig. 2 exemplifies concentration-time curves after a volunteer simultaneously swallowed an enteric coated tablet and a solution of VPA-D-4. The delay time – time spent in the stomach, where this tablet is insoluble – was different for different volunteers. Other retard formulations gave a slower rise in blood VPA. Finally, one marketed formulation was so perfectly retarded that only part of the drug was released during the transit time in the intestine. The *in vivo* and the *in vitro* results agreed for every formulation tested (unpublished work).

The important advantage gained by the simultaneous application of the drug under study and the solution of labelled VPA was the reliability of the information gained by only a very few experiments. This aspect has been discussed in detail by d'A. Heck et al. for Imipramine tablets [11]: less than 10% of the normal number of volunteers was required to obtain equally reliable data if labelled drug were applied simultaneously in the bioavailability study.

Pharmacokinetics of verapamil after i.v. and oral application

Fig. 3 shows comparable peak concentrations with the stated doses of i.v. verapamil and oral deuterated verapamil. This drug shows a high first-pass effect (almost 80% metabolized during first pass through the liver), such that constant biochemical conditions are especially important ([12]; likewise important where there is enzyme induction). Only with this methodology could absolute bioavailability be measured, since small physiological changes (e.g. liver blood flow) or changes of the metabolic capacity of the liver alter the amount of drug reaching the systemic circulation [4, 12].

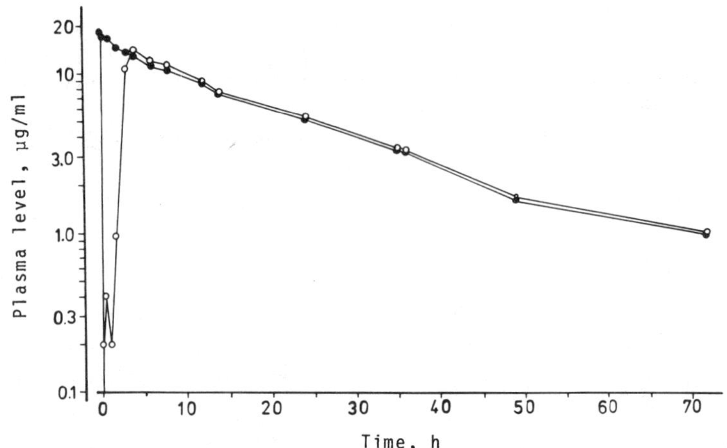

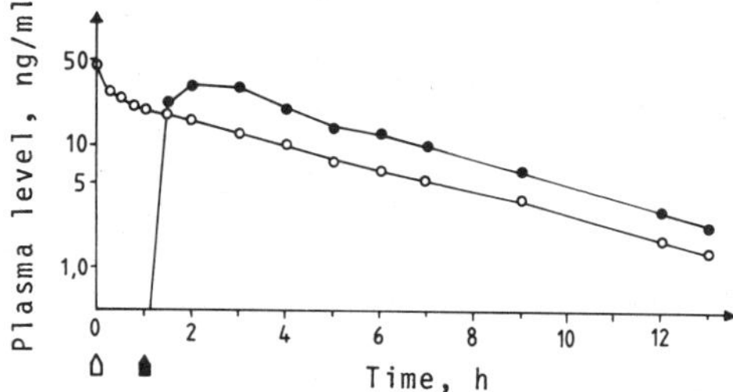

Fig. 2 *(above)*. VPA levels in a volunteer after swallowing 150 mg of VPA Na salt as an enteric coated tablet and the same dose of $[^2H_4]$VPA Na salt in solution (*open circles* and *solid circles* respectively). The relative bioavailability determined in this experiment was 95% for VPA.

Fig. 3. Racemic verapamil levels in a volunteer after 10 mg veramil i.v. and 80 mg $[^2H_3]$veramil as a solution orally (*solid* and *open circles* respectively) *as arrowed*.

Pharmacokinetics of pseudoracemates of verapamil

Fig. 4 shows the partial mass spectra of two of the labelled verapamil molecules we use. The quanternary carbon atom makes the molecule chiral. In the racemate studies $[^2H_3]$verapamil (verapamil-D-3) was used; for the studies with the labelled stereoisomers the drug molecules with two deuteriums (position shown in Fig.) were synthesized by Dr. E. Schmidt and Mrs. Vogelgesang. The chiral starting material was a gift of the Knoll AG, Ludwigshafen.

In a volunteer dosed orally with the two stereoisomers, one with deuterium label, there were different concentrations (Fig. 5) due to the ~3 times larger volume of distribution for the L-verapamil and to the stereospecific first-pass metabolism (experiments by B.M. Vogelgesang, M. Eichelbaum & E.K. Schmidt; the L-form is preferentially metabolized).

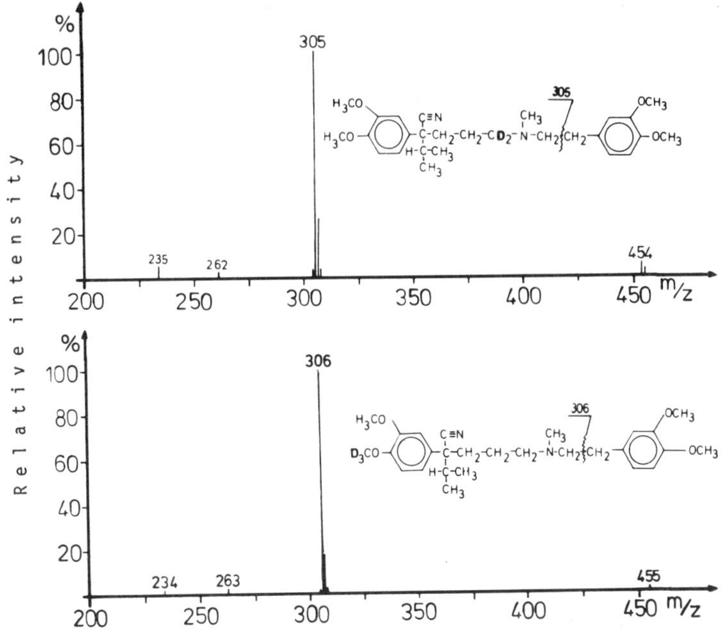

Fig. 4 *(above)*. Partial 20 eV EI mass spectra of deuterated verapamil (+) (*top*; [^{2}H$_2$], mol. wt. 456, no M+) and racemic (*bottom*; [^{2}H$_3$], mol. wt. 457, no M+), recorded with a LKB 2091 mass spectrometer.

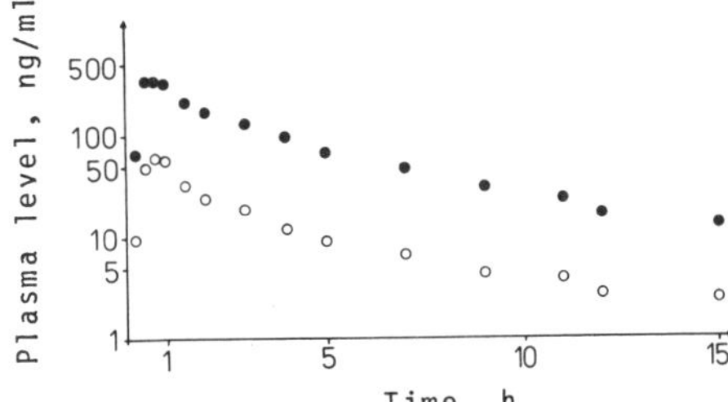

Fig. 5. Levels of unlabelled L-(-)-verapamil *(open circles)* and [^{2}H$_2$](+)-verapamil (*solid circles*) in a volunteer after swallowing 86 mg of each hydrochloride in 100 ml carbonated water.

Kinetics of carbamazepine in patients during steady state

Fig. 6 shows that the kinetics of a drug can be measured in patients at steady state without interrupting therapy. The plots are *calculated* (solid line = unlabelled drug plus once-administered labelled drug). Note that the steady state is not disturbed by the experiment. Individual curves for different patients on carbamazepine therapy have been published [e.g. 13]. During continuous treatment, carbamazepine induces its own metabolism, and hence plasma concentrations decrease. In Fig. 6 the curve for the labelled drug

Fig. 6. Calculated plasma-level time courses for a patient in steady state receiving (at 8 a.m.) one pulse dose of the labelled drug instead of his regular dose (dose times *arrowed*). ······, labelled drug; ---, unlabelled; ——, summation (total drug). Calculation assumptions (representative for VPA monotherapy): 70 kg human, 600 mg VPA-Na 8 a.m. & 8 p.m.; 1st order kinetics; distribution vol. 0.15 l/kg; ½-life 10 h.

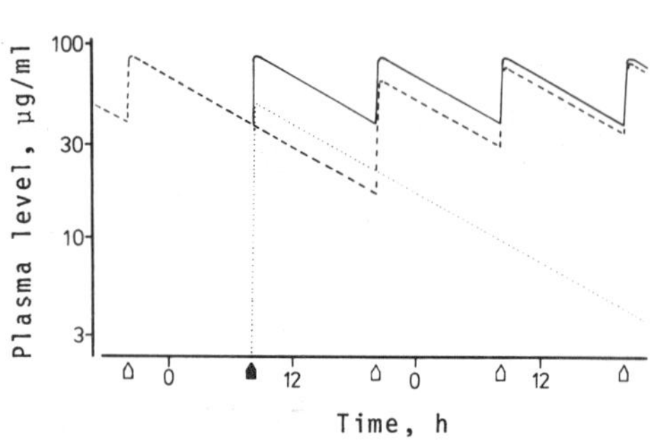

would have a lower maximum and decline more steeply. The half-life for carbamazepine is ~28 h after a single dose, ~12 h under mono-therapy and only ~8 h under combined antiepileptic therapy [13].

This is an example which is representative for several antiepi-leptic drugs. Pharmacokinetic parameters from single-dose studies in healthy volunteers are essential, but are often unrepresentative of patients under long-term treatment. This is especially true if enzyme-inducing drugs are included in the regimen.

CONCLUDING COMMENTS

The examples shown from our work as well as many studies from other groups have firmly established the advantages to be gained from pharmacokinetic studies using suitable stable isotope labelled drugs. The steady-state situation can be measured in patients. Time-dependent changes in drug disposition are minimized. Fewer subjects are required for bioavailability studies. Studies normally conducted on two occasions can be performed only once. These advan-tages translate into less inconvenience for patients and volunteers together with less variability in pharmacokinetic parameters.

Acknowledgements

Experiences and results described have been gained in the clinical pharmacology group of the Department of Internal Medicine of the University of Bonn. We thank our medical colleagues, our technicians and our students. Our mass spectrometer was a donation from the Dr. Robert Pfleger-Stiftung, Bamberg; the computer system came from grant PTB 8303, Bundesministerium für Forschung und Tech-nologie.

References

1. Millard, B.J. (1978) *Quantitative Mass Spectrometry*, Heyden, London, 160 pp.
2. Draffan, G.H. (1978) in *Stable Isotopes, Applications*........ (Baillie, T.A., ed.), Macmillan, London, pp. 27–42.
3. Baba, S., Kato, S., Morishita, S. & Sone, H. (1978) *J. Med. Chem. 21*, 525–529.
4. Eichelbaum, M., von Unruh, G.E. & Somogyi, A. (1982) *Clin. Pharmacokin. 7*, 490–507.
5. Hege, H.G. & Weymann, J. (1982) in *Stable Isotopes* [*Proc. 4th Int. Conf.*, Juelich, 1981] (Schmidt, H.-L., Foerstel, H. & Heinzinger, K., eds.), G. Fischer, Stuttgart, pp. 679–684.
6. von Unruh, G.E., Jancik, B.C. & Hoffmann, F. (1980) *Biomed. Mass Spectrom. 7*, 164–167.
7. Wolf, D., Cohen, H., Meshorer, A., Wasserman, I. & Samuel, D. (1979) in *Stable Isotopes [Proc. 3rd Int. Conf.]* (Klein, E.R. & Klein, P.D., eds.), Academic Press, New York, pp. 353–360.
8. Horning, M.G., Haegele, K.D. & Brendel, K. (1978) in *Stable Isotopes, Applications in Pharmacology, Toxicology & Clinical Research* (Baillie, T.A., ed.), Macmillan, London, pp. 55–64.
9. Gugler, R. & von Unruh, G.E. (1980) *Clin. Pharmacokin. 5*, 67–83.
10. Hoffmann, F., von Unruh, G.E. & Jancik, B.C. (1981) *Br. J. Clin. Pharmacol. 19*, 383–385.
11. d'A. Heck, H., Buttril, S.E., Flynn, N.W., Dyer, R.L., Anbar, M., Cairns, T., Dighe, S. & Cabana, B.E. (1979) *J. Pharmacokin. Biopharm. 7*, 233–248.
12. Eichelbaum, M., Somogyi, A., von Unruh, G.E. & Dengler, H.J. (1981) *Eur. J. Clin. Pharmacol. 19*, 133–137 (see also 127–131).
13. Eichelbaum, M., Köthe, K.W., Hoffmann, F. & von Unruh, G.E. (1982) *Eur. J. Clin. Pharmacol. 23*, 241–244.

#A-4

ISOTOPE-LABELLED MATERIALS AS INTERNAL STANDARDS

Robin Whelpton

Department of Pharmacology & Therapeutics
The London Hospital Medical College
Turner Street, London E1 2AD, U.K.

The use of isotope-labelled drugs as internal standards, or to elucidate some analytical problem, is a very powerful technique. The approach is often quicker and easier than alternative methods and sometimes is the only solution to the problem. Chromatographic separation followed by mass spectrographic (MS) detection of test and isotopic labelled materials is generally considered to be the definitive analytical technique but is not without its difficulties and pitfalls.

Experiences gained in the development of a GC-MS assay of trifluoperazine are discussed. This involved the use of stable and radioactive isotope labelled compound. The potential problems due to isotope effects are exemplified by the decomposition of [^{3}H]fluphenazine and references to the literature. A role for 'enantiomeric dilution' analysis of chiral compounds, with its similarity to the stable isotope approach, is proposed.

The concept of two molecules which are chemically identical but which can be distinguished by some physical technique is obviously very appealing to the analyst. Molecules containing stable isotopes are differentiated by MS (usually using selected ion monitoring, SIM) and radioactive molecules quantified by measuring the emissions (usually by liquid scintillation counting). Normally the sample has to be divided and total and radioactive drug determined separately — a disadvantage of the radioactive approach compared to MS where the same aliquot serves for determining both labelled and non-labelled compound. The nuclides suitable for incorporating in drug or biologically interesting molecules include ^{2}H (deuterium, D), ^{13}C, ^{15}N and ^{18}O (stable isotopes), and ^{3}H (tritium, T), ^{14}C, ^{32}P, ^{35}S and ^{131}I (radioisotopes). [An accompanying art. (G.E. von Unruh) is relevant and #A-6 in Vol. 7; consult Index, e.g. 'Stable....', in earlier vols.–*Ed.*]

AVAILABILITY AND PREPARATION OF LABELLED MOLECULES

Several radioactive 'tagged' drugs are commercially available. Their number is increasing steadily - probably in response to growing interest in RIA and receptor-binding studies. Pharmaceutical companies are another source. Alternatively, the compound can be synthesized; this calls for specialization and usually 'custom-synthesis'. However, stable isotope material can often be prepared from readily available starting materials. High-purity D-labelled reagents such as acetic anhydride, acetyl chloride, iodomethane and lithium aluminium hydride are relatively inexpensive and available from laboratory chemical suppliers. (An extensive list of stable isotope labelled materials, including some drugs, is available from Merck, Sharp & Dohme or their agents.)

For SIM the analyte and internal standard (i.s.) need to differ in mass by about 3 units. This reduces the interference resulting from natural isotopes, which would otherwise limit the sensitivity of the assay. Natural carbon contains ~1% ^{13}C, but the proportion of 'heavier' isotopes is greater for S, Cl and Br. Replacement of methyl by CD_3 –is a convenient way of increasing the mass by 3 units. This was the approach we adopted to produce the i.s. for our trifluoperazine assay [1]. *N*-desmethyltrifluoperazine was prepared from 2-trifluoromethylphenothiazine according to Scheme 1. The first step was carried out using sodium hydride in dimethylformamide (DMF) rather than by the published method using sodium in liquid ammonia [2]. Methylation with deutero-iodomethane gave the required compound which initially was purified by TLC but in a subsequent preparation was crystallized as the dimaleate.

Total synthesis may be avoided by modifying the drug to be assayed. For example, it would be possible to demethylate chlorpromazine via its amine oxide [3] and re-methylate with labelled iodomethane. Normorphine is obtained by treating morphine with cyanogen bromide [4]. Intermediates can be obtained by metabolism, or i.s. samples for drug metabolites, e.g. 7-hydroxychlorpromazine, obtained merely by metabolism of labelled drug [5]. A potential problem when starting with the drug to be assayed and suitably modifying it is reduced isotopic purity due to contamination by starting-material.

USE OF INTERNAL STANDARD

The use of isotopic internal standards is a development of isotope dilution analysis in which a known quantity of radioactive compound is added to a 'pool' of chemically identical but unlabelled compound. A portion of the 'pool' is assayed and the new specific activity calculated, from which the amount of material originally present can be obtained. By adding a constant amount of labelled i.s. to all samples, including calibration standards, a calibration curve for the unknown/i.s. response ratio *vs.* concentration can be

2-trifluoromethylphenothiazine + 1,3-bromochloropropane

$\downarrow$ NaH/DMF

3-(2-trifluoromethylphenothiazinyl-10)-propyl chloride

$\downarrow$ piperazine

N-desmethyltrifluoperazine

$\downarrow$ CD_3I

Scheme 1.
Preparation
of an i.s.

$[^2H_3$-*methyl*]trifluoperazine

constructed. Although specific activity is not calculated as such, the assay is in essence by an isotope dilution approach.

The i.s. is added at the earliest possible stage – e.g. to the plasma sample or tissue homogenate – and taken through the entire sample-preparation procedure. It is assumed (or is shown) that the isotopic material mixes homogeneously and is chemically identical to the non-labelled compound being assayed. With SIM of stable isotopes, the quantification of unknown and i.s. can be regarded as two separate, parallel assays. For trifluoperazine the respective molecular ions, m/z 407 and 410, were monitored. The amount of i.s. that can be added is related to the abundance of fragments with the same m/z as the test compound, in this case 407. Thus, the higher the isotopic purity, the greater can be the difference in concentrations of unknown and i.s. Using an i.s. with an isotopic purity of ~99%, trifluoperazine could be assayed in the presence of a 100-fold excess of i.s. The i.s. acts as a carrier, and the calibration curve is constructed over a narrow range of *total* concentrations – in this case between 127 and 63 ng per tube.

The approach with a radioactive i.s. is somewhat different. The minimum amount of material is used because the assay cannot differentiate between 'cold' and labelled drug – hence the need for high specific activity material. To this end, nuclides with short half-lives are an advantage. The sample is usually divided for separate assay of total and radioactive material. With HPLC the same aliquot may be used, the radioactivity being determined in a collected fraction or with a flow-through scintillation counter. If the specific activity of a radioactive drug is not high enough for use as an i.s., it may have a role in elucidating analytical problems or determining absolute recoveries. $[^3H]$trifluoperazine was used to estimate the recovery of added material. Radioactive trifluoperazine was also used to show that difficulties encountered in attempting to develop a GC-NPD method were largely due to adsorptive losses from aqueous solutions (Table 1). Silanizing the glassware reduced the adsorption but increased the formation of emulsions, so that smaller aliquots had to be taken and there was no overall gain.

Table 1. Adsorption of trifluoperazine dihydrochloride (100 ng/ml) from aqueous solutions. Values are % remaining at the stated time.

Material	5 min	15 min	30 min	60 min	18 h
Soda glass	72.7	67.8	52.5	41.6	15.5
Pyrex glass	78.2	70.1	60.3	52.9	24.0
Pyrex glass, silanized	87.4	84.7	76.4	75.3	73.6
PTFE	84.9	67.9	58.3	47.3	24.8

Note on the glassware: it was routinely pre-treated with chromic acid and rinsed.

ISOTOPIC EFFECTS AND OTHER PROBLEMS

So far it has been assumed that labelled and non-labelled compounds are chemically identical. This is not the case, but with the correct choice of nuclide and site of labelling, isotopic effects can be minimized. Deuterium-labelled compounds elute from GC columns slightly earlier than the corresponding non-labelled material, and almost complete separation by HPLC has been reported for one pair of compounds [6]. These effects, once noted, should present little difficulty.

Potential difficulties arise when bonds linking the isotope are broken - e.g. during decomposition or chemical reaction - as these are more stable. Substituting a heavier atom reduces the vibrational energy of the bond and more energy is required to break it. The phenomenon is more marked for ^{1}H, ^{2}H and ^{3}H, as the mass varies by 2- or 3-fold [7]. The effect is well known to organic chemists who deliberately label molecules such that the bond linking the isotope will be involved in the chemical reaction. The reaction is qualitatively the same but its rate is reduced compared with that for unlabelled material. When radioactive compounds are used there is usually an increase in specific activity, assuming that not all the molecules are labelled and that the reaction does not go to completion.

At this point it is worth considering what happens to the specific activity of a radioactive sample during one half-time. If the proportion of radioactive compound is almost negligible (say < 1%) the total remains constant but the amount of radioactive molecules falls to one-half - i.e. the specific activity is reduced to half in one half-time. However, if all the molecules contain a radioactive atom the *total* declines at the same rate as the radioactive decay but the specific activity is unchanged. When a significant proportion of molecules are labelled the change in specific activity is between the two extremes: thus if, initially, 50% are labelled the specific activity falls by one-third, not one-half, in the first half-time.   When purifying some tritiated fluphenazine for use as an internal standard, isotopic effects in decomposition were noted. It was supplied in August 1976 at 18.8 Ci/mmol (after TLC purification and UV quantitation).   In April 1982 TLC showed extensive decomposition, and an aliquot was purified and quantitated by HPLC-UV: the specific activity of the collected material was 28.9 Ci/mmol. To show that the difference was not due to a different method of purification, the material was also purified by TLC (August, 1982): the specific activity as ascertained by HPLC-UV was 27.8 Ci/mmol. Non-radioactive fluphenazine decomposed more rapidly than radioactive, until all the molecules contained one tritium atom (28.9 Ci/mmol represents 100% with one ^{3}H). As far as I know, this is the first time that isotopic effects have been described with regard to decomposition of radioactive compounds.

GC-MS is probably the most specific technique available; but, even so, selectivity problems were encountered during the development of the trifluoperazine assay.   On OV-17 columns an impurity with a molecular fragment of m/z 410 chromatographed at almost the same retention time as deuterium-labelled trifluoperazine. By monitoring other fragments, m/z = 411, 412, etc., the impurity was shown not to be due to contamination of the extraction tubes with labelled drug. The impurity and [^{2}H$_3$]trifluoperazine were resolved on OV-225 columns.

Differences between stable and radioactive internal standards are summarized in Table 2.

ENANTIOMERIC DILUTION ANALYSIS

With the recent 'explosion' of papers detailing the development of 'chiral' phases for HPLC and GC, it is easy to believe that soon there will be chromatographic methods to resolve all chiral drugs. For drugs or natural products which are marketed as one enantiomer, enantiomeric dilution analysis will be a possibility. The rationale is very similar to the use of stable isotope internal standards. Provided that the enantiomeric purity is high, one enantiomer could be used at relatively high concentrations as a carrier/i.s. in the assay of the other isomer. The i.s. would have to be kept away from other chiral molecules with which it might interact: thus, it might not be appropriate to add it to the biological sample. Within certain constraints, enantiomeric dilution analysis could be as powerful as GC-MS but much less costly.

References

1. Whelpton, R., Curry, S.H. & Watkins, G.M. (1982) *J. Chromatog.* *228*, 321-326.
2. Sherlock, M.H. & Sperber, N. (1961) *U.S. Patent 2,985,659.*

Table 2. Comparison of the use of stable and radioactive isotopic internal standards.

Aspect	Stable	Radioactive
Preparation	standard methods	more specialized
Proportion of i.s. used	greater than unknown, e.g. $\times 100$	negligible compared with unknown, e.g. $< 1\%$
Detection	MS of i.s. and unknown together	Separate detection of i.s. radioactivity
Problems	isotope effects; hydrogen exchange	isotope effects; hydrogen exchange; radiolysis; adsorption, etc.

References, continued

3. Curry, S.H. & Evans, S. (1976) *J. Pharm. Pharmacol.* *28*, 467–468.
4. Lednicer, D. & Mitscher, L.A. (1977) *Organic Chemistry of Drug Synthesis*, Wiley, New York; see p. 291.
5. Alfredsson, G., Wode-Helgodt, B. & Sedvall, G. (1976) *Psychopharmacol.* *48*, 123–131.
6. de Ridder, J.J. & van Hal, H.J.M. (1976) *J. Chromatog.* *121*, 96–99.
7. Ander, P. & Sonnessa, A.J. (1965) *Principles of Chemistry,* Collier-Macmillan, London; see pp. 747–750.

#NC(A)

NOTES and COMMENTS relating to

Sample handling and usfulness of isotopes

Comments related to particular contributions

#A-1, #A-3 & #A-4, p. 61
#NC(A)-2, p. 62
(See p. 375 for points relevant to #A-2)

#NC(A)-1

A Note on

APPROACHES FOR DETERMINING ETHYLENEDIAMINE AND ITS
METABOLITES: USE OF [^{14}C]- AND [D$_4$]-ETHYLENEDIAMINE

John Caldwell and Ian A. Cotgreave

Department of Pharmacology
St. Mary's Hospital Medical School
London W2 1PG, U.K.

Ethylenediamine (1,2-diaminoethane) is a small, highly basic molecule of some industrial importance. It is used in the manufacture of various polymers, resins and dyestuffs, is present in cutting oils and wetting agents, and is used in pharmaceutical formulation. Our interest in ethylenediamine stems from its presence in the widely used bronchodilator drug, aminophylline, which consists of theophylline and ethylenediamine in a molar ratio of 2:1. The ethylenediamine component, unlike the majority of excipients used in pharmaceutical formulation, is not devoid of biological activity [see 1], and is thus worthy of some attention. There is very little information in the literature on the disposition of ethylenediamine in animals or man, and we have therefore made a detailed study of the metabolic fate of this molecule in the rat and in human volunteers.

Ethylenediamine and its metabolites pose great problems of detection to the analyst, since they lack suitable physicochemical characteristics. Ethylenediamine has little of no UV absorbance, gives a poor FID response in GC, and cannot be extracted into organic solvents from aqueous solution. Taken together, these properties render the measurement of ethylenediamine difficult. Accordingly, we have developed an assay for ethylenediamine in body fluids [2] in which it is converted into its *N,N'*-di-*m*-toluoyl derivative in the biological matrix, by treatment with *m*-toluoyl chloride. This derivative, unlike the parent molecule, is readily extracted from aqueous solution into dichloromethane. After removal of the solvent *in vacuo* and reconstitution of the residue, the derivative is then assayed by HPLC-UV on a reversed-phase C-18 column with 40% (v/v) aqueous acetonitrile as mobile phase. The internal standard (i.s.), cadaverine (1,5-diaminopentane), is added to the biological fluid before addition of *m*-toluoyl chloride, and is likewise converted into its

di-*m*-toluoyl derivative.   Ethylenediamine in plasma is assayed directly as described, but for urine a preliminary extraction at pH 14
is required to remove bases which otherwise interfere.

The availability of these assay methods has permitted for the
first time the investigation of the pharmacokinetics of ethylenediamine in man. Aminophylline was given to volunteers i.v. and orally
(on separate occasions) and blood samples taken at regular intervals.
Plasma was separated and assayed for ethylenediamine as described [2].
After i.v. injection, ethylenediamine was very rapidly distributed
through a relatively small volume of distribution, and was rapidly
eliminated from the plasma, with a beta-phase half-life of ~1 h. In
blood samples obtained after oral administration of aminophylline,
ethylenediamine levels were low, indicating poor bioavailability [3].

Urine from the subjects given aminophylline was also assayed
for ethylenediamine, showing very low recovery of unchanged compound.
Thus, following oral administration only 3% of the dose was excreted
as ethylenediamine in the urine, whereas following i.v. injection
18% was excreted [4]. The urinary content of ethylenediamine rose
markedly following acid hydrolysis, suggesting the presence of an
acid-labile conjugate(s). HPLC of the dichloromethane extract of a
m-toluoyl chloride-treated urine revealed, besides the ethylenediamine derivative, another UV-absorbing compound. It had the same
retention time as authentic *N*-acetyl-*N'*-*m*-toluoyl-ethylenediamine,
and its identity has been confirmed by GC-MS(EI).

FATE OF ETHYLENEDIAMINE IN THE RAT

Following i.p. injection of $[^{14}C]$ethylenediamine into rats,
the major pathways of elimination of ^{14}C were the urine (47% of
dose) and the expired air (18%) as $^{14}CO_2$ [5]. The urinary metabolites
were separated by TLC and by HPLC, the latter following treatment of
the urine with *m*-toluoyl chloride, and identified by comparing their
chromatographic characteristics with those of authentic standards.
The major compounds present in the urine were ethylenediamine itself
(9% of dose), *N*-acetylethylenediamine (31%) and *N,N'*-diacetylethyl-
enediamine (2%; TLC only). Besides these ethylenediamine-related
compounds, both TLC and HPLC revealed the presence of a small amount
of $[^{14}C]$hippuric acid in the urine.

It seems reasonable to ascribe the formation of both $[^{14}C]$hippu-
ric acid and $^{14}CO_2$ to the conversion of ethylenediamine to glycine
by the action of diamine oxidase. However, the ubiquity of glycine
as an endogenous amino acid renders its analysis as a metabolite of
a xenobiotic very difficult, and it has been necessary to test the
'deamination hypothesis' for ethylenediamine metabolism in a number
of ways.

The principal fate of benzoic acid in rats is conjugation with glycine to give hippuric acid. Co-administration of benzoic acid (orally) with ethylenediamine (i.p.) altered the fate of the latter. Urinary ^{14}C excretion rose to 64% of dose, whilst the % dose exhaled as $^{14}CO_2$ fell to 7%. The increased urinary radioactivity was due solely to increased formation of $[^{14}C]$hippuric acid, the acetylation of ethylenediamine being unaffected by the benzoic acid given too.

One important biosynthetic role of glycine is its incorporation into haem; of the 2 carbon atoms present, only that bearing the carboxyl and amino functions is incorporated. Rats were given $[^{14}C]$-ethylenediamine and 7 days later were sacrificed and exsanguinated. The specific activity of erythrocyte haem was measured, presumably reflecting incorporation of glycine derived from ethylenediamine. The haem specific activity was approximately halved in rats which had received benzoic acid together with $[^{14}C]$ethylenediamine. In these animals, radioactivity was also found incorporated into plasma proteins, and again this was reduced by about one-half by the co-administration of benzoic acid. However, this latter result is incomplete evidence of glycine formation from ethylenediamine, which could be transformed to various other amino acids that are also incorporated into proteins.

Further evidence for *in vivo* deamination of ethylenediamine to yield glycine has been provided by use of the diamine oxidase inhibitor, aminoguanidine. Co-administration of this compound with $[^{14}C]$-ethylenediamine resulted in a fall in the amount of exhaled $^{14}CO_2$, with little change in urinary ^{14}C excretion. When aminoguanidine, benzoic acid and $[^{14}C]$ethylenediamine were all given together, the amount of $[^{14}C]$hippuric acid in the urine was slightly reduced compared with animals not given benzoic acid.

Taken together, these various results support the concept that in addition to undergoing extensive *N*-acetylation, ethylenediamine is deaminated in the body to give glycine. This is then incorporated into pathways of biosynthesis and intermediary metabolism. Little has been published on the quantitative aspects of the fate of glycine *in vivo*. So that inferences drawn from the above experiments could be confirmed, rats were given $[^{14}C]$glycine alone and together with benzoic acid and/or aminoguanidine. With glycine alone, 4% of the ^{14}C dose was excreted in the urine in 24 h, mainly as $[^{14}C]$hippuric acid, and 41% in the expired air as $^{14}CO_2$. The pattern of metabolism and excretion of glycine showed co-administration influences similar to those with ethylenediamine, as did the incorporation of glycine-derived ^{14}C into erythrocyte haem and plasma protein.

In vitro studies

Although the above experiments give good evidence for the conversion of ethylenediamine into glycine in the rat, they are all indirect.   Since glycine incorporation is rapid, extensive and by diverse pathways, it is virtually impossible to obtain direct evidence in the whole animal.   For this reason, we have also examined the conversion of ethylenediamine to glycine in rat-liver homogenates.   Incubation led to an enzyme- and time-dependent disappearance of $[^{14}C]$ethylenediamine, accompanied by the appearance of $[^{14}C]$-glycine. This was shown both by ion-exchange chromatography with no pre-conversion, and by HPLC preceded by treatment of the incubation mixture with benzoyl chloride to form N-benzoyl-glycine (hippuric acid).   That the glycine was available for hippuric acid biosynthesis, as had been shown *in vivo* (see above), was demonstrated by fortifying the incubation mixture with the co-factors for amino acid conjugation - ATP, CoA.SH and Mg^{2+} ions - and adding D_5-benzoic acid. HPLC showed the formation of $[^{14}C]$hippuric acid, and GC-MS of this material showed the presence of D_5-hippuric acid.

FATE OF ETHYLENEDIAMINE IN MAN

The finding of deamination to glycine is daunting for studies in man; to use radiolabelled forms of ethylenediamine in man would be impermissible since these would very likely result in radioisotope incorporation into various macromolecules.   Accordingly, for human studies a combination of 'cold' methods, entailing chemical derivatizations, and deuterium-labelled ethylenediamine have been used.

The identification of N-acetylethylenediamine in the urine of volunteers receiving aminophylline (orally or i.v.) has already been mentioned.   The identity of this metabolite was confirmed by the administration of an equimolar mixture of protonated and $[D_4]$ethylenediamine.   Plasma and urine samples were taken at various times following drug administration, and were treated with m-toluoyl chloride.   Extracts of these derivatized samples were examined by GC-MS, and spectra were obtained of N,N'-di-m-toluoylethylene diamine and of N-acetyl-N'-m-toluoylethylenediamine.   In both of these spectra, all ions diagnostic of the compounds were found in pairs, separated by 2, 3 or 4 amu, arising from the mixture of protonated and deuterated forms. The amount of N-acetylethylenediamine present, assayed by HPLC (see above), was very similar to the total of ethylenediamine conjugates, measured following acid hydrolysis.   This indicates that N,N'-diacetylethylenediamine, if present at all, is only a very minor metabolite (<2% of the dose) in man.

The acetylation of ethylenediamine was also examined in a small (n = 12) panel of healthy volunteers, and was found to exhibit considerable inter-individual variation. However, when these volunteers were phenotyped with respect to their genetically-determined capacity to acetylate sulfamethazine, no correlation was seen between the acetylation of this aromatic amine and that of the aliphatic diamine. It thus seems that ethylenediamine should be regarded as one of the so-called 'monomorphic' substrates for N-acetylation.

Since radioisotopes are precluded for studying the fate of ethylenediamine in man, it is extremely difficult to show the conversion of ethylenediamine to glycine in volunteers. Indeed $[D_4]$ethylenediamine will be deaminated to yield $[D_2]$glycine, which will presumably be incorporated in part into urinary hippuric acid; yet the size of the endogenous pool(s) with which the ethylenediamine-derived glycine will mix is such that the enrichment will be virtually impossible to detect. Thus, the normal excretion of hippuric acid is 1-2 g/day (equivalent to 420-840 mg glycine), while even if all of the dose of ethylenediamine is deaminated, this will produce only ~20 mg of glycine to add to the endogenous pool(s). The formation of glycine from ethylenediamine in man has therefore been examined *in vitro* in human liver homogenate.

$[^{14}C]$Ethylenediamine was incubated with whole homogenate of human liver essentially as outlined above for the rat. The disappearance of ethylenediamine and the formation of glycine was shown similarly, with benzoyl chloride treatment of the incubation mixture before HPLC. The activity of human liver in this regard was markedly lower ($<15\%$) than that of rat liver. The enzymic synthesis of hippuric acid could not be demonstrated as for rat liver (see above), probably due to the very low activity of human liver homogenate.

CONCLUDING COMMENTS

Taken together, these various experimental approaches have provided important information about the metabolism of ethylenediamine, a compound of some toxicological significance about whose metabolism little was previously known. These approaches may also be of interest for the more general problem of the incorporation of residues of foreign compounds into endogenous pathways of metabolism and biosynthesis. Although compounds able to participate in intermediary metabolism in such ways are the cause of considerable concern, the literature details few experimental approaches to such problems.

Acknowledgements

This work was supported by a grant from Napp Laboratories Ltd. We are grateful to Dr. P.D. Farmer for assistance with mass spectrometry. We thank Prof. R.L. Smith for his encouragement.

References

1. Cotgreave, I.A. & Caldwell, J. (1983) *J. Pharm. Pharmacol.* *35*, 774-779.
2. Cotgreave, I.A. & Caldwell, J. (1983) *Biopharm. Drug Disp.* *4*, 53-62.
3. Cotgreave, I.A. & Caldwell, J. (1983) *J. Pharm. Pharmacol.* *35*, 378-382.

4. Caldwell, J. & Cotgreave, I.A. (1983) *Br. J. Clin. Prac.,
    Suppl. 23,* 22-25.
5. Caldwell, J. & Cotgreave, I.A. (1983) *Br. J. Pharmacol. 78,*
    62P.

#NC(A)-2

A Note on

SELECTIVE SAMPLE PREPARATION TECHNIQUES FOR TRACE ANALYSIS

Reed C. Williams

Clinical & Instrument Systems Division
E.I. du Pont de Nemours
Wilmington, DE 19898, U.S.A.

The isolation and determination of very small quantities of materials in complex mixtures is a major problem for analytical chemists in many fields.   Historically [cf. R.P. Maickel, #A-1 -*Ed.*], solvent extraction has been the principal technique used for sample preparation; now solid-phase materials such as bonded silicas and small-particle resins are becoming accepted as convenient and efficient extraction sorbents. Two types of these solid-phase materials are now discussed and compared: lipophilic sorbents such as C-18 bonded silica, and anion-exchange resins.

EQUIPMENT AND PACKING MATERIALS

Samples were extracted with the Du Pont PREP^{TM} automated sample processor. The PREP^{TM} (Clinical & Instrument Systems Division, Du Pont) is a centrifugally based, microprocessor-controlled instrument [1].   Extraction takes place within prepared cartridges which fit into a 12-place reversing rotor in the sample processor. An extraction cartridge consisting of a resin column, effluent cup, sample recovery cup and a cap is shown in Fig. 1. Sample aliquots are buffered and placed in the reservoir of the resin column; the remainder of the extraction sequence is done automatically. The sample is extracted onto the sorbent within the column, rinsed with a wash solvent, eluted into the sample recovery cup with an elution solvent, and evaporated to dryness. The entire process takes 10-30 min depending on the chosen extraction program. The dried extract can then be re-constituted for the chosen mode of analysis.

Since the same automated procedure is used for all samples, the extraction variables will be the sorbent, the sample buffer, the wash solvent and the elution solvent. Two types of lipophilic sorbent were used for this study: the Type OD cartridge contains a C-18 bonded silica (~20% w/w of octadecyl hydrocarbon); Type W is packed

with a lipophilic cross-linked polystyrene resin. The Type AS cartridge contains a strong anion-exchanger - a polystyrene resin with
quaternary amine groups, compatible with both aqueous and organic
solvents. Although these are proprietary Du Pont extraction sorbents,
the comparisons and results reported here should be applicable to
other similar lipophilic and anion-exchange extraction sorbents.

APPLICATIONS OF LIPOPHILIC SORBENTS

Lipophilic solid-phase sorbents are the most convenient and
widely used of the new extraction materials [2-4]. [See #C-4 in
Vol. 10, this series, for applicability to pesticide residues in
crops and soils.- *Ed*.]. Commercially available types range from cross-
linked polystyrenes to C-18 bonded silica. The latter, widely used,
can behave as a very deactivated silica for extracting very polar
compounds, but will ordinarily act as a lipophilic sorbent for most
other compounds including those discussed here.

These bonded silicas are easy to use. First they must be acti-
vated by rinsing with a water-miscible solvent such as methanol or
acetone, suitably 1 ml, and rinsed with 1 ml of water. Then, in the
example shown in Fig. 2, 1 ml of blood serum containing 50 ng of
theophylline was buffered to pH 5 with 1 ml of 0.1 M KH_2PO_4 and put
through the cartridge. After washing with 1 ml water, elution with
2 ml acetone and drying down gave a residue which was taken up in
mobile phase and injected onto a C-8 HPLC column with a UV detector;
Fig. 2 gives other conditions. Theophylline was readily detectable.

Theophylline has strong UV absorption and is easily detected;
serum is relatively free of potential interferences. With urine
(1 ml), however, interfering components vitiate the resolution of
theophylline (Fig. 2, right). Urine is too complex for the C-18
packing to give a clean extract under the conditions used, as also
found (Fig. 3, right) when urine was run with a Type W cartridge. The
latter worked well (Fig. 3) with serum (1 ml) spiked with 50 ng of
diazepam and its metabolite nordiazepam and buffered to pH 9 with 0.1 M
K_2HPO_4 (elution with 2 ml of acetone).

Evidently lipophilic sorbents are convenient and can effectively
extract drugs at ng/ml levels from biological samples. However,
they may extract too many components and be insufficiently selective.
Selectivity for acidic or basic analytes is offered by ion-exchange
sorbents, the usefulness of which is no longer confined to very
polar and water-soluble compounds - as now exemplified.

APPLICATIONS OF AN ANION-EXCHANGE SORBENT

For runs with the PREP^{TM} Type AS cartridge, an aqueous solution
was made up with the 5 compounds shown in Table 1, ranging from a
water-soluble acidic compound to an almost insoluble basic compound.

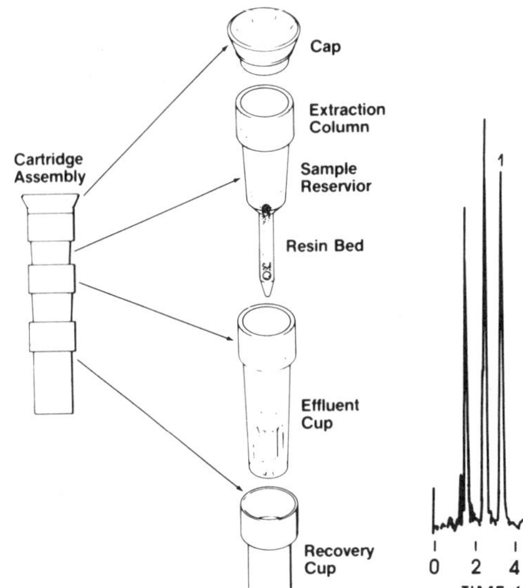

Fig. 1. Extraction
cartridge assembly
(PREP^{TM}).

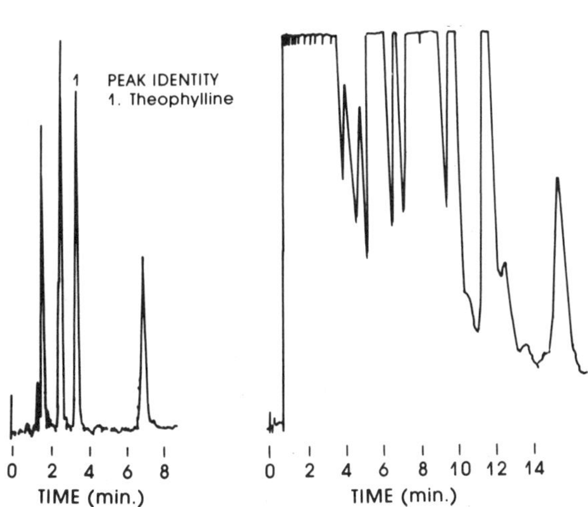

Fig. 2. HPLC examination of eluates from
Type OD cartridges. *Left:* extract from
serum (1 ml) containing theophylline (50 ng).
Right: extract from urine (1 ml).
HPLC on C-8 column (Zorbax), with $CH_3CN/$
10 mM KH_2PO_4 (10:90 by vol.), 2.5 ml/min;
detection at 274 nm.

Fig. 3. HPLC examination
of eluates from Type W
cartridges. *Left:* extract
from serum (1 ml) spiked
with nordiazepam (1.),
diazepam (2.) and an i.s.
(3.). *Right:* extract
from urine (1 ml).
HPLC with CH_3CN/20 mM
KH_2PO_4 pH 3.0 (45:55 by
vol.), with 254 nm detec-
tion; otherwise as in
Fig. 2.

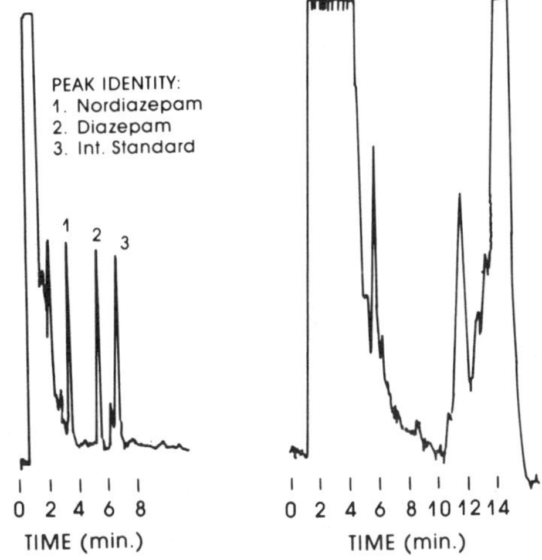

Table 1. Model compounds loaded onto and recovered from Type AS cartridge. Eluting solvents: (1) methanol, (2) methanol with 2 g/l boric acid (pKa = 9), (3) methanol/acetic acid (pKa = 4.75), 90:10 by vol.; (4) methanol/formic acid (pKa = 3.75), 70:30 by vol. Recoveries with S.D. values are based on 6 observations.

Compound	Concn., µg/ml	pKa	Water solubility	Eluent	Recovery
Pyrilamine	50	–	insoluble	(1)	70 ±2%
Oestriol	10	10	insoluble	(2)	97 ±2.1%
Phenobarbital	20	7	1 g/l	(3)	99 ±1.2%
Salicylic acid	20	3.5	2 g/l	(4)	88 ±1.9%
4-Hydroxy-3-methoxymandelic acid (VMA)	50	3.5	freely soluble	(4)	92 ±1.0%

The sample (1 ml) was made alkaline (1 ml 0.1 M NaOH) and loaded; elution was with a series of solvents, as given in the heading to Table 1, in the automated processor. Each eluate fraction (2 ml) was dried down where applicable, and analyzed for the 5 drugs, with the recovery results shown in Table 1. Initial passage of water had shown that all the drugs had been retained, including pyrilamine (it was eluted by methanol); most water-insoluble neutral and basic drugs are in fact retained, by lipophilic attraction to the organic portion of the resin. The slightly acidic oestriol was eluted by methanol with a boric acid modifier. Phenobarbital and two acids of pK 3.5 were eluted respectively by methanol/acetic acid and methanol /formic acid. Organic elution solvents had to be used in this study because 4 of the drugs were relatively water-insoluble. Volatility of elution solvents and acid modifiers facilitates evaporation to dryness and concentration of samples for analysis.

The pKa of acidic modifiers (heading to Table 1) reflects that of eluted drugs: the modifier should be at least as acidic as the drug, to achieve elution. Moreover, these anion-exchange separations evidently offer selectivity with proper choice of wash and elution solvents. Thus, for salicylic acid a methanol/acetic wash solvent in place of water would wash off all neutral and slightly acidic components with pKa above ~4.75. Methanol/formic acid elution would bring off salicylic acid while leaving more acidic compounds (pKa >3) on the resin. The extraction would thus show preference for acidic compounds with pKa close to 4. This selectivity is comparable to that of the acid/base liquid back-extraction techniques. For urine spiked with salicylic acid, RP-HPLC interferences were minimized by use of an AS rather than a C-18 cartridge [elution with solvent (4) - Table 1 heading - after washing with (3)].

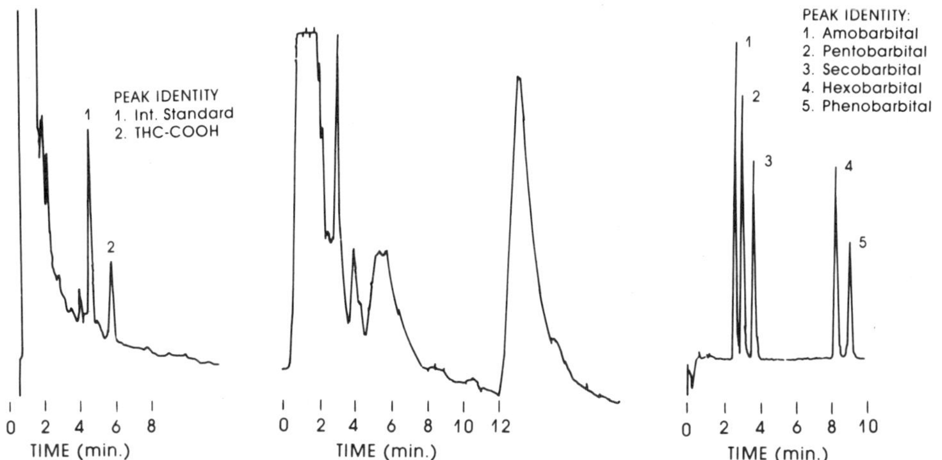

Fig. 4. GC chromatograms of THC metabolite extracts: *left*, 10 ml urine, Type A cartridge; *right*, 1 ml urine, C-18 silica cartridge. GC column: 3% OV-17, 6 ft. × 2 mm (gas-flow 30 ml/min), at 255° with injector at 260°; detection by FID at 275°. Derivatization performed as in [5].

Fig. 5. GC of barbiturates: 0.25 ml urine; Type AS cartridge. GC as on left, but 155° (4 min) → 250° @ 5°/min; injector at 265°; NPD at 275°.

One example of an application concerns the primary metabolite of tetrahydrocannabinol (THC), 11-nor-Δ^9-THC-9-COOH, of which urine may contain 20-200 ng/ml. GC-FID is a sensitive means of analysis but requires a very selective extraction to remove the metabolite from interfering components. This is effected with a Type A cartridge which contains the same anion-exchange resin as Type AS but is dry-packed. The resin is transferred from the cartridge to a tube containing 10 ml of base-hydrolyzed urine and then, carrying the metabolite, back to the cartridge which is then washed with 2 ml methanol and eluted, with 2 ml of ethyl acetate/methanol/acetic acid (90:10:4 by vol.). The dried-down residue was examined by GC [5] (Fig. 4, left), in comparison with that from extracting 1 ml of urine onto a CV-18 cartridge and eluting with acetone (Fig. 4, right). Evidently the anion-exchange extract gives a much cleaner background.

A further example concerns barbiturates (Fig. 5). Urine (0.25 ml) buffered with 1 ml of pH 12 phosphate buffer was extracted onto a Type AS cartridge, which was washed with methanol and eluted with hexane/ethyl acetate/acetic acid (63:9:8 by vol.). The extracts were evaporated to dryness, derivatized, and then analyzed on a GC with a N-P detector [5]. Fig. 5 shows a typical chromatogram of an extract of spiked urine. The extracts are very clean with no interferences.

CONCLUSION

Both the lipophilic and the anion-exchange sorbents can be used
for the extraction of trace components in complex mixtures. However,
the anion-exchange sorbents potentially have greater selectivity for
acidic analytes; those that are compatible with both aqueous and
and organic solvents can be used to extract both water-soluble and
insoluble acidic analytes. Proper choice of acidic modifiers in
wash and elution solvents can result in very selective extractions
for analytes of interest.

References

1. Williams, R.C. & Viola, J.L. (1979) *J. Chromatog. 185,* 505-513.
2. St. Onge, L.M., Dolor, E., Anglim, M.A. & Least, C.J. (1979)
    Clin. Chem. 25, 1373-1379.
3. Krahn, M.K., Collier, T.K. & Malins, D.C. (1982) *J. Chromatog.
    236,* 441-452.
4. Balkon, J., Donnelly, B. & Prendes, D. (1982) *J. Forensic Sci.
    27,*   23-31.
5. Whiting, J.D. & Manders, W.W. (1982) *J. Anal. Toxicol. 6,*
    49-52.

#NC(A)-3

A Note on

SIMPLIFIED APPROACHES TO THE HPLC DETERMINATION OF DRUGS IN BIOLOGICAL SAMPLES

G.S. Clarke, L.K. Liu and M.L. Robinson

E.R. Squibb & Sons
International Development Laboratories
Moreton, Merseyside L46 1QW, U.K.

Requirement	*Simple assays for drugs in biological fluids.*
End-step	*RP-HPLC on a C-18 column, with appropriate detection.*
Sample preparation	*Minimal, e.g. merely filtration or centrifugation. For pre-concentration or purification, a 'loop column' is advantageous.*
Comments	*Procedures are mentioned for cephalosporins, metoprolol and nadolol, especially in urine.*

The simplest approach to sample preparation for HPLC is centrifugation and/or filtration of the sample, followed by direct injection onto the column. This is particularly suitable for drugs which are not bound to the matrix, and where concentrations are well above the limit of detection. An example of this approach is the analysis of cephradine in urine (1-100 µg/ml), as reported from this laboratory in an earlier volume of this series [1]. Another example of direct injection (sometimes with pre-dilution) is a reported assay for cefmenoxime in urine [2].

For metoprolol in urine (0.5-100 µg/ml), the centrifuged sample with mobile phase added is injected directly onto the column (ODS-Hypersil). The mobile phase is water/acetonitrile/triethylamine, 80:20:1 by vol., adjusted to pH 3.0 using orthophosphoric acid. A fluorimetric detector is used (excitation, 220 nm; emission, 297 nm) as a well-known way to achieve selective detection with little interference from endogenous material.

If a concentration or purification step is required, a short column of a chemically bonded stationary phase, mounted in the

Fig. 1. Schematic diagram of the HPLC system with a loop column for purification and enrichment.

Mode of use:
The loop (4.2 mm i.d. column) contains Lichroprep Si 60 ODS. The sample is injected, followed by a flush with water, to waste.
The exit then becomes the entry point for back-flushing, onto the analytical column, with mobile phase (~500 µl in a typical assay).
Direct pumping of mobile phase onto the column then commences.

A novel aspect is the back-flushing onto the analytical column, rather than forward flushing from the enrichment column. (Suitable 6-port valve: Rheodyne 7125.)

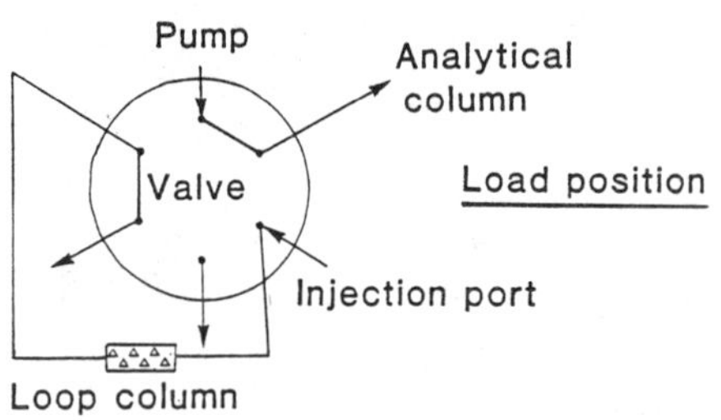

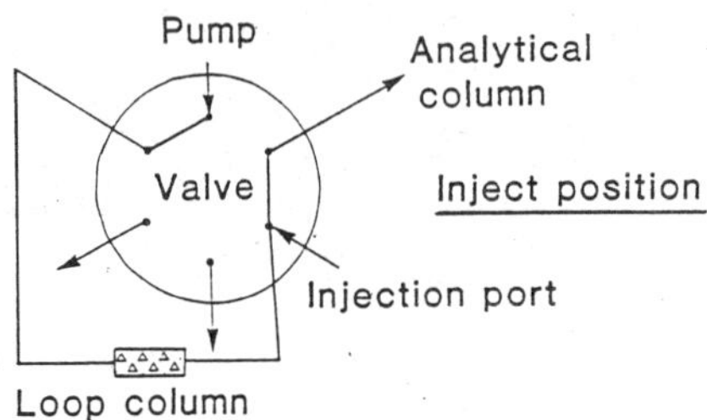

sample loop of the high-pressure injection valve, may be used, as in previous trace-enrichment applications [3, 4]. In our current method for the beta-blocker nadolol in urine, the sample is centrifuged and up to 500 µl injected onto a Lichroprep Si 60 ODS column mounted in the injector loop as in Fig. 1; the procedure is given in the legend. The mobile phase is water/acetonitrile/methanol/conc. ammonia, 400: 300:300:0.5 by vol. The flow rate is 1 ml/min, and detection is at 230 nm. [Cf. Vol. 7, this series, #B-3 – R.W. Frei & U.A.Th. Brinkman.–*Ed.*]

Mere protein precipitation has sufficed in assaying serum for cephradine [1], cefmenoxime [2] (supernatant then concentrated) and ceftrioxone [5]; but drug may be co-precipitated or trapped, and so lost.

References

1. Forster, T.C., Heyes, W.F., Jones, G.T. & Wood, P.R. (1978) in *Blood Drugs and Other Analytical Challenges* (Reid, E., ed.) [Vol. 7, this series], Horwood, Chichester, pp. 341–344.
2. Noonan, I.A., Gambertoglio, J.G., Barriere, S.L., Conte, J.E. & Lin, E.T. (1983) *J. Chromatog. 273*, 458–463.
3. Burns, D.A. (1981) *Pharm. Tech. 5*, 53–60.
4. Koch, D.D. & Kissinger, P.T. (1980) *Anal. Chem. 52*, 27–29.
5. Ascalone, V. & Dal Bo, L. (1983) *J. Chromatog. 273*, 357–366.

Comments on material in #A

Comments on #**A-1**, R.P. Maickel – SEPARATION SCIENCE

S.H. Curry, *reiterating perennial problems* [cf. NC(F)-3, Vol. 10, this series – *Ed.*].– When you homogenize a tissue, how do you know that you have separated out the drug in quantitative yield? and how do you set up a series of added-to-sample standards for a tissue? *Reply* (R.P. Maickel): the problems are indeed considerable ('spiking' of tissues with a fine-gauge needle has been tried); yet it can be reckoned that 85% yields are achievable on homogenizing tissues. *Question by* U.A. Th. Brinkman.– Since many 'difficult' compounds in the pharmaceutical and biomedical fields are metabolites and so are rather polar, will not LC-MS become more important? Its cost can be minimized through using the same MS for GC and LC, with easy switching. *Reply.–* I agree concerning the nature of the problem and the way the pharmaceutical industry should go. But for most U.S. Universities MS is too expensive to run and maintain, and so other solutions have to be found.

Separating peptides from biological fluids *(answer to query by* L.A. Sternson) may depend on molecular size (protein precipitation is feasible if mol. wt. below ~3000) and on whether biological activity has to be retained (denaturation acceptable?). For chromatography, e.g. of encephalin (*answer to* J.A.F. de Silva), RP-HPLC with gradient elution has worked well, coupled to post-column detection with either fluorescamine or o-phthalaldehyde reagent. Distinct peptide fractions have been obtained; but we have yet to prove the homogeneity of each fraction.

Comments on #**A-3** & **A-4**, G.E. von Unruh, STABLE ISOTOPES
[cf. final p. of #F]   & R. Whelpton, ISOTOPIC INTERNAL STANDARDS

G.E. von Unruh, *replying to* H. de Bree: in the bioavailability comparison with verapamil, which has a wide therapeutic range, unlabelled and labelled drug were given in equal amount, resulting in a doubled total dose. D. Newell (*to whom the reply was "no"*, *likewise from* R. Whelpton) wondered about experience of using ^{13}C-NMR as a non-invasive technique for pharmacokinetic monitoring *in vivo*.

Comment by R.P. Maickel.– In analyzing biological samples for a drug, endogenous trace constituents may 'collect' on a GC or HPLC column: those that originate in today's sample may well appear in the recordings of tomorrow's assays! *Concerning adsorptive loss of drug* (R. Schmid).- R. Whelpton's finding that loss occurred even with silanized glassware is not surprising. It seems that nitrogen-containing compounds, especially secondary amines, associate strongly with residual -OH groups in the glass, which cannot be totally removed by silanization. We overcome this problem by adding, in the analysis, a suitable non-interfering amino compound, thereby competitively blocking free -OH groups in the glass. Our experience with tricyclics has shown applicability to GC also. (The problem, and attempted remedies, feature in #A-1 also, and throughout this series, e.g. in #E, Vol. 12.- *Ed.*)

Comment on #NC(A)-2, Reed C. Williams - SAMPLE PREPARATION

Point made by Senior Editor.- Disclosure of the nature of the polymeric material would be welcomed; one wonders whether it is akin to XAD-2, to which contributions in earlier vols. allude. *In reply to a question*: the solid sorbents indeed extract as efficiently if the serum level is ng/ml rather than µg/ml (with equally good recovery), e.g. in the case of theophylline and diazepam (Reed Williams).

SOLID-PHASE SAMPLE PREPARATION: *Forum contribution (no publication text)* by P.A. Harris (Analytichem, Harbor City, CA)

An example of the advantageous use of bonded silica sorbents is the isolation of the carboxylic acid metabolite of tetrahydrocannabinol from urine, hitherto achieved by iso-octane extraction followed by GC. Sorbent extraction followed by HPLC gives improved selectivity and sensitivity (5 ng/ml urine detectable) with good recovery (89%). In the sorbent extraction of tricyclic antidepressants from serum, transfer to the HPLC column can be done quantitatively with an automatic syringeless injection device, allowing selective detection at 254 nm; 20 ng/ml is detectable, and recoveries exceed 95%. *Remark by* R. Schmid: with Analytichem cyano cartridges (especially 40 µm) we have never achieved near-100% retention in extracting tricyclics from serum.

Annotations by Senior Editor: PRE-HPLC SAMPLE TREATMENT

ODS-silica (as a short column) can extract lipophilic small molecules from plasma, not retaining plasma proteins if pre-treated with human plasma; analytes responsive to this no-deproteinization approach include doxorubicin, methotrexate, procainamide, propranolol and theophylline [1]. For various drugs (e.g. cyclobenzaprine, diflunisal, indomethacin) in plasma or urine, a clear supernatant conducive to RP-column life is obtainable by successively adding $ZnSO_4$, $Ba(OH)_2$ and acetonitrile (at least 2 vol.; alternatively methanol) [2].

1. Yoshida, H., Morita, I., Masujima, T. & Imai,H. (1982) *Chem. Pharm. Bull.*30, 3827-3830.
2. Ng, L.L. (1983) *J. Chromatog.* 257, 345-353.

Section #B

STRATEGIES FOR HPLC (OTHER THAN DETECTION) AND FOR TLC

#B-1

DEVELOPMENT AND APPLICATION OF AN HPLC PRE-CONCENTRATION DEVICE SUITABLE FOR LARGE AMOUNTS OF BIOLOGICAL SAMPLE

H.M. Ruijten, P.H. van Amsterdam and H. de Bree

Duphar Research Laboratories
P.O. Box 2, Weesp, The Netherlands

A cone-shaped HPLC pre-column has been developed to cope with the problems that arise during the work-up of large volumes of biological sample. The device offers large loading capacity at the front of the pre-column. It is constructed of stainless steel and PCTFE. Its pressure resistance (up to 40 MPa) allows full integration into HPLC systems. It is observed that there is scarcely any decrease in performance on going from a small to a large sample volume using the cone-shaped pre-column. Its possibilities in biomedical research are discussed, with metabolic examples.

For compounds in biological matrices, quantitative and qualitative analysis is not easy. In general the hydrophilic analyte is in low, and irrelevant substances are present in large, amount. The introduction of reverse-phase (RP) HPLC tackled the polarity problem and also helped in approaching the low-concentration problem. Up to now several neat systems have been described and introduced to pre-concentrate analytes from ml-amounts of sample for quantitation [1-3]. (Trace enrichment has been considered by R.W. Frei earlier in this series: #D-5 in Vol. 7 and #B-3 in Vol. 10.- *Ed.*) But enrichment aimed at isolation and identification remained troublesome. Accordingly we developed a system for pre-concentration and isolation of analytes from biological matrices with integral use of HPLC.

The system comprises pre-columns mounted on a valve, allowing on-line connection with a separation column after pre-concentration. The pre-columns are cone-shaped to comply with the need for large capacity at the beginning of the pre-column and to allow a neat uni-diameter coupling. Being of s.s. construction, the device is pressure-proof up to 40 MPa. The system accepts all usual brands of analytical column by use of adapter inserts. Besides serving for pre-concentration, pre-columns packed with RP material may be used

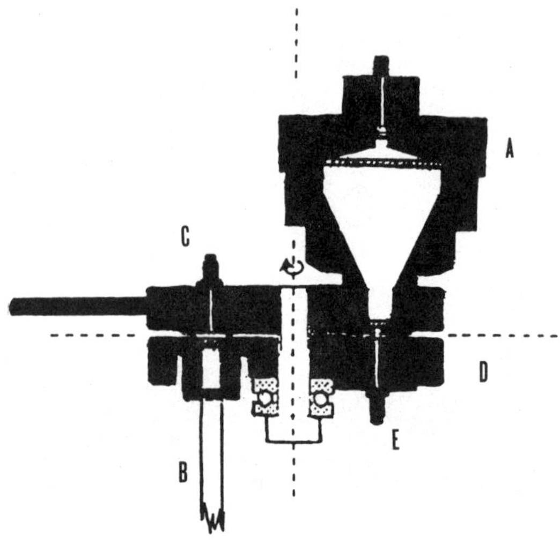

Fig. 1. The cone-shaped pre-column (pre-cone), assembled. A, pre-cone; B, separation column; C, capillary connection (when pre-cone not in use); D, valve; E, waste exit. The upper part of D can rotate with respect to the lower part (PTFE gliding gaskets). All filters (s.s. frits) are borne by gaskets of PCTFE (polychlorotrifluoro-ethylene).

to deproteinize samples. The proteins are not retained on the usual stationary phases and are easily washed off with with twice the pre-cone's water capacity, while the compounds of interest are retained on the column. On a nitrile phase some proteins do have some retention in water, and they will denature as soon as an organic modifier is introduced in the elution program.

If a nitrile phase has to be used, this problem is easily dealt with by supplementing the aqueous component of the eluent with 10% (v/v) dimethylsulphoxide. Thereby the proteins also elute with V_o.

DESCRIPTION OF THE DEVICE[*]

The construction is shown in Fig. 1. To obtain a good distribution of the sample at the front of the pre-cone (A), the upper section is filled with a material which is inert in the chosen chromatographic system. In the RP mode we used regular glass beads of 100 µm diameter. The upper part of valve D can rotate with respect to the lower part so as to connect or disconnect pre-cone and separation column. We have 3 sizes of cone-shaped pre-column in use:

(a) preparative: 26 ml inner vol.; 45 mm long; 39 mm diam.
(b) semi-preparative: 6 "   "   " ; 27 "   " ; 24 ."   "
(c) analytical: 1.5 "   "   " ; 13 "   " ; 14 "   " .

The amount of biological sample that can be pre-loaded onto the pre-cones described, expressed as vol. of urine, is ~2 l for the preparative, 500 ml for the semi-preparative and 50 ml for the analytical pre-cone. As will be shown, in the modes of operation described the pre-cones hardly affect the chromatographic performance.

---

[*] now commercially available [from Duphar (dept. Bedrÿfs-beheer), Weesp]

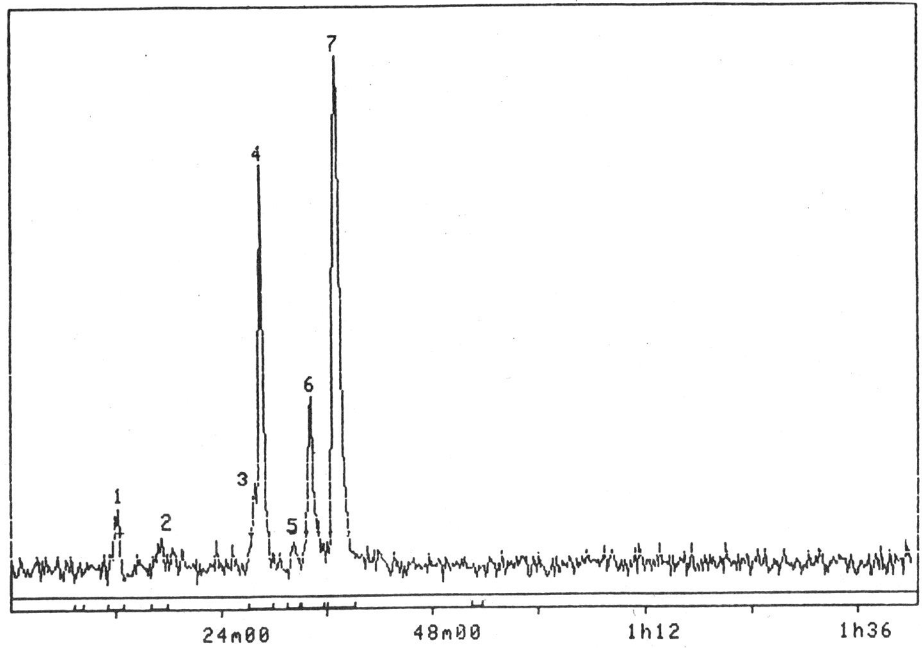

Fig. 2. Urinary metabolite pattern of clovoxamine in the rat.
Load vol. 1 ml. On-line radioactivity detection. Column 500 X
i.d. 9 mm, packed with Nucleosil 7C8. A combination of gradient
and isocratic elution was used, from water to methanol.

HPLC EQUIPMENT

To perform a gradient elution program we used two Waters M 6000 A
pumps together with an M 720 microprocessor. Fluorimetric detection
was done with a Perkin-Elmer LC-204 A; on-line radiochemical detection with a Berthold BVF 22550 module or an Isomess IM 2000. Small
volumes of sample were injected by means of a syringe-loading sample
injector, and larger volumes by merely pumping the sample via the
HPLC pump onto the pre-cone.

APPLICATIONS IN BIOMEDICAL RESEARCH

Metabolism *in vivo*

To establish metabolite patterns, samples of various kinds of
biological fluid have been examined by HPLC. In general a 500 X i.d.
9 mm separation column was used for such studies. A segmented gradient profile provides high resolution. Proteinaceous samples (plasma, organ homogenates) can be deproteinized using the analytical
pre-cone. Fig. 2 shows such a metabolite pattern for the drug
clovoxamine (structure overleaf) in urine.

Clovoxamine

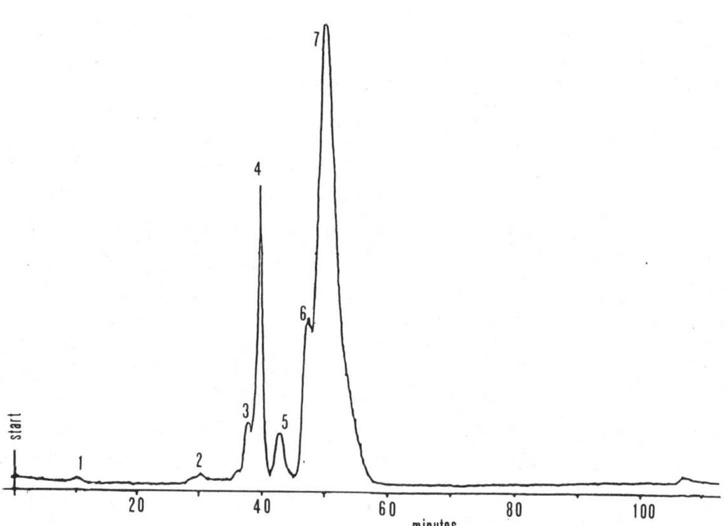

After these highly selective HPLC patterns for metabolites have
been established, they are used directly in pilot work on isolation.
The only change necessary for scaling up the amount of sample (up to
litre amounts) is the use of a pre-cone of appropriate capacity. The
same column and chomatographic conditions are used; only time factors and flow are increased with the increase of retention volume.
Fig. 3 exemplifies such an isolation run guided by the pattern in
Fig. 2.   The fractions thus obtained are generally not yet pure
enough for NMR or other spectral techniques, but by using heart-
cutting in one or two re-runs and finally performing the TLC runs, pure
compounds are obtained.   Thereby not only the 'main' metabolites are
isolated but the whole metabolic pattern can be elucidated [4].   (In
Vol. 12, #A-2, W. Dieterle discussed HPLC metabolite isolation.-*Ed.*)

Fig. 3.   Isolation of clovoxamine metabolites from a pool of rat
urine.   Amount concentrated: 250 ml, using the preparative pre-cone.
Chromatography as in Fig. 2 but corrected for use of the pre-cone.

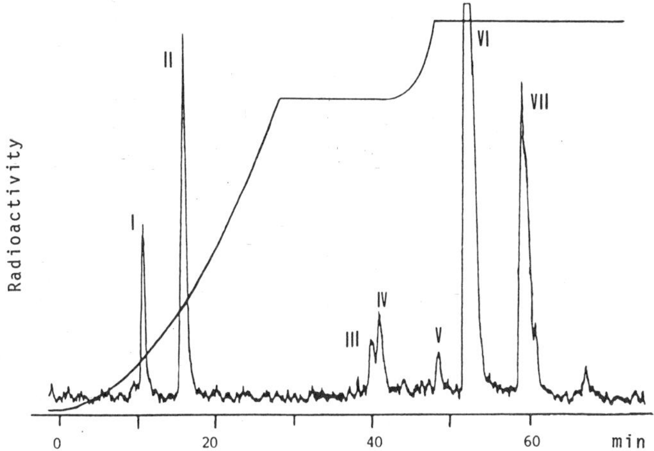

Fig. 4. Metabolite pattern of secoverine incubated with a liver homogenate. Column: Nucleosil 7C8, 500 × 9 mm. A combination of gradient and isocratic elution from water to methanol was used. Injection: 1 ml. The analytical pre-cone was used for deproteinizing.

In vitro investigations

To obtain insight into the metabolism of a drug at an early stage of development, or in the search for active drug metabolites, investigations are done in liver-homogenate supernatants. Such an investigation in our laboratory, concerning the drug secoverine, is described elsewhere in this vol. [#NC(E)-1]. We ran an exploratory chromatogram on 1 ml homogenate supernatant after incubation in the presence of secoverine (Fig. 4); the analytical pre-cone was used for deproteinizing. When scaling up from 1 ml to 650 ml, using the preparative pre-cone, we encountered no problems such as clogging or other chromatographic disturbances (Fig. 5). Peaks I and II seen in the pilot run are not in evidence in the isolation run. This is due to their very low capacity factor in the RP system. The other peaks were recovered quantitatively and the relevant ones were identified. As in the experiment to which Fig. 3 relates, here it is possible to isolate clean products in a quick and elegant way for further structural analysis.

Body-fluid analysis

Using the analytical pre-cone, up to 50 ml urine or plasma can be introduced onto an analytical column. Accordingly, the theoretical detection limits may be substantially lowered, the only limiting factor being detector sensitivity. Fig. 6 exemplifies the straightforward analysis of whole blood for pamoic acid.

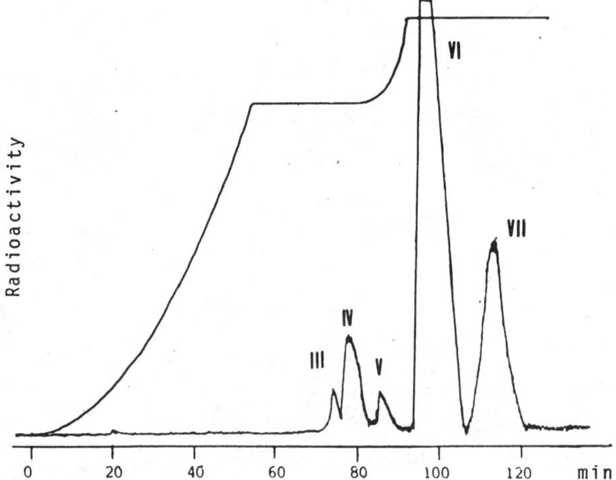

Fig. 5, *above.*
Isolation chromatogram of
650 ml liver homogenate
containing 10 mg secoverine
after concentration on the
preparative pre-cone.
Chromatographic conditions
as for Fig. 4.

Fig. 6. Pamoic acid in
whole blood. Column:
Nucleosil 7C8, 250 x 6.2 mm;
same packing in the analytical pre-cone. Injection:
1 ml blood containing 120 ng
pamoic acid. Fluorimetric
detection (ex. 368 nm; em,
518 nm).

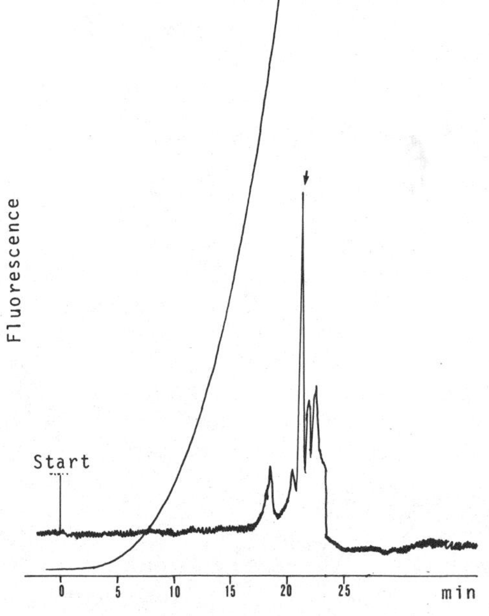

References

1. Roth, W., Beschke, K., Jauch, R., Zimmer, A. & Koss, F.W. (1980)
    J. Chromatog. 222, 13-22.
2. de Jong, G.J. & Zeeman, J. (1982) *Chromatographia 15*, 453-458.
3. Werkhoven-Goewie, C.E., Brinkman, U.A.Th. & Frei, R.W. (1981)
    Anal. Chem. 53, 2072-2080.
4. Ruijten, H.M., de Bree, H., Borst, A.J.M., de Lange, N.,
    Scherpenisse, P.M., Vincent, W.R. & Post, L.C. (1984)
    Drug Metab. Disp. 12, 82-92.

#B-2

MULTI-SOLVENT OPTIMIZATION OF HPLC SEPARATIONS

G.B. Cox

Du Pont (UK), Wedgwood Way
Stevenage SG1 4QN, U.K.

The development of analytical methods for HPLC has in the past been based on experience, intuition and serendipidity. With the high efficiency of present-day columns, the main scope for improving separations lies in adjusting the mobile phase for selectivity. Optimization has been fostered (Snyder, Kirkland) by a 'Selectivity Triangle' based on solution interactions – acidic, basic and dipole. Thereby optimal resolution of components for a particular column is achievable by choosing solvents from a relatively small range.

With a biological matrix, solvent selectivity effects may enable the elution of components of interest to be matched with gaps in the elution pattern of the matrix. This is exemplified by the assay of anticonvulsant drugs in plasma.

The development of analytical methods using HPLC has long been a complex operation, due in great part to the large number of variables which can profoundly influence the separation ultimately attained. This has resulted in long method-development times and in methods which may be quite far from the optimum, being the first result deemed to be just acceptable. The problems here are two-fold: the method may be unstable in that very small changes in solvent composition or pH or changes in batch of column packing can deleteriously influence the separation, or a new factor such as interference from the sample matrix or a hitherto unrecognized metabolite may become apparent, resulting in complete re-assessment of the developed method. In either case, in the absence of any defined method-development protocol, the analyst is faced with complete redevelopment of the separation. The consequent long development times (especially when a separation is not, in fact, possible and much time is spent in proving this) have focussed attention on techniques of method optimization in attempts to make the whole process of method development more efficient.

From the basic resolution equation, the resolution between two components in a separation is a function of column efficiency, capacity factor and selectivity. Resolution increases with the square root of column efficiency (defined as no. of theoretical plates). Once a reasonably high efficiency is attained, further increases in resolution by increases in plate no. are limited by column length, back-pressure, etc. Variation in capacity factor is also restricted in its usefulness in improving resolution. As k' increases beyond 2 its effect on resolution becomes quite small. At values below this it is often true that change in k' can have a dramatic effect on R_s. The problem frequently encountered, however, is that increasing k' values for poorly retained components also increases those of the longer retained components. This results in prolonged analysis times and reduced sensitivity for the resulting method. The foregoing implies that selectivity is the most useful parameter for investigation.

The selectivity of a chromatographic system is a function of the nature of both mobile and stationary phases. In general, especially in reversed-phase (RP) systems, it is the mobile phase which has by far the greater influence on selectivity. The selectivity of mobile phases was first studied quantitatively by Snyder [1] who investigated the interactions existing in solution between solvent and solute molecules. By utilizing a number of carefully selected test solutes he was able to measure parameters describing three main interactions: proton donor, proton acceptor and dipole. These were plotted on a triangular graph. The results not only demonstrated that similar solvents fell in relatively small areas within the selectivity triangle as might be expected, but also illustrated in what way and by how much these solvents and solvent groups differ from each other in terms of selectivity. This facilitates the choice of selectivity-influencing solvents for a particular separation system. The strategy is to choose solvents which are as far apart in Snyder's selectivity triangle as possible but which are also compatible with the chromatography; there is little point in trying to use an ether or hydrocarbon to influence selectivity in aqueous RP chromatography due to their immiscibility.

The chromatographic compatibility criterion restricts the choice of solvents such that it is simple to arrive at the best selection of 3 chromatographically strong solvents which influence selectivity. These are used with a weak 'base' solvent which modifies the strength of the solvent to adjust the separation time to a reasonable value. In RP-HPLC the selectivity-adjusting solvents are methanol, acetonitrile and tetrahydrofuran, and the base solvent is water. Achieving acceptable peak shape for many compounds with ionizable functionality frequently calls for modification of the aqueous phase by adjusting pH and ionic strength. In polar bonded phase (NP) chromatography the selectivity-modifying solvents are an ether, usually methyl t-butyl ether, chloroform and dichloromethane. The fourth, solvent-strength modifying, solvent, is either non-polar, e.g. hexane or trichloro-trifluoroethane (Freon^{TM} TF 113) or very polar, e.g. methanol. The choice is dictated by the polarity of the analytes. For adsorption chromatography the situation is somewhat different owing to the localization effects which occur on the surface of silica with some solutes and solvents. In order to exploit these, the selectivity-adjusting solvent set usually chosen is the same as for NP except that acetonitrile, a solvent with good localizing properties, is substituted for chloroform. Then Freon^{TM} has to be used in place of hexane to allow complete miscibility of solvents.

ESTABLISHING CONDITIONS TO ACHIEVE SELECTIVITY

Once the solvent set is chosen for an experiment, it remains to choose the appropriate compositions and to investigate the selectivity effects experimentally. The choice of mobile phase composition is simply made by considering the time-span required for the separation and the probable number of analyte components. Keeping in mind the above discussion of resolution as affected by k', the solvent strength should be adjusted such that the peaks are retained beyond a value of 2 if possible. Probably the most convenient procedure by which the composition to give the desired k' can be ascertained is to run a scouting gradient and use this data to predict it, either mathematically using Snyder's gradient-elution equations [2] or by trial-and-error.

Once the composition of the first solvent pair is established the compositions of the other two pairs may be calculated from the known solvent-strength data [3]. Whilst they may need slight correction to establish exactly equivalent k' values in the three solvent pairs (for the RP mode: methanol/water, acetonitrile/water and tet-rahydrofuran/water), these derived values set the compositions at the apices of a constant solvent strength selectivity triangle.

The selectivity effects have been shown by Kirkland and co-workers [4] to be readily investigated by a statistically based mixture-design experiment set which involves separations with 7 different solvent mixtures. These mixtures are shown in Fig. 1 to be those occurring at the three corners of the constant solvent strength selectivity triangle, the mid-points of its sides, and a point at its centre. The values of the compositions are simply determined by mixing the solvents established at the apices in appropriate proportions. The results of the 7 experiments generally yield sufficient data for prediction of the composition to give the best possible separation. The prediction may be made either by calculation from a knowledge of how the k' of each component varies with solvent composition and subsequent estimation of resolution of each pair of compounds at all possible compositions - 'overlapping resolution mapping' [4] - or from careful inspection of the 7 chromatograms and visual interpolation between points.

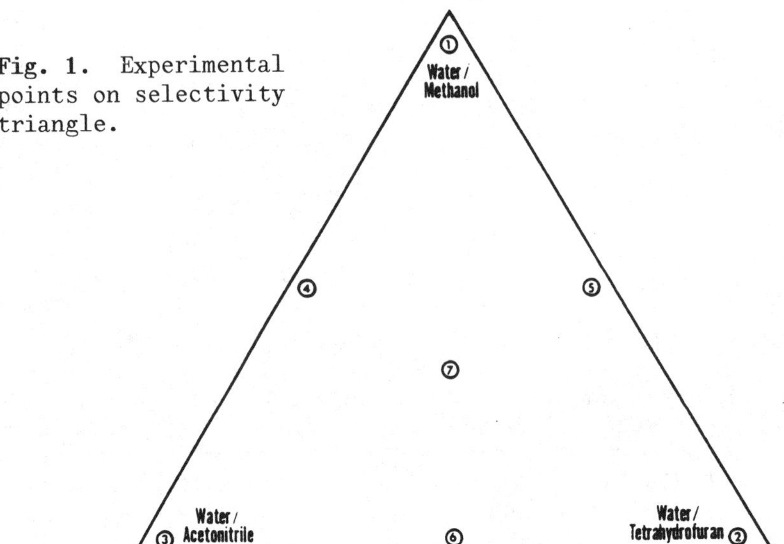

Fig. 1. Experimental points on selectivity triangle.

There are two basic applications of this technique in analysis of body fluids. One is to optimize the separation of analytes to ensure maximum resolution. The other is to match the chromatographic behaviour of the analytes to that of the matrix in which they are to be analyzed. Thereby it is possible to ensure minimal interference of co-extractives from the matrix with the analysis. As one moves to analyses of higher and higher sensitivity in attempts to reach the minimum detection limits necessary in many bioanalyses, this technique becomes of increasing importance.

ILLUSTRATIVE OPTIMIZATIONS

Instrumentation and general procedures

Mobile phases were dynamically mixed from methanol, acetonitrile, tetrahydrofuran and water containing 0.5% (v/v) orthophosphoric acid to give pH 2.2. HPLC was performed in a Du Pont Sentinel automated instrument with a Zorbax C-8 column, 150 x (i.d.) 4.6 mm, at 50°. The flow-rate was 3 ml/min.

Extractions of plasma were carried out using a Du Pont 'Prep' automated sample processor. Type W (XAD-2) cartridges were employed for the extractions, using 1 ml aliquots of plasma buffered to pH 4.8 by sodium dihydrogen phosphate. Each cartridge was washed with 1 ml of distilled water and was extracted with 1 ml of acetone. The acetone extracts were evaporated to dryness at 50° and the residues were dissolved in 100 µl of 40% (v/v) methanol/0.5% phosphoric acid.

A mixture of 5 anticonvulsant drugs and, as internal standard (i.s.), ethyl 5-*p*-tolylbarbiturate were analyzed using the standard Sentinel automated method-development software with the k' for the final component in the samples defined at a value of 10.

Plasma extracts from the sample processor were analyzed under the same solvent conditions as those defined by the Sentinel program, using injection aliquots of 10 μl with detection at 220 nm (sensitivity 0.016 AUFSD). A sample of trimethoprim was analyzed under identical conditions.

Anticonvulsant separation

Following the gradient 'scout', a solvent composition of 40.7% methanol in 0.5% aqueous phosphoric acid was established as resulting in the required k' for the final component, in this case carbamazepine. Solvent conditions were found for the other two corners of the triangle to give equivalent k' values for the final peak, and the remaining 4 sets of compositions were calculated. These are shown in Table 1. The k' values of the anticonvulsants and their i.s. at the 7 sets of experimental conditions are shown in Table 2.

Several conclusions can be drawn from the data. As expected, the separation was not especially difficult, adequate resolution being achieved using methanol-water (Fig. 2). It is noteworthy that phenytoin and carbamazepine are not resolved in acetonitrile (Fig. 3) and the primidone peak moves closer to the solvent front. The most dramatic solvent-selectivity effects are observed on addition of tetrahydrofuran to the mobile phase. At point 3 the elution order changes completely. Here carbamazepine epoxide elutes first (in contrast with points 1 and 2 where it eluted third). Carbamazepine moves from the end of the chromatogram to third place, and phenobarbitone moves to a larger k' (Fig. 4). Point 4 also allows good resolution of all components, whilst the remainder show poorer resolution between primidone and carbamazepine epoxide which also elute

Table 1. Solvent compositions for experimental set (as %, v/v).

Point no.	% Water	% Methanol	% Acetonitrile	% Tetrahydrofuran
1	59.3	40.7		
2	74.7		25.3	
3	74.7			25.3
4	66.9	20.4	12.7	
5	74.6		12.7	12.7
6	66.9	20.4		12.7
7	69.6	13.6	8.4	8.4

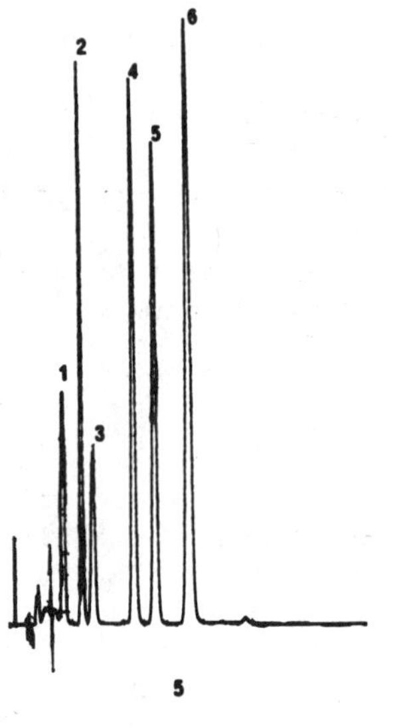

Fig. 2 *(left),* **Fig.** 3 *(bottom left)* & **Fig. 4** *(below).* HPLC of a mixture of primidone (peak 1 in Figs. 2 & 3), phenobarbitone (2), carbamazepine epoxide (3), i.s. (4), phenytoin (5) and carbamazepine (6). In Fig. 4 the respective peak nos. are 2, 4, 1, 6, 5 and 3.

The mobile phase, made up to 100% with water at pH 2.2, contained solvent as specified (v/v): **Fig.** 2, methanol, 40.7% – point 1; **Fig.** 3, acetonitrile, 25.3% – 2; **Fig.** 4, tetrahydrofuran, 25.3% – 3. (**1**, **2** & **3** signify experimental points.)

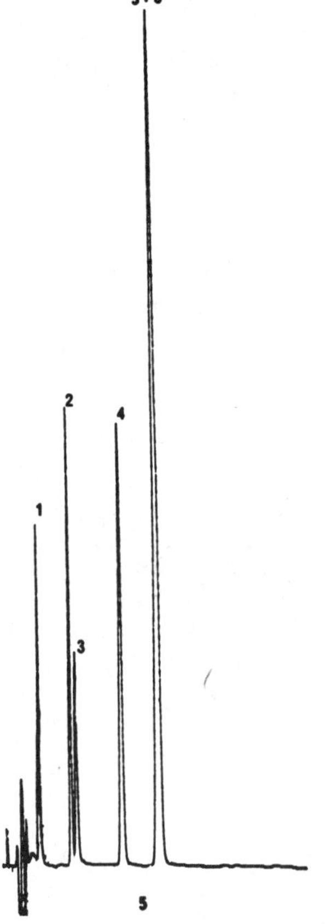

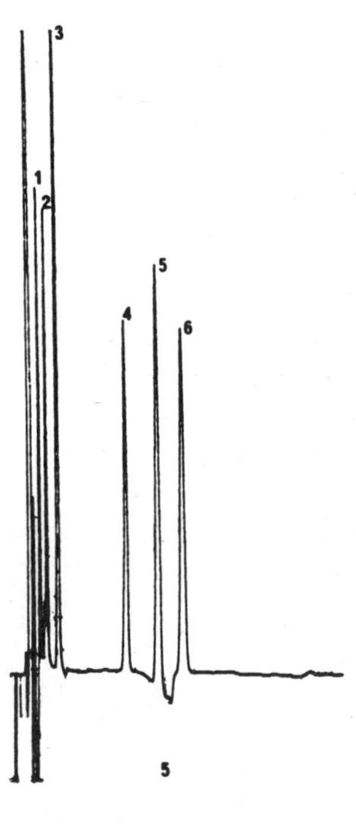

Table 2. HPLC of anticonvulsants: k' values, points 1-7 (see text).

Compound	1	2	3	4	5	6	7
Primidone	2.2	1.7	1.3	2.4	1.4	1.5	1.8
Phenobarbitone	3.2	4.1	6.6	4.7	5.0	4.3	4.5
Carbamazepine epoxide	3.5	4.4	0.88	4.9	1.5	1.3	1.8
Internal standard	6.4	7.0	10.3	9.4	8.6	7.6	8.2
Phenytoin	7.5	10.5	8.7	12.4	9.4	8.2	9.6
Carbamazepine	9.6	10.5	2.0	13.6	3.6	3.2	4.5

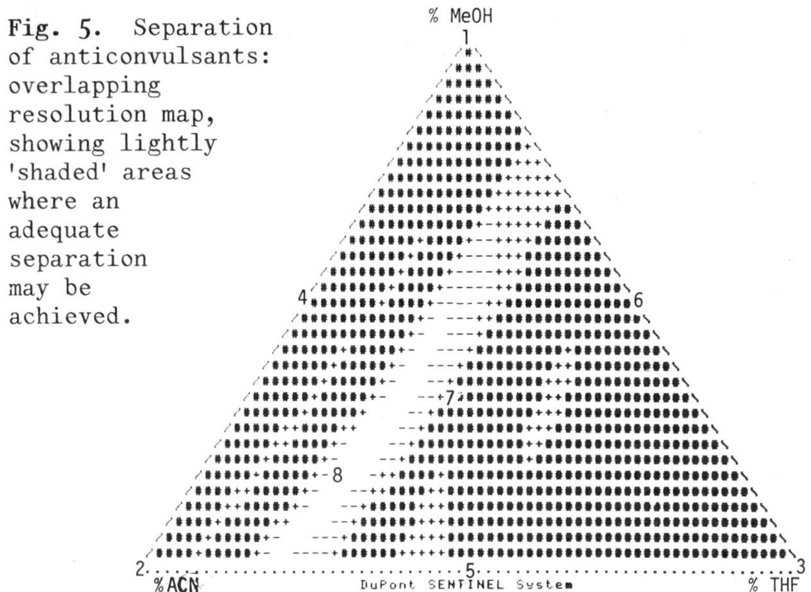

Fig. 5. Separation of anticonvulsants: overlapping resolution map, showing lightly 'shaded' areas where an adequate separation may be achieved.

relatively close to the solvent front. The overlapping resolution map (Fig. 5) shows a quaternary mobile-phase mixture as giving the optimum separation. This optimum is defined by the system as that which gives the maximum selectivity between all components and tends to arrive at a solution which tries to achieve equal spacing of peaks through the chromatogram. This was found to be at a solvent composition of 6.4% methanol, 16.2% acetonitrile, 5.1% tetrahydrofuran and aqueous solvent comprising the remainder; Fig. 6 shows an HPLC run.

The resolution map in fact shows several areas (lightly shaded) where an adequate separation may be achieved. In passing from one to another, the separation moves through areas of poor separation as peaks cross over. Yet all these areas represent better separation than those experimental results (points 1 & 4) noted as adequate.

Fig. 6. Separation of anticonvulsants: experimental point 8, representing optimum solvent composition: methanol, 6.4%; acetonitrile 16.2%; tetrahydrofuran, 5.1%; water (pH 2.2) to 100%.
Peaks:
1, primidone;
2, carbamazepine epoxide;
3, phenobarbitone;
4, carbamazepine;
5, i.s.;
6, phenytoin.

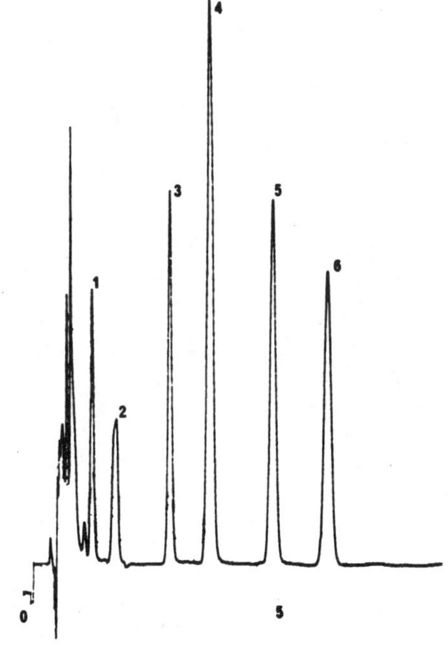

The value of these data is that if interference is encountered from the matrix or if another component has to be added, it is relatively easy to re-define suitable separation conditions by moving to one of the other suitable areas of the triangle.

Plasma extraction relating to trimethoprim assay

The aim was to establish mobile-phase conditions which would allow the determination of trimethoprim without interference from plasma co-extractants. For the experimental set defined in Table 1, the trimethoprim k' values were: point 1, 7.2; 2, 8.1; 3, 11.1; 4, 5.7; 5, 8.9; 6, 5.4; 7, 6.4. The extracts from 1 ml aliquots of plasma were also analyzed under the same set of conditions (high sensitivity, low-UV detection). Fig. 7 shows analyses at the three corners of the selectivity triangle. Again, dramatic differences are seen in the chromatograms; most of the peaks move with changes in solvent composition. The negative peak seen in Fig. 7c is due to the sample not containing tetrahydrofuran, having been dissolved in methanolic solvent; it could be eliminated by dissolution in aqueous tetrahydrofuran if use of this set of conditions were desired. Use of acetonitrile/water as mobile phase allows elution of trimethoprim in a section of the plasma chromatogram free of interference.

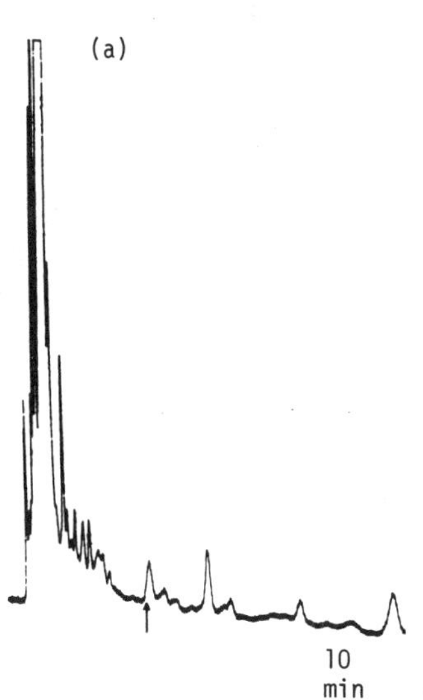

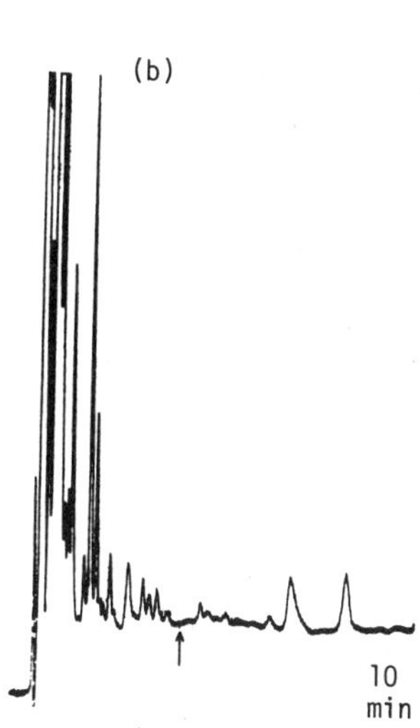

Fig. 7. HPLC of extract from plasma blank (*arrow* shows the retention position for trimethoprim). (a), Point 1: 40.7% methanol.
(b), Point 2: 25.3% acetonitrile.
(c), Point 3: 25.3% tetrahydrofuran.

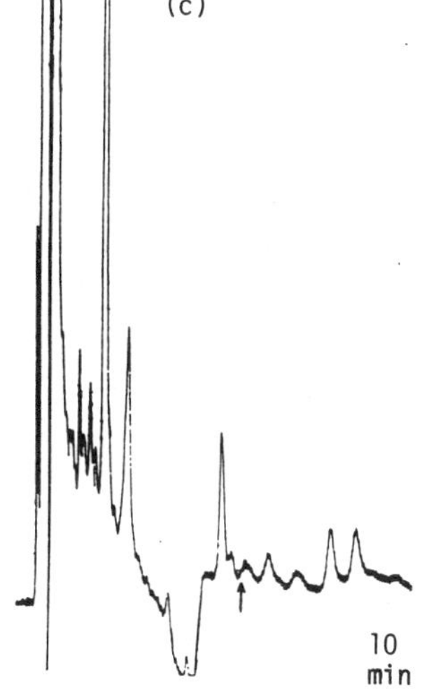

CONCLUSION

Application of an established multi-solvent optimization scheme to the separation of anticonvulsant drugs allows delineation of a number of sets of conditions yielding different selectivities and elution orders of the components besides furnishing optimum separation. The same approach used for investigation of elution of an intended analyte and, on the other hand, of components extracted from its matrix allows identification of spaces in the matrix chromatogram into which the analyte may be 'fitted' to allow high-sensitivity analysis with minimal interference. [Mobile-phase optimization also features in the following article, #B-3, by E.P. Lankmayr & W. Wegsheider.- *Ed.*]

References

1. Snyder, L.R. (1978) *J. Chromatog. Sci. 16*, 223-234.
2. Snyder, L.R. (1980) in *High Performance Liquid Chromatography* Vol. 1 (Horvath, C., ed.), Academic Press, New York, pp. 207-331.
3. Snyder, L.R., Dolan, J.W. & Grant, J.R. (1979) *J. Chromatog. 165*, 3-30.
4. Glajch, J.L., Kirkland, J.J., Squire, K.M. & Minor, J.M. (1980) *J. Chromatog. 199*, 57-79.

#B-3

OPTIMIZATION STRATEGIES FOR CHROMATOGRAPHIC ANALYSIS OF MULTI-COMPONENT MIXTURES

E.P. Lankmayr and W. Wegscheider

Institute for Analytical Chemistry, Micro-
and Radiochemistry
Technical University
A-8010 Graz, Austria

In column-chromatographic analysis of biological samples, complex chromatograms with long elution times are often encountered. Once the general separation conditions have been chosen, e.g. type and mode of use of chromatographic column and mode of detection, a number of continuously tunable variables can be used to optimize the overall analytical procedure. Necessarily a criterion has to be defined that permits an objective judgement of the resulting chromatograms. For this purpose chromatographic response functions (CRFs) used with formal strategies of experimental optimization have been developed. The strategies involve either entirely empirical search and mapping techniques or semi-empirical functions that account for specific molecular interactions. In any case the optimization procedure has to be chosen in accordance with the complexity of the separation problem, and the CRF must reflect fundamental analytical performance characteristics such as precision, accuracy and analysis time. Example in APPENDIX: thyronine I-enantiomers.

Since the very beginning of chromatography major efforts have been devoted to developing optimum separation conditions. By and large, optimization to establish an acceptable separation within a reasonable time is done by trial-and-error. Potential interactions of various experimental factors can lead to very complex situations that make it difficult to tune those parameters that are most important for an optimum separation. Consequently, a systematic approach for optimizing chromatographic conditions is preferable. Various optimization techniques exist. The most commonly used quality criteria and search strategies are discussed below.

QUALITY CRITERIA FOR CHROMATOGRAPHIC SEPARATIONS

In order to characterize chromatographic separations it is a necessary, but not sufficient, condition that the optimization criterion be related to chromatographic resolution. For the quantitation of separation quality an objective criterion – a 'chromatographic response function' (CRF) – has to be defined. The following is a representative collation from the many published.[*]–

Function (1)   (2)   (3)   (4)

$$\sup_{R}\{\min\ \alpha_{ij}\} \qquad \varepsilon_R\{R_{s,\min} > \rho_0\} \qquad \sum_j \ln f_j/g_j \qquad 1/t\ \pi_j\ f_i/(g_j + 2n_j)$$

ref. [1]   [2]   [3]   [4]

The first response function was proposed by Laub & Purnell [1, 5]. Separation is improved by maximizing selectivity $\alpha_{ij} = k_i/k_j$ for the most difficult to resolve pair of solutes i and j. This criterion has proved successful for optimizing the composition of binary stationary phases in GC and has been extended to HPLC separations [6, 7].

Function (2) is used mainly along with 'overlapping resolution mapping'. This approach models the retention times (k' values) by an empirical function, and calculates the resolution for each pair of adjacent peaks, finally indicating a region in space where a pre-selected minimal resolution ρ_0 is achieved for all pairs [2, 8].

The first concept, however, that considered the separation of all sample components giving a detector signal was introduced by Morgan & Deming [3]. Their generalization of the concept of peak separation [9] resulted in CRF no. (3). The estimation of f and g is shown schematically in Fig. 1A (see legend for definitions). In multi-component analysis these quantities are determined for all solute pairs and used for quantitation. The numerical value of CRF (3) ranges from minus infinity in the case of severe overlap of any pair of solutes to zero in the case of baseline separation of all detected compounds.

Taking account of more general analytical aspects, CRF (4) has been defined by us [4]. This is a conceptual extension of CRF (3) that additionally deals with analysis time t and background noise n[†] (Fig. 1B). By definition, this CRF is concerned with achieving a maximum separation in the shortest time possible. On the other hand it approaches a minimum of zero either in cases of severe peak overlap $(f_i/g_i \to 0)$ – i.e. when a previously detected signal disappears – or in cases when the background noise becomes a limiting factor

---

[*] Some symbols are defined in the text that follows, or in the Fig. 1 legend. R_s = chromatographic resolution; ρ_0 = pre-selected minimum value of R_s. [†] Noise is expressed in terms of baseline amplitude.

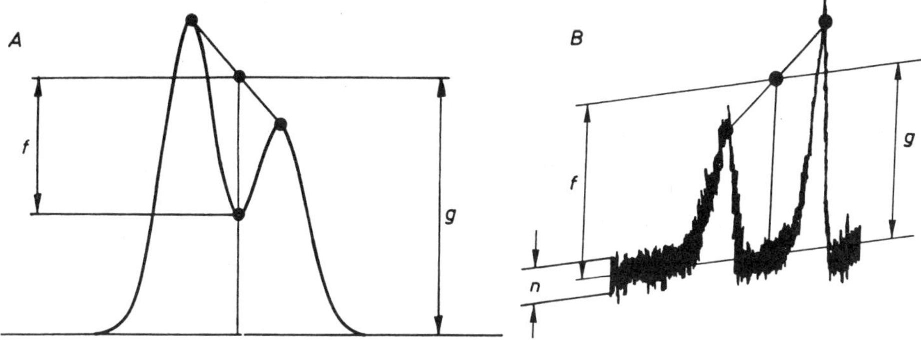

Fig. 1. Definition of peak separation [3, 4].
A: determination of f and g in presence of overlap but negligible noise.
B: trace-analytical situation with noise present.

$(n_i \rightarrow g_i)$, especially for precision in trace analysis even though no or negligible overlap may occur (Fig. 1B). The CRF may also approach zero when the time required to elute the last peak tends to infinity.

Summarizing the features of the different CRFs, it can be stated that those designated (1) and (2) above are measures of separation based on chromatographic theory, whereas peak separation as used in (3) and (4) is based on the properties of the output of the chromatographic system and, as such, also accounts for variable signal strength. Hence the correct estimation of chromatographic resolution from the chromatogram is not biased by peak asymmetry. An objective evaluation of the relationships between CRFs and their performance characteristics in terms of precision, accuracy and analysis time in multi-component systems revealed that up to now CRF (4) is the only function that accounts for all these basic analytical requirements [10]. CRFs (1) and (2) are sensitive only to the pair of solutes that is hardest to resolve, thus balancing one poor separation either against a possibly large number of excellent separations or even against a possibly large number of similarly bad separations. Accuracy is directly addressed only by those CRFs that are sensitive to the detection process itself, automatically considering relative signal heights and peak asymmetries.

OPTIMUM LOCALIZATION AND SELECTIVITY EFFECTS

Research on formal strategies for pin-pointing optimum separation conditions has so far come up with a number of alternatives.

The 'window diagram' and 'overlapping resolution mapping' techniques represent two possible strategies. Another frequently applied search strategy, likewise needing only modest computational effort, is a variable SIMPLEX algorithm as proposed by Nelder & Mead [11]. It must be emphasized, however, that an optimization strategy is never restricted to a particular optimization criterion and *vice versa*.

Empirical optimization ultimately resulting in fully automated instrumentation is useful only when variables are at hand that permit to tune effectively separation selectivities as well as retention times. The most common variables are mobile phase composition [2, 12-16], pH and ionic modifiers [4, 7, 17-19] and stationary phases [20]. This is by no means an exhaustive list of the variables useful in optimization. Additionally there seems to be quite a potential for incorporating secondary chemical equilibrium effects [21] and the variation of separation temperatures [22, 23].

With any strategy the operator has to select those experimental variables that are most conducive to good separation. Also, lower and upper boundaries have to be defined for related variables which by definition directly affect each other (e.g. solvent composition) as well as for independent variables which have little or no influence on any other experimental parameter (e.g. pH, temperature). The actual applicable range is governed by theoretical and practical aspects of stability, miscibility and analyte solubilities as well as mobile-phase and stationary-phase components. Thus a multi-dimensional space of variables is defined and has to be considered in any strategy aimed at establishing optimum conditions.

A semi-empirical approach has been proposed for optimizing the separation of a 9-component mixture of weak acids, weak bases and zwitterionic compounds [7, 17]. In this case a 4-level 2-factor experimental design was employed for specification of 16 mobile-phase compositions consisting of combinations of 4 pH values and 4 concentrations of octylamine hydrochloride as ion-interacting reagent (IIR). Retention data for all 9 solutes with these mobile phases were used to construct a multi-factor window diagram (Fig. 2).

The optimum was found with a 45 min separation at pH 3.7 and 0.75 mM octylamine hydrochloride. Fig. 2 demonstrates that the window-diagram approach is a very efficient mapping method for chromatographic methods if sufficient mechanistic information is available for a proper description of a response surface. Drastic limitations can be observed in fully empirical optimization procedures where extensive numbers of experiments may be required to map the space of variables without on-line identification of the sample constituents. Moreover, information based on α does not take into account the important influence of the capacity factor (k') values at a given column plate number. Resolution of two components always

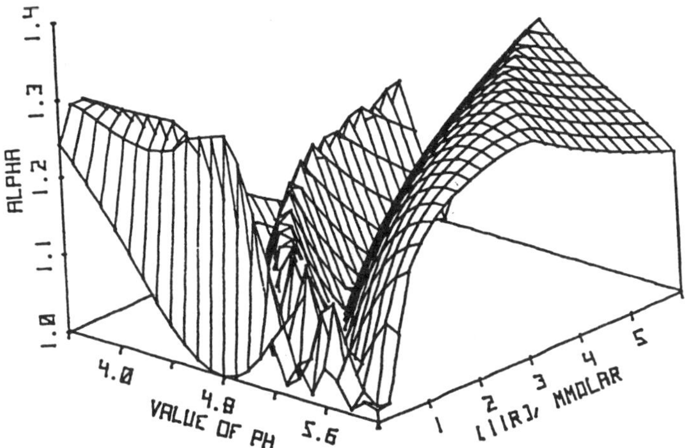

Fig. 2.   'Minimum plot' for a 9-component mixture ('ALPHA', α) [7].

is a function of their k' values and is insufficiently described by
α values alone.

For optimization of ternary or quaternary mobile-phase compositions the so-called overlapping resolution mapping technique has
become quite useful.   Its basic idea is centered around Snyder's
solvent selectivity triangle concept [24] and a statistical mixture
of experimental sets [8].   Retention data measured for each compound
at several solvent-composition ratios are fitted to a second-order
polynomial to ultimately obtain resolution maps for each pair of
of adjacent peaks. This method requires the definition of a limiting
resolution value which has to be met by all solute pairs in a pre-
selected k' window.   Consequently an experienced operator and/or
adequate chromatographic pre-information are desirable. The characterization of the response surface by a second-order polynomial
represents an additional drawback since only variables that yield
relatively smooth response curves can furnish realistic results.

Finally the use of SIMPLEX algorithm as search strategy should
be discussed. Its ability to localize an optimum without any need
to characterize the entire response surface has made it quite popular
[3, 4, 19, 25]. The algorithm directs the adjustment of experimental
conditions away from those which give a poor response towards conditions with a more favourable response. This provides an efficient
search procedure and can also be easily integrated into an automated

system. It should be noted, however, that direct-search algorithms, such as the SIMPLEX, should be used on low modal response surfaces only. In cases where several optima are to be observed, SIMPLEX searching can turn out to be tedious and time-consuming, with convergence to any of the optima present. This would be the case especially in the optimization of more complex separations where generally several optima can be observed (2 from each pairwise reversal of order).

In the future, separation optimization without intervention in the calculation program will also have to incorporate sequential search strategies that account automatically for complexities in the separation problem. In view of multi-dimensional detection systems and improved data-reduction techniques, a new definition of corresponding quality criteria will become necessary.

References

1. Laub, R.J. & Purnell, J.H. (1975) *J. Chromatog.* *112*, 71-79.
2. Glajch, J.J., Kirkland, J.J., Squire, K.M. & Minor, J.M. (1980) *J. Chromatog.* *199*, 57-79.
3. Morgan, S.L. & Deming, S.N. (1975) *J. Chromatog.* *112*, 267-285.
4. Wegscheider, W., Lankmayr, E.P. & Budna, K.W. (1982) *Chromatographia* *15*, 498-504.
5. Laub, R.J. & Purnell, J.H. (1976) *Anal. Chem.* *48*, 799-803 and 1720-1724.
6. Deming, S.N. & Turoff, M.L.H. (1978) *Anal. Chem.* *50*, 546-548.
7. Sachok, B., Kong, R.C. & Deming, S.N. (1980) *J. Chromatog.* *199*, 317-325.
8. Rautela, G.S., Snee, R.D. & Miller, W.K. (1979) *Clin. Chem.* *25*, 1954-1964.
9. Kaiser, R.E. (1960) *Gas Chromatographie*, Geest & Portig, Leipzig, p. 33.
10. Wegscheider, W., Lankmayr, E.P. & Otto, M. (1983) *Anal. Chim. Acta* *150*, 87-103.
11. Nelder, J.A. & Mead, R. (1965) *Comput. J.* *7*, 308-313.
12. Bakalyar, S.R., McIlwrick, R. & Roggendorf, E. (1977) *J. Chromatog.* *142*, 353-365.
13. Schoenmakers, P.J., Billiet, J.A.H. & deGalan, L. (1979) *J. Chromatog.* *185*, 179-195.
14. Antle, P.E. (1982) *Chromatographia* *15*, 277-281.
15. Snyder, L.R., Glajch, J.L. & Kirkland, J.J. (1982) *J. Chromatog.* *218*, 299-326.
16. Glajch, J.L., Kirkland, J.J. & Snyder, L.R. (1982) *J. Chromatog.* *238*, 269-280.
17. Kong, R.C., Sachok, B. & Deming, S.N. (1980) *J. Chromatog.* *199*, 307-316.
18. Lindberg, W., Johansson, E. & Johansson, K. (1981) *J. Chromatog.* *238*, 201-212.

19. Berridge, J.C. (1982) *J. Chromatog. 244*, 1-14.
20. Goldberg, J.C. (1982) *Anal. Chem. 54*, 342-345.
21. Karger, B.L., Le Page, J.N. & Tanaka, N. (1980) in *High Performance Liquid Chromatography, Advances and Perspectives*, Vol. 1 (Horváth, C., ed.), Academic Press, New York, pp. 113-206.
22. Gant, J.R., Dolan, J.W. & Snyder, L.R. (1979) *J. Chromatog. 185*, 153-177.
23. Hammers, W.E., Theeuwes, A.G.M., Brederode, W.C. & De Ligny, C.L. (1982) *J. Chromatog. 234*, 321-336.
24. Snyder, L.R. (1978) *J. Chromatog. Sci. 16*, 223-234.
25. Watson, M.W. & Carr, P. (1979) *Anal. Chem. 51*, 1835-1842.

APPENDIX: An illustrative optimization

The following example concerns the determination of enantiomeric iodinated thyronines in human blood plasma and urine samples. In order to achieve selective and sensitive detection of the analytes in such complex sample matrices, a catalytic reaction detector has been developed [26] (Fig. 3). It operates on the basis of the I^--catalyzed reaction of chloramine-T and tetrabase. In a small column filled with zinc powder the iodine bound in thyronine is effectively reduced to free iodide anions. The experimental set-up also incorporates a pre-column derivatization of the thyroid hormones with *tert*-butyloxy-L-leucine-*N*-hydroxysuccinimide ester in order to obtain diastereomeric derivatives which can be resolved on commercially available reverse-phase columns [27].

The factors to be optimized in the chromatographic separation process included the pH, methanol concentration and total buffer concentration in the mobile phase. For optimum localization in the space of variables with boundaries specified prior to analysis, i.e. methanol concentration from 40 to 90% (v/v), pH from 2.5 to 8 and total buffer concentration from 5 to 25 mmol, CRF no. 4 was coupled to a SIMPLEX search procedure. The first mobile phase was chosen randomly with 50% methanol content and a 11.8 mM buffer pH 3.0. The variation of the CRF during progress of the optimization runs and the resulting chromatogram after vertex 12 are shown in Fig. 4.

An interesting aspect, frequently encountered in laboratory practice, concerns the adaption of a separation system optimized with a specific separation column onto a similar column which may differ in efficiency and/or surface chemistry. The LiChrosorb RP 18 column (Merck, Darmstadt, GFR) used for the first set of experiments was replaced by a Spherisorb 5 ODS column (Phase Separations, Queensferry, U.K.) which represents a comparable column type. The new run was started under the optimum conditions as determined with the LiChrosorb column. Owing to different column efficiency and potential changes in chromatographic selectivity, the SIMPLEX terminated

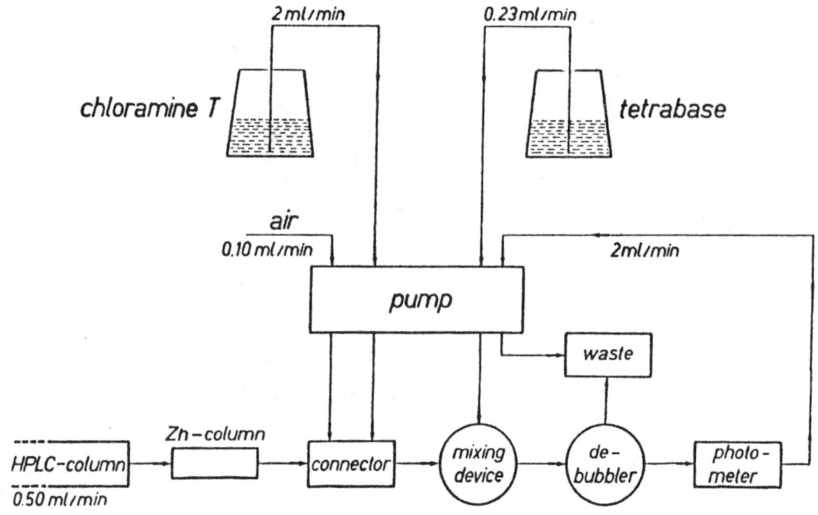

Fig. 3. Configuration of the catalytic reaction detector.

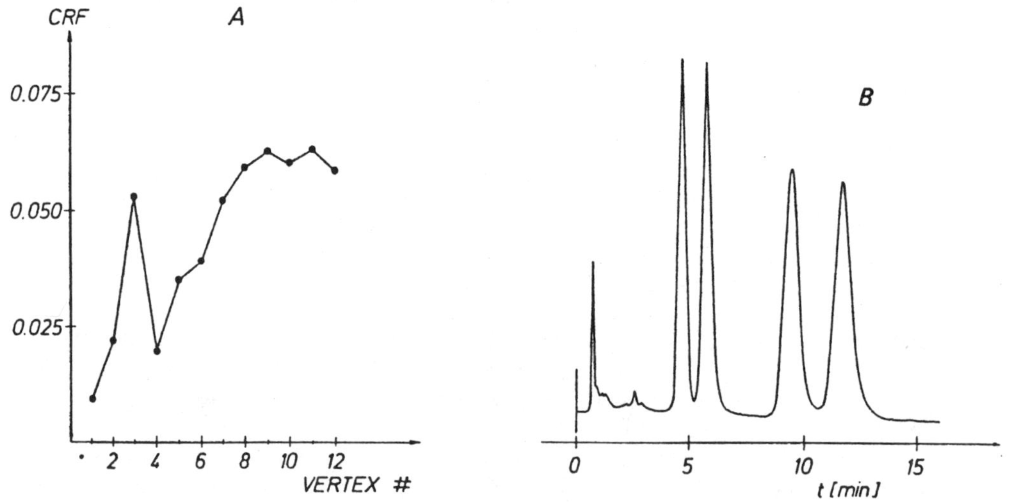

Fig. 4. *A:* Chromatographic response function as SIMPLEX proceeds.
B: Chromatographic separation of isomeric thyronines after vertex 12 (T = thyronine).
Column: LiChrosorb RP-18, 7 µm, 150 x i.d. 3.2 mm. Mobile phase: 18.8 mM phosphate;citrate buffer, pH 3.2, with 59.2% (v/v) methanol; 1 ml/min.
Elution order: L-T$_3$, D-T$_3$, L-T$_4$, D-T$_4$.

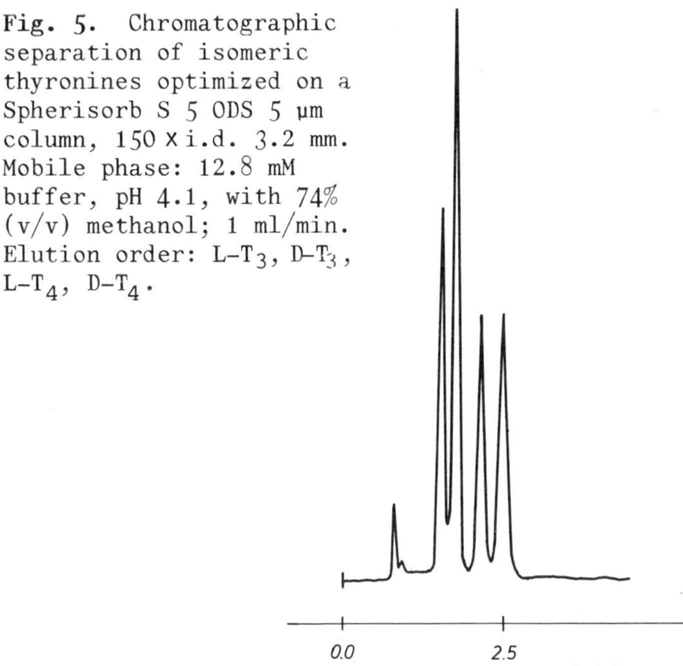

Fig. 5. Chromatographic separation of isomeric thyronines optimized on a Spherisorb S 5 ODS 5 µm column, 150 x i.d. 3.2 mm. Mobile phase: 12.8 mM buffer, pH 4.1, with 74% (v/v) methanol; 1 ml/min. Elution order: $L-T_3$, $D-T_3$, $L-T_4$, $D-T_4$.

at different experimental conditions. The new optimum was localized at pH 4.1, 12.8 mM buffer concentration and a methanol concentration of 74%. The optimization converged to an optimum defined by the same elution order, but analysis time decreased by a factor of more than 4 (Fig. 5), thus establishing conditions suitable for application in extended clinical studies [28, 29].

In conclusion it should also be emphasized that the optimization procedure described requires neither information concerning the number and nature of the analytes nor any prior information on their chromatographic behaviour. As with any optimization strategy, however, the operator must select those experimental variables that influence selectivity and retention most significantly.

References for Appendix

26. Lankmayr, E.P., Maichin, B. & Knapp, G. (1981) *J. Chromatog.* *224*, 239-248.
27. Lankmayr, E.P., Budna, K.W. & Nachtmann, F. (1980) *J. Chromatog.* *198*, 471-479.
28. Leb, G. Budna, K.W., Sommersgutter, M., Goebel, R., Lankmayr, E.P., Passath, A. & Knapp, G. (1981) *Acta Med. Austriaca 8*, 71-74.
29. Leb, G., Lankmayr, E.P., Goebel, R., Pristautz, H., Nachtmann, F. & Knapp, G. (1981) *Klin. Wochenschr. 59*, 861-863.

#B-4

REVERSED-PHASE TLC ON ALKYL-BONDED AND PARAFFIN-COATED SILICA GEL

I.D. Wilson

Department of Drug Metabolism
Hoechst UK
Walton Manor
Walton, Milton Keynes MK7 7AJ, U.K.

Reversed-phase (RP) TLC is well suited to the chromatography of polar compounds. Advantages of RP-TLC compared to normal-phase (NP) TLC include:- high recoveries of material from the plate, ease of development of solvent systems, and in some cases greatly reduced decomposition of sensitive compounds.

Two types of RP-TLC plate are available, chemically modified (or bonded) and physically coated. In the former an alkyl chain, most commonly C-18, is attached to a silica gel support using the appropriate organosilane. Coated plates are prepared by dipping (or developing) a pre-coated silica-gel TLC plate in a solution of the stationary phase (e.g. liquid paraffin) in a volatile solvent. On evaporation of the solvent the silica gel receives an even coating of the stationary phase, and can then be used in the reversed-phase. Both types of RP-TLC plate have proved to be suitable for the chromatography of a wide range of organic compounds.

The RP-TLC technique has been known for many years [e.g. 1] but until recently has been little used. The increased popularity of the method probably owes much to the wider understanding of chromatography in the RP mode, and the commercial availability of alkyl-bonded plates. Both of these advances probably resulted from the development of RP-HPLC as a major chromatographic technique.

RP-TLC PLATES: GENERAL PROPERTIES

One of the manufacturers, Merck, produces C-2, C-8 and C-18 RP-TLC plates so as to cover a range of polarities. Macherey Nagel &

Co. (MN) achieve the same result by regulating the extent of silanization of their C–18 bonded plates to give "50, 75 and 100% coverage". Whatman also produce a C-18 modified plate. A property shared by all these plates is that they are intensely hydrophobic, limiting the amount of water which can be used in the mobile phase. To some extent this problem may be overcome for the Whatman plates by the addition of salt to the solvent system, allowing much larger amounts of water to be used in the mobile phase.

More recently a number of manufacturers have introduced non-hydrophobic bonded RP-TLC plates.* These include a C-18 bonded plate from Merck and a C-12 bonded plate from Antec. Both types are suitable for use with 100% water as solvent. Similarly paraffin-coated RP-TLC plates are non-hydrophobic and these may also be used with 100% water. We have examined the properties of the various C-18 C-18 bonded plates using the ecdysteroids (insect growth regulators) as model compounds [2, 3]. The relationship between R_f and the methanol content of the mobile phase for ecdysone is shown in Fig. 1 for fully silanized C-18 bonded plates. There is little difference among different makes of plate (except for load, *below*). We find that Merck C-8 and MN "75%" C-18 give similar results to their longer chain, or more completely silanized C-18, counterparts. It is only with the Merck C-2 and MN "50%" C-18 plates that significant differences from the C-18 (or "100%" C-18) plates are seen. Both the C-2 and MN "50%" C-18 plates require less organic modifier in the mobile phase to achieve the same R_f as their C-18 or "100%" C-18 equivalents.

A useful feature of RP-TLC plates (both bonded and coated) is the wide range of solvent compositions over which there is a linear relationship between R_f and methanol concentration (Fig. 1). This kind of relationship is not restricted to the ecdysones, but applies to all the compounds which we have investigated. This property allows facile adjustment of the mobile phase composition to achieve the desired R_f. [The following amplification is complemented by #NC(B)–1.]

SOLVENT SYSTEMS

A wide range of organic solvents are suitable for RP-TLC including propan-2-ol (IPA), acetonitrile (ACN), acetone (AC), ethanol (EtOH), methanol (MeOH), tetrahydrofuran (THF) and dioxan. Indeed, any water-miscible organic solvent should be suitable for RP-TLC (in contrast to RP-HPLC where the UV properties of the solvent greatly restrict the choice). In our investigations on the RP-TLC of the ecdysteroids [2] we established the following eluotropic series: IPA >ACN > AC >EtOH > MeOH. As in RP-HPLC, different organic modifiers can be used to give different separations because of changes in the selectivity of the system (e.g. see [4]). The changes in R_f for a range of 11 ecdysteroids relative to ecdysone (= 100%) are shown in

* tabulated by U.A.Th. Brinkman in #NC(B)–1; cf. footnote at end of art.

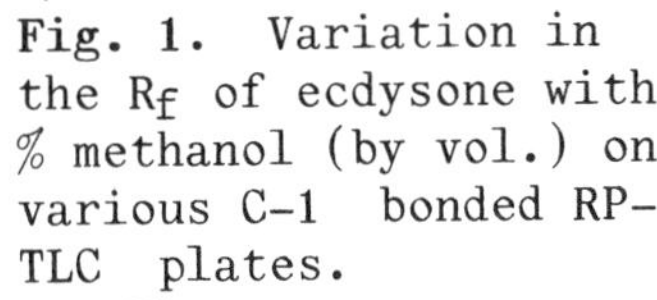

Fig. 1. Variation in
the R_f of ecdysone with
% methanol (by vol.) on
various C-1 bonded RP-
TLC plates.
▲ , Whatman;
■ , Merck;
● , MN "100%";
□ , Merck (non-
hydrophobic).

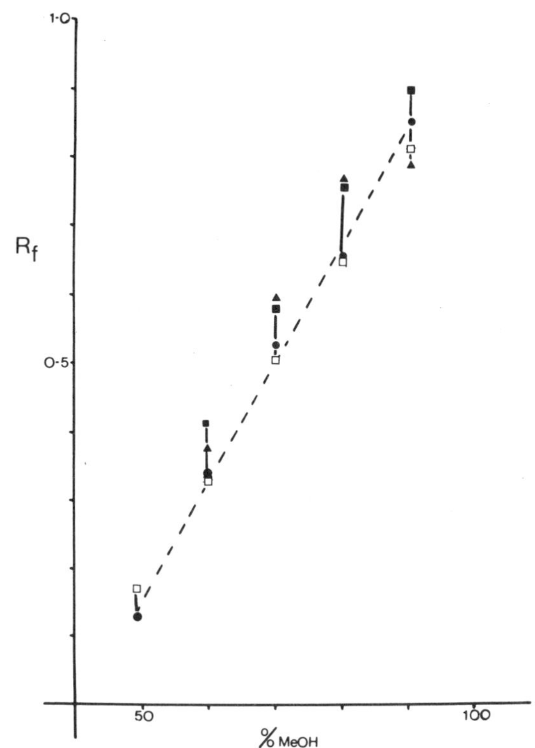

Fig. 2, using simple binary mixtures of IPA, MeOH or ACN with water.
With the appropriate choice of organic modifier most separations
should be possible.*

RECOVERY OF COMPOUNDS FROM RP-TLC PLATES

Another useful feature of RP-TLC which we have exploited is the
relatively high recoveries (compared to NP-TLC) of substances from
the plate following chromatography. The ecdysteroids, for instance,
are subject to high adsorptive losses on NP-TLC (~50%) whilst much
higher recoveries (up to 100%) are possible from RP-TLC plates, both
bonded and coated [2, 3]. This property may be of particular use
for the isolation of relatively polar substances such as drug metabolites. A point worth noting is the poor quality of some bonded
RP-TLC plates; they have to be extensively washed prior to chromatography in order to remove impurities that co-elute when the sample
is recovered from the plate [2]. Not only are higher recoveries of
polar compounds achieved with RP-TLC, but we have observed that
substances which decompose when chromatographed on silica gel in the
normal phase are not degraded when subjected to RP-TLC [5].

---

* Recently we have found that ion-pair reagents enable the hydroxybenzoic acids to be chromatographed.

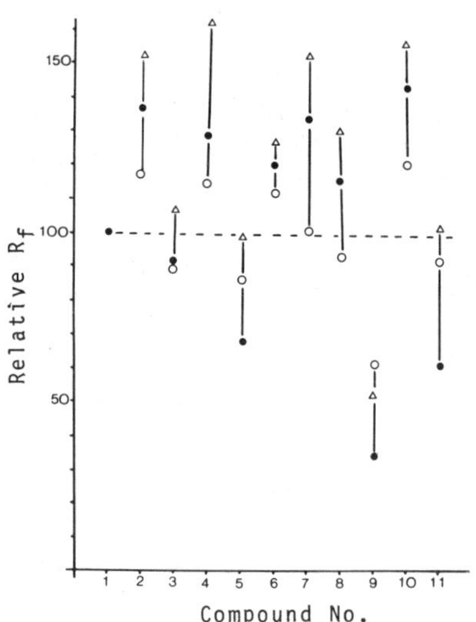

Fig. 2. Solvent selectivity, evidenced by R_f values for a range of ecdysteroids (on Merck C-18 plates), as influenced by the choice of organic modifier: ●, IPA; △, MeOH; ○, ACN. All R_f values are normalized to that of ecdysone (Cpd. **1**, = 100%; shown ----). **2**: 20-hydroxyecdysone; **3**: 2-deoxy-20-hydroxyecdysone; **4**: inokosterone; **5**: muristerone; **6**: makisterone A; **7**: cyasterone; **8**: poststerone; **9**: ponasterone A; **10**: polypodine B; **11**: ajugasterone C.

APPLICATION OF RP-TLC TO THE ASSAY OF ISOXEPAC IN URINE

We have used RP-TLC on C-18 bonded plates in an assay designed to determine patient compliance during treatment with the non-steroidal anti-inflammatory drugs (NSAID's) isoxepac, indomethacin and aspirin. RP-TLC was chosen because an adequate separation of these three NSAID's proved very difficult using NP-TLC on silica gel, but was easily achieved on C-18 bonded plates. In this assay 1 ml of the patient's urine was acidified to pH 2 using HCl, and extracted into 5 ml of chloroform. An internal standard (i.s.) was added to each sample to check both the extraction and the chromatography. The extract was then blown to dryness under a nitrogen stream, taken up in a small volume of methanol, and applied to the plate. [In Vol. 12, this series, I.D. Wilson et al. discussed isoxepac in respect of conjugate features, investigated with use of HPLC.-*E.R., Ed.*]

We examined both Merck and MN "100%" C-18 plates for use in this assay. Fig. 3 shows clear differences between the two types of plate, loaded with identical amounts of the same samples: whilst the Merck plate gives excellent results with i.s. and isoxepac each readily apparent, the MN plate is obviously overloaded, and uninterpretable. No sample capacity differences were seen with standards.

We used the hydrophobic type of RP-TLC plate for the isoxepac compliance assay, necessitating extraction of the sample in order to obtain it in a form suitable for application to the plate. Were we to develop such an assay now, we would use non-hydrophobic plates (either bonded or coated), and direct application of urine.

We have found it possible to apply urine and bile to the non-hydrophobic RP-TLC plates without difficulty. With plasma it is

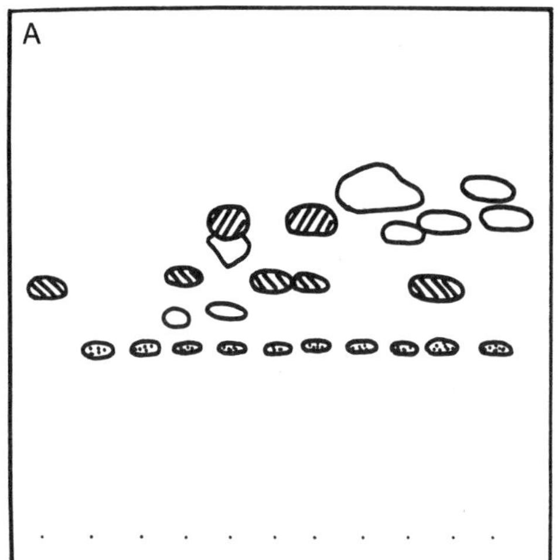 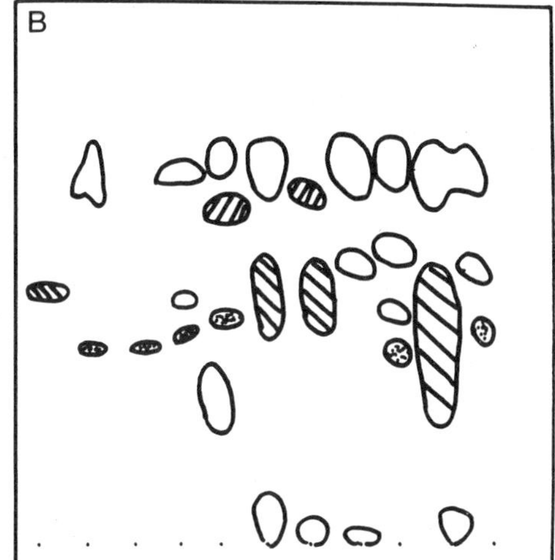

Fig. 3. RP-TLC of urine samples from the compliance assay for
isoxepac, with C-18 plates: A = Merck,   B = MN. Isoxepac, ⊘
(in each case an isoxepac standard is present on the left); i.s. ⊘
Visualization was based on native fluorescence; excit. at 366 nm.

necessary to first precipitate proteins (with an equal vol. of ACN).
If the proteins are not precipitated the samples chromatograph as
badly distorted streaks [5].   A comparison of the Merck C-18 hydrophobic and non-hydrophobic plates in terms of R_f vs. methanol content of the mobile has demonstrated that they were similar (cf.
Fig. 1).   Transfer of separations from one type to the other should
not be difficult.

PARAFFIN–COATED RP-TLC PLATES

There are a number of advantages to be gained by using 'home-
made' coated plates rather than commercially supplied bonded RP-TLC
plates. Firstly, they can be prepared very simply in the laboratory
as required, eliminating the need for large stocks. Our method is
to make up 100 ml of a solution of 7% paraffin ('NUGOL'; v/v) in
dichloromethane, which is then placed in a 25 x 25 x 5 cm glass TLC
tank. The required number of pre-coated silica-gel TLC plates are
placed in the tank, and the solvent allowed to migrate to the top of
the plates in the usual way. When the solvent front reaches the top
the plates are removed and allowed to dry in a fume cupboard. This
simple process converts silica-gel TLC plates into RP-TLC plates.

Secondly, the low cost of the materials used means that coated
plates are relatively cheap compared to their chemically modified
alternatives. A third advantage of coated plates is that they may
be produced in any size (e.g. preparative size) and on any type of

backing (e.g. plastic or metal foil). This is important as currently the range of chemically modified plates is mostly confined to glass-backed plates, in a limited range of sizes. Lastly, as mentioned earlier, paraffin-coated plates are non-hydrophobic and can be used with entirely aqueous samples or solvent systems.

COMPARISON OF BONDED AND COATED RP-TLC PLATES

We have compared the chromatography of a wide range of organic compounds on paraffin-coated silica gel plates and Merck non-hydrophobic C-18 RP-TLC plates. The compounds investigated included ecdysteroids, aminophenols, NSAID's, antipyrine and a number of mono- and dihydroxybenzoic acids [6]. We found both types of plate to be suitable for all of these compounds, with the exception of the hydroxybenzoic acids (which gave only poorly defined streaks). Provided that some adjustment is made to the solvent (the bonded plates requiring rather more methanol in the mobile phase), the two types of plate appear interchangeable [6].

Our experiences suggest that the choice between bonded and coated plates does not depend on their chromatographic properties but rather on factors such as economy and convenience. The coated plates are cheap, but must be prepared before use, whilst the bonded plates are convenient but rather costly.*

CONCLUSION

Given the popularity and ease of use of NP-TLC on silica gel, it is unlikely that reversed-phase chromatography will achieve the dominant position in TLC that it occupies in HPLC. However, RP-TLC represents a useful extension of the technique of TLC, especially for polar compounds. It is likely, therefore, that the number of applications of RP-TLC will continue to increase as the technique becomes more widely known, and its advantages appreciated.

References

1. Smith, I., ed. (1969) *Chromatographic and Electrophoretic Techniques*, Vol. 1, 3rd edn., Heinemann, London, 1080 pp.
2. Wilson, I.D., Scalia, S.& Morgan, E.D. (1981) *J. Chromatog. 212*, 211-219.
3. Wilson, I.D., Bielby, C.R. & Morgan, E.D. (1982) *J. Chromatog. 242*, 202-206.
4. Brinkman, U.A.Th. & de Vries, G. (1983) *J. Chromatog. 265*, 105-110.
5. Wilson, I.D. (1983) *J. Pharm. Biomed. Anal. 1*, 219-222.
6. Wilson, I.D. (1984) *J. Chromatog.*, in press.

* Bonded plates may be made 'in-house' by silanizing silica gel TLC plates (2% v/v organosilane in toluene [2]). The process is expensive in reagents and solvents, and hazardous because of the reactivity of the silanes; purchase of bonded plates is preferable.

#NC(B)

NOTES and COMMENTS relating to

Strategies for HPLC (other than detection) and for TLC

Comments related to particular contributions

#B-1, #B-2 & B-3, p. 107
#B-4, #NC(B)-1 & #NC(B)-2, p. 108

#NC(B)-1

A Note on

CHEMICALLY BONDED STATIONARY PHASES IN (HP)TLC

U.A. Th. Brinkman and G. de Vries

Department of Analytical Chemistry
Free University
de Boelelaan 1083
1081 HV Amsterdam, The Netherlands

In modern TLC, including its high-performance (HP) mode, silica is the stationary phase material preferred by a large majority of practitioners, and TLC on chemically bonded phases occupies a rather modest position. This is in marked contrast with the situation in HPLC where much work is done with chemically bonded phases, especially C-18 and C-8 modified silicas. Still, in recent years the widespread use of these apolar stationary phases in what often is called reversed-phase (RP) HPLC has promoted interest in their utilization in TLC, and a recent review on RP-TLC features over 90 references [1].

Today, a wide variety of good quality pre-coated plates for RP-TLC is commercially available, as is evident from the summary in Table 1 [2]. Complementing the range of plates with an apolar coating, plates with an amino (Merck) and a phenyl (Whatman) phase have recently been marketed. A large majority of the pre-coated plates appear to be of normal TLC quality ($d_p \geq 10$ μm), the exceptions being the RP-coated HPTLC-quality plates ($d_p = 5\text{-}7$ μm) and, possibly, the MN-type layers ($d_p = 5\text{-}10$ μm).

Satisfactory overall separation efficiency is combined with compatibility with mobile phases containing up to at least 80% water for the newly introduced RP-8- and RP-18-coated TLC plates, and with the KC_{18} and $Si\text{-}C_{18}$ plates provided that ~3% of NaCl is added to the mobile phase. The Nano-Sil C18-50 plates can be used up to ~70% of water. The Merck RP-coated HPTLC plates and the Nano-Sil C18-75 and -100 plates can be used over only a limited range of mobile phase compositions (<30-40% water). With the OPTi-UP C_{12} plates, migration does not pose a serious problem at any mobile phase composition, but often resolution is poorer than with the other plates. Amongst various other types of plates tested, the plots of migration time

Table 1. Apolar chemically bonded RP-TLC plates on the market.

Manufacturer	Plate designation	C chain	Miscellaneous features
Antec	OPTi-UP C_{12}	12	Inorganic binder
Baker	Si-C_{18}	18	± pre-scored
Macherey, Nagel	NANO-SIL C18-100, -75, -50	18	
Merck	Kieselgel silanisiert	2	
	RP-2, -8, -18	2, 8	Acid-resistant F_{254} TLC
	(HPTLC)	18	plates have improved
	RP-8, -18 (TLC)	8, 18	water-compatibility
Whatman	KC$_{2,\,8,\,18}$	2, 8, 18	± pre-adsorbent strip

vs. mobile phase composition are notably diverse, seemingly indicative of fundamental differences in the manufacturing process [2].

The advantage with the RP-18- and RP-8-coated TLC plates of un-
limited use without added NaCl is partly offset by the shorter migration time observed with the KC_{18} plates in the often used 30-70%
methanol range. Obviously replacing methanol by acetonitrile will
decrease development time markedly. Acetone and dioxane also are
promising alternatives as the organic modifier in RP-TLC [3].

All types of pre-coated plates display identical separation
sequences with apolar test analytes such as phthalate esters and
chloroanilines. Considerable and often inexplicable differences
occur, however, in the separation of, e.g., chlorophenols and aminophenols [2]. No general conclusion can be reached here as regards
the preferred type of RP-TLC plate. One can merely state that the
varied, if unpredictable, behaviour of the pre-coated plates adds to
the technique's potential as a tool for creating separations. Yet,
when using RP-TLC for a rapid evaluation of mobile phase systems for
use in RP-HPLC one should proceed with caution. Generally speaking,
correlation of RP-TLC and RP-HPLC data will present no problems when
dealing with apolar solutes. With polar solutes, however, differences in interactions with residual silanol groups and binders, and
effects caused by the addition of buffer constituents or ion-pairing
agents may well ruin the expected similarity between RP-TLC and RP-
HPLC retention data. RP-HPLC prediction is pursued in #NC(B)-3.

References

1. Brinkman, U.A.Th. & de Vries, G. (1982) *J. High Resol.
    Chromatog. & Chromatog. Comm. 5*, 476-482.
2. Brinkman, U.A.Th. & de Vries, G. (1983) *J. Chromatog. 258*,
    43-55.
3. Brinkman, U.A.Th. & de Vries, G. (1983) *J. Chromatog. 265*,
    105-110.

#NC(B)-2

A Note on

TLC AS A PILOT TECHNIQUE FOR REVERSED-PHASE HPLC

N. Lammers, J. Zeeman, G.J. de Jong and [*]U.A.Th. Brinkman

Research Laboratories
Duphar, P.O. Box 2
1380 AA Weesp
The Netherlands

[*]Department of Analytical
Chemistry, Free University
De Boelelaan 1083
1081 HV Amsterdam
The Netherlands

Studies now presented have shown that the RP-HPLC behaviour of test solutes correlates well with TLC on alkyl-bonded silica if the mobile phase contains only water and an organic modifier, but less well if it contains an ion-pair reagent and buffer. We have investigated conditions for getting an ion-pair effect in RP-TLC.

In RP-HPLC - the mode used for ~90% of our HPLC separations - mobile-phase parameters to be considered in the time-consuming development of new separation procedures include the nature and percentage of organic modifier, and type and concentration of the buffer and of the ion-pair reagent. Several authors [1-7] have recently reported on the close similarity in retention data between HPLC and TLC.

Now that several manufacturers offer TLC plates of silica with chemically bonded alkyl groups [8; and #B-4, NC(B)-1, this vol.], TLC has warranted trial as a pilot technique for RP-TLC, since varying the mobile phase composition is easier and quicker in TLC than in HPLC. Our study has been based on some HPLC separation procedures recently developed in our laboratories.

MATERIALS AND PROCEDURES

The following TLC plates were used: KC_{18} (Whatman, Springfield Mill, Maidstone, U.K.) [termed KC18 below], Nano-SIL C_{18}-50 (Macherey, Nagel & Co., Düren, W. Germany) [MN50], and RPS Uniplates (Analtech, Newark, DE, U.S.A.) [RPS]. The layers were impregnated with the ion-pair reagents by dipping the plates for 10 min in a methanolic solution of the reagent, whose concentration was the same as that in the mobile phase. The plates were developed in a double-trough developing tank (Camag, Muttenz, Switzerland).

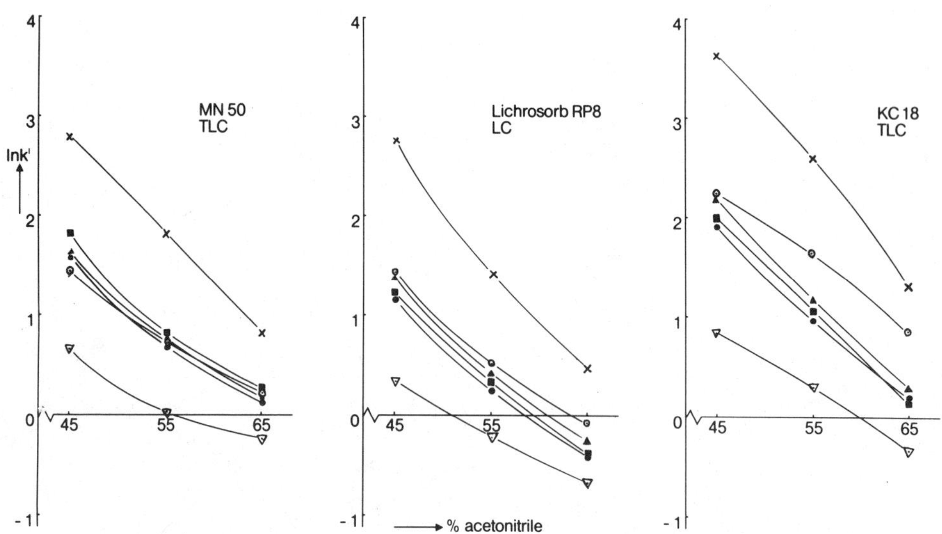

Fig. 1. Dependence of ln k' of diflubenzuron and its impurities on the mobile-phase acetonitrile content (along with water containing 10% dioxane). TLC with MN50 & KC18 plates; HPLC on Lichrosorb RP-8.

The HPLC stationary phases were 5 μm LiChrosorb RP-8 (Merck, Darmstadt) and 7 μm Zorbax C_8 (DuPont, Wilmington DE). Detection was by 254 nm UV (both TLC and HPLC) or iodine vapour (RPS TLC). All chromatographic data are expressed in terms of ln k' where k'= capacity factor. In TLC, ln k' is directly proportional to $R_M = \log (1/R_F - 1)$.

TRIAL OF MOBILE PHASES WITHOUT BUFFERS AND ION-PAIR REAGENTS

If the mobile phase contains only water and organic modifier(s) the similarity between TLC and HPLC is very close (relevant refs. in [8]). We have now confirmed this for some systems with mixtures of acetonitrile/water, dioxane/water and acetonitrile/dioxane/water as mobile phases. With the latter mixture we have studied the influence of varying acetonitrile content on the retention of the urea insecticide diflubenzuron and its impurities (Fig. 1). The two TLC plates were chosen for the comparison with an HPLC stationary phase because of their compatibility with mobile phases containing a high proportion of water [9]. The retention order of the test solutes on KC18 and LiChrosorb RP-8 is seen to be the same, whereas with MN50 the retention order is somewhat different. Plots of ln k'$_{HPLC}$ *vs.* ln k'$_{TLC}$ show correlation coefficients 0.991 for KC18 and 0.983 for MN50.

TRIAL OF ION-PAIR SYSTEMS

For separating basic and acidic compounds, RP ion-pair chromatography is often recommended. Such systems are relatively compli-

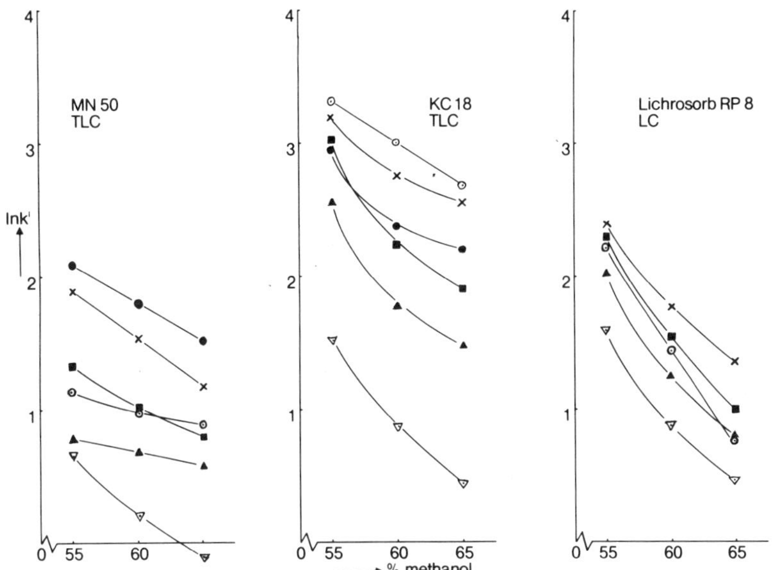

Fig. 2. Dependence of ln k' of fluvoxamine and its impurities and degradation products on the mobile-phase content of methanol (along with 0.04 M phosphate buffer pH 3 containing 5 mM Na heptylsulphonate). Stationary phase as in Fig. 1.

cated and optimization is often time-consuming. Hence it is very important to compare TLC and HPLC. The first example (Fig. 2) concerns the new antidepressant fluvoxamine, which contains a primary amino group, and its impurities and degradation products; the methanol content of the mobile phase was varied. The retention order is evidently very similar for MN50 and LiChrosorb RP-8, but with KC18 the relative retention of one impurity (o) deviates markedly.

Fig. 3 shows the influence of the concentration of the ion-pair reagent, sodium heptylsulphonate. Because the retention behaviour of the solutes is very different for the two TLC plates, a second HPLC stationary phase, Zorbax C_8, was included in this study. Evidently a different retention order occurs with each chromatographic system. Besides, the dependence of the retention of the ion-pairing test solutes (Fig. 3, legend) on the heptylsulphonate concentration is much more pronounced with Zorbax C_8 than with any other stationary phase. Moreover, it is very remarkable that the retention of the impurity (o) on KC18 remains about constant over the whole concentration range of heptylsulphonate. It seems that the ion-pair effect is absent for this compound on KC18. In other words, the value of TLC as a pilot technique for HPLC is highly debatable with the present test system. Evidently the use of stationary phases from different manufacturers enhances the complexity of the problem.

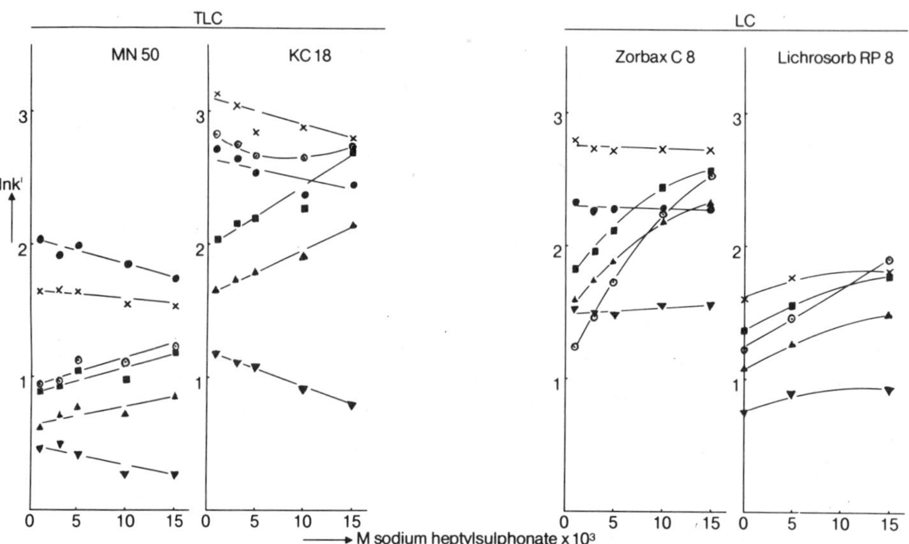

Fig. 3. Dependence of ln k' of fluvoxamine and its impurities and degradation products on the mobile-phase content of Na heptylsulphonate, viz. its molarity (as on axis) in the mixture of methanol and 0.04 M phosphate buffer pH 3 (60:40 by vol.). Stationary phase: MN50, KC18 (TLC), Zorbax C_8 and LiChrosorb RP-8 (HPLC). Ion-pairing compounds: ■, ▲, o.

Another parameter of interest is the length of the alkyl chain of the ion-pair reagent. Fig. 4 shows the effect of varying chain length on the retention of the fluvoxamine solutes. The test system was TLC on MN50 either non-impregnated or pre-impregnated with sodium heptylsulphonate (see earlier). With no impregnation an increase in chain length from C-3 to C-8 distinctly increases retention for the ion-pairing test solutes, as is to be expected. However, a noticeable decrease occurs upon further increasing the chain length to C-12. In contrast, with impregnated plates a continuous increase in retention is observed from C-3 through to C-12 for the ion-pairing test solutes, whereas the behaviour of the remaining compounds is essentially the same in both cases. Apparently, with higher mol. wt. ion-pair reagents impregnation of the plates is obligatory (see also refs. [10-12]).

For morphine and some of its degradation products, no ion-pair effect was observed even after dipping of KC18 and MN50 plates in a methanolic solution of sodium dodecylsulphate. This is in contrast with data for HPLC, where SDS addition to the mobile phase causes considerable retardation of the basic test compounds. With RPS plates, however, impregnation with SDS caused a 10-30% decrease of the R_F values for the pertinent compounds. Indeed, the retention order in the HPLC system and on the impregnated RPS plates was essent-

Fig. 4. Dependence of ln k'
of fluvoxamine and its
impurities and degradation
products on the chain length
of the ion-pair agent
(0.015 M in 0.04 M phosphate
buffer pH 3; mobile phase
methanol/buffer, 60:40 by
vol.). Stationary phase:
MN50, non-impregnated (a)
and impregnated (b).
Ion-pairing compounds: ■,
▲, ○.

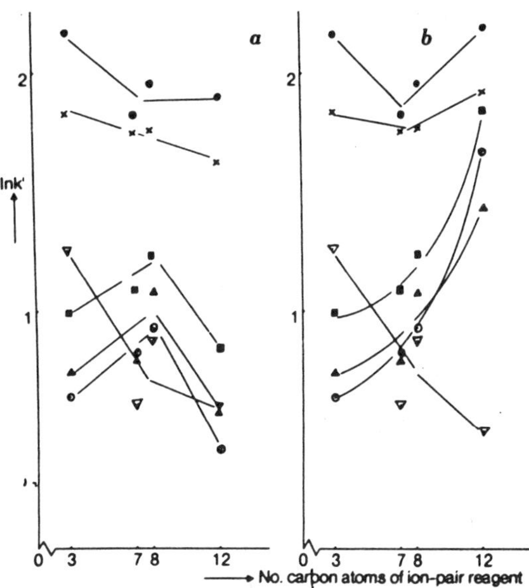

ially the same. Still, the large differences between the capacity
factors in TLC (−1.5 < ln k' < 0.5) and HPLC (0.7 < ln k' < 4) seriously
detract from the usefulness of TLC as a pilot technique for optimization of HPLC. The improved behaviour of the RPS plates as compared
with the KC18 and MN50 plates can perhaps be explained by the different nature of the stationary phases involved. Whereas KC18 and
MN50 have chemically bonded phases, the RPS plates contain a physically held long-chain hydrocarbon. This difference may well influence the ease of dissolution of the SDS and the sorption mechanism
of all the compounds.

As a final example, ion-pair chromatography was performed on a
test mixture containing an acidic compound; hexylamine was selected
as ion-pair reagent. Using KC18 and MN50 plates the ion-pair effect
was observed only after impregnation of the TLC plates with the
amine. In other words, contrary to the results with the acidic
ion-pair reagents, impregnation now is already required with rather
short alkyl chains. As with the earlier examples, impregnation of
the TLC plate did not affect the R_F values of the other test solutes
in the mixture.

CONCLUSION

For RP chromatographic systems in which the mobile phase contains only water and organic modifiers, TLC will generally be a useful pilot technique for HPLC. For systems involving an ion-pair
reagent (and a buffer), the situation is less favourable. It is
more difficult to cause an ion-pair effect in TLC than it is in HPLC,

and impregnation of the TLC plate by a dipping technique often is a
pre-requisite for acceptable performance. Nevertheless, differences
in retention order and/or magnitude of the ion-pair effect in TLC as
compared to HPLC may be expected. The nature of the thin-layer
material, the type and alkyl chain length of the counter-ion and the
nature of the solute (cf. the comparison of fluvoxamine and morphine)
can all play a role here. Besides, one should also realize that
chemically bonded phases for use in HPLC obtained from different
manufacturers often have different characteristics, which may cause
dramatic differences in retention (see also [13-16]).

Further research is obviously necessary in order to get a clear
insight into the role of the various parameters in RP ion-pair TLC.
In such work, for comparison with HPLC the chromatographic system –
and especially the chemically bonded phase – should be carefully
selected.

References

1. Issaq, H.J., Shaikh, B., Pontzer, N.J. & Barr, E.W. (1978)
    J. Liq. Chromatog. *1*, 133-149.
2. Jork, H. & Wimmer, H. (1981) *GIT Fachz. Lab.* 7, 566-573.
3. Bieganowska, M. & Soczewiński, E. (1981) *J. Chromatog.* *205*,
    451-456.
4. Golkiewicz, W.(1981) *Chromatographia 14*, 629-632.
5. Okumura, T. (1981) *J. Liq. Chromatog.* *4*, 1035-1064.
6. Guinchard, C., Masson, J.D., Truong, T.T. & Porthault, M. (1982)
    J. Liq. Chromatog. *5*, 1123-1140.
7. Issaq, H.J., Klose, J.R. & Reitz, G.A. (1982) *J. Liq. Chromatog.*
    5, 1069-1080.
8. Brinkman, U.A. Th. & de Vries, G. (1982) *J. High Resol.
    Chromatog. Chromatog. Comm.* *5*, 476-482.
9. Brinkman, U.A. Th., de Vries, G. & Cuperus, R. (1980) *J.
    Chromatog.* *198*, 421-428.
10. Lepri, L., Desideri, P.G. & Heimler, D. (1978) *J. Chromatog.*
    153, 77-82.
11. Volkmann, D. (1979) *as for 8.*, 2, 729-732.
12. Gonnet, C., Marichy, M. & Naghizadeh, H. (1980) in *Recent
    Developments in Chromatography and Electrophoresis,* Vol. 10
    (Frigerio, A.& McCamish, M., eds.), Elsevier, Amsterdam, pp. 11-19.
13. Ogan, K. & Katz, E. (1980) *J. Chromatog.* *188*, 115-127.
14. Amos, R. (1981) *J. Chromatog.* *204*, 469-478.
15. Panthananickal, A. & Marnett, L.J. (1981) *J. Chromatog.* *206*,
    253-265.
16. Goldberg, A.P. (1982) *Anal. Chem.* *54*, 343-345.

Comments on material in #B

Comments on #B-1, H.M. Ruyten et al. - HPLC PRE-CONCENTRATOR

Question by G.S. Clarke.- What are the nature and life-span of the column material? *Reply by* H. de Bree.- The material is the same as in the analytical column except that its particle size is larger; it serves for 10 runs each with 500 ml of urine.   *Answer to* I.D. Wilson: each unit costs ~£2500. *Comment by* U.A.Th. Brinkman.- I gained the impression that breakthrough/% recovery was not felt to be important; in examples shown, two compounds appeared to be completely lost. *Response.-* We first run a profile, then we know which compounds we are looking for; where structure identification is the aim, losses are acceptable. If compounds actually disappear, the remedy might be a different material or a change in pH.

Comments on #B-2, G.B. Cox, & #B-3, E.P. Lankmayr- HPLC OPTIMIZATION

Comment by R. Schmid.- As an investigator not knowledgable in HPLC 'software' I see much scope for optimization approaches applied to defined separation problems in clinical analytical chemistry. A. Gulaid, *addressing* G.B. Cox.- Since EC detection requires an electrolyte, is consideration of this a follow-on from the optimization exercise? *Reply.-* In an example shown, pH 2.2 phosphoric acid was present; hence one can use any buffer and pH with the optimization system.

Points relating to #B-3.- J.C. Berridge has two relevant publications besides the one cited: (1982) *Anal. Proc.18,* 472-475; (1984) *Analyst 109,* 291-293. G.B. Cox (cf. #B-2) comments that #B-3 exemplifies the overlapping resolution mapping technique which he described. It should be noted (he adds), incidentally, that polynomials of higher order, e.g. special or full cubic, could be used with the technique. Even so, these equations will not handle discontinuous functions although it is difficult to envisage chromatographic parameters which would give rise to such response curves.

G.B. Cox also makes a general point.- I am increasingly concerned with the current usage of the term 'optimization'. There are unfortunately many parameters which influence a separation and which could therefore be optimized. In most cases only one or two of them are treated in available optimization schemes. In my article (#B-2), for example, only solvent selectivity is optimized whilst #B-3 describes optimization of pH, buffer concentration and solvent strength. Neither deals with optimization of temperature or flow-rate. It appears that some form of nomenclature describing what has been optimized would be helpful. In addition, work designed to demonstrate which parameters have profound effects on the final separation and which may be safely ignored would be timely to prevent optimization routines becoming as tedious and time-consuming as the largely random method-development processes which they were designed to replace.

Comments on #B-4, I.D. Wilson – RP-TLC, and #NC(B)-1, U.A.Th. Brinkman
     [also N. Lammers et al.,#NC(B)-2] – RP-TLC, & RP-HPLC DESIGN

Response to V. Goldberg.- Different batches of the same make of plate have not shown much variation; but only one batch is used at a time.* U.A.Th. Brinkman, *responding to* C.D. Bevan.- The anomalous effect of adding NaCl (to 3%) is still found with washed plates; so it would be unhelpful to pre-test plates for Na^+ or Cl^-. The idea of choosing NaCl came from an American paper; no explanation of why it works is offered.

Remarks by L.A. Sternson.- One source of the differences observed in retention behaviour between RP-TLC and RP-HPLC may arise from the fact that in HPLC the organic component of the mobile phase adsorbs in Langmuir fashion onto the hydrophobic surface, modifying its characteristics. Thus, equilibrated HPLC systems present surfaces different in properties from the phase used in TLC which has not been pre-equilibrated with the developing solvent. *Remarks by* E.P. Lankmayr.- Equilibration of RP-HPLC columns for ion pairing is well known to be rather time-consuming, indicating that the equilibrium is reached rather slowly. Accordingly, it seems obvious that an RP-HPLC column and an RP-TLC plate could not give comparable data; a TLC plate will primarily generate a solvent gradient and thus will not be in the same equilibrium as an HPLC column.

Citations by Senior Editor

Irwin, W.J., Hempenstall, J.M. & Li Wan Po, A. (1984) *Lab. Pract. 33*, 74-77.-"Mechanical damage in reversed-phase HPLC columns". The upper few mm of the column may become channelled, resulting in progressively broadening peaks which finally become multiple. The phenomenon can be obviated by using radially compressed columns.
Bonicamp, J.M. (1982) *J. Chromatog. Sci. 20*, 389-390. "Removal of oxidants from ethyl acetate." To remove contaminants that may cause TLC streaking, the solvent (which can be tested with KI) can best be treated with a 5% solution of hydroxylamine hydrochloride, then filtered. *Editor's reminder:* The Index entry 'Solvents' in earlier vols. is a lead-in to various tips on chromatographic and sample-preparation solvents, as also considered in #A-1, this vol.

------

* *Related discussion points* (G.B. Cox, U.A.Th. Brinkman).- 'ODS' HPLC separations vary with the manufacturer. Plates and columns of the same make may differ in the silica or bonding process, if indeed both are available.

Section #C

HPLC DETECTION, AND DETERMINATION OF METAL-COMPLEX DRUGS

#C-1

COMPARATIVE PERFORMANCE OF DIFFERENT HPLC SYSTEMS WITH ELECTROCHEMICAL DETECTION

David Perrett

Department of Medicine
St. Bartholomew's Hospital Medical College
London EC1A 7BE

For suitable compounds, HPLC with electrochemical detection offers excellent sensitivity with relatively good selectivity; but accurate and reliable operation, particularly at levels of a few pmol, relies upon both correct selection and good matching of equipment and its operation. Often the detection of trace compounds requires the instrumentation to give satisfactory results at signal-to-noise ratios of below 10 at fmol levels. The ability of some commercially available detectors, a number of different pumps, and various electrodes to meet these requirements has been investigated in connection with measuring the levels in plasma of catecholamines and a number of thiols including thiol drugs.

The importance of reducing noise in HPLC systems operating with electrochemical (EC) detectors is well recognized. The absolute level of noise, in any one system, is often limited by the available equipment, although there are other sources of noise that depend less on the equipment. Table 1 lists some of the causes of noise that we have encountered with three different systems used in our laboratory. Some useful although not guaranteed solutions for these problems are also given in the Table, with references to more detailed information. [#A-3 by C.R. Jones in Vol. 12 is pertinent.–*Ed.*]

High specificity is obtainable since only electrochemically active species are detected and then only above certain applied potentials. Specificity, and also noise levels, are also influenced by the correct choice of working electrode. The specific detection of thiols at mercury [1] or gold-mercury amalgam [2, 3] electrodes is well known, but other electrode materials can offer useful advantages of specificity over the common carbon-based electrodes. Recently

Table 1. Some sources of noise with electrochemical detectors.

Type	Source	'Solution'
Elect- rical	Detector sensititivy too high	Reduce if possible
	Operating potential too high	Reduce to lowest value with requisite sensitivity
	Electrode type and/or design	Choose appropriate electrode
	Electrode poorly prepared	Re-pack or polish electrode
	Electrostatic effects	Common ground system [4]
	Time constant of amplifier	Select better setting
	Spikes from other lab. equipment	Isolate detector from source
Liquid	Grade of buffers/reagents employed	Use Analytical Reagent quality, and re-circulate
	Leaks in system	Seal leaks
	Bubbles in buffers or the flow cell	De-gas and re-circulate
	Closed electrical circuit in buffer	Let solvent drip into buffer at outlet
Others	Pulsations from the pump	Select better pump or add more pulse dampeners
	Metal fittings in the systems	Remove metal sinters and/or add EDTA to buffers
	Temperature fluctuations at electrode	Thermostat cell and/or shield from draughts

a copper electrode has been used to detect amino acids [5], and gold electrodes for thiols [6, 7]. The growing popularity of HPLC with EC detection is apparent not only from the many applications recently published, particularly for catecholamines [8, 9; cf. #F-1, this vol.], but also by the growing number of EC detector manufacturers.

EQUIPMENT

The EC detectors regularly used were the 4 and 4A types from Bioanalytical Systems Inc. (W. Lafayette, Indiana) and a type LCA 15 from EDT Research Ltd. (London NW10 7LKU). The EC detector cells employed with the former were a BAS TL3 carbon-paste cell, and TL4 cells equipped with glassy-carbon, gold, silver and platinum working electrodes and a glassy-carbon auxiliary electrode. The EDT system was supplied with an LCA13 glassy-carbon electrode. Other detectors were kindly loaned for this study. All studies were performed in the oxidative mode. The various pumps employed are shown in Table 2. The detector outputs were recorded using either Servoscribe recorders or a HP3390A integrator (Hewlett Packard) which was used in the plot mode for the studies on detector noise levels.

Table 2. Pump-associated noise in electrochemical (EC) detection. The column was 3 μm ODS-Hypersil, 100 x i.d. 4.6 mm, run at ambient temperature with 0.1 M citrate/phosphate pH 6 buffer containing 10% (v/v) methanol. Detection was with a BAS glassy-carbon electrode at +0.4 V, time constant 1 sec; the HP3390A recorder was used in the plot mode. F = fast noise, S = slow noise (see text); values are the peak-to-peak noise, in pA.

Flow, ml/min	Altex 110A[†]		Knauer 64		ACS 300/01		S-P8770		Waters 6000	
	F	S	F	S	F	S	F	S	F	S
0.5	10	41	*	*	10	33	15	37	16	32
1.0	17	40	10	48	10	36	15	38	44	73
1.5	20	37	12	54	12	44	17	67	51	83
2.0	25	53	15	66	12	62	20	59	31	83
2.5	30	82	20	95	10	55	15	75	46	96

[†] fitted with an additional pulse-damper   * not determined

ROLE OF THE PUMP IN NOISE REDUCTION

EC detector cells can act rather like pressure transducers, generating a small current with small changes in pressure/flow-rate. For minimum noise the liquid system should be well damped. This has led some workers to claim that only syringe pumps are suitable for high-sensitivity applications [4]. Pulsations in HPLC pumps are reduced by various means - electronic correction (e.g. Spectra-Physics, Altex), rapid stroke (e.g. ACS 300, Knauer) or special gears (Waters 6000) - but the EC detector tests these approaches severely.

Overall system noise consists of at least two components, fast and short-term; fast noise is smaller in peak-to-peak amplitude and leads to a general blurring of the baseline. I have defined slow noise as that occurring on a cycle possibly greater than 1 min but giving deflections that could at worst be mistaken for small peaks and are certain to lead to errors in quantitation. Baseline drift caused by long-term changes in detector response has not been considered here since, in high-sensitivity analytical work, drift and the disturbances caused by the solvent front are often so large that the detector needs frequent baseline correction. To some extent the measurement of system noise is rather subjective. A published discussion [10] on the evaluation of noise in HPLC UV detectors is also largely applicable to EC detectors.

Table 2 shows the values of the noise components for 5 well-known HPLC pumps when coupled directly to an ODS-silica column

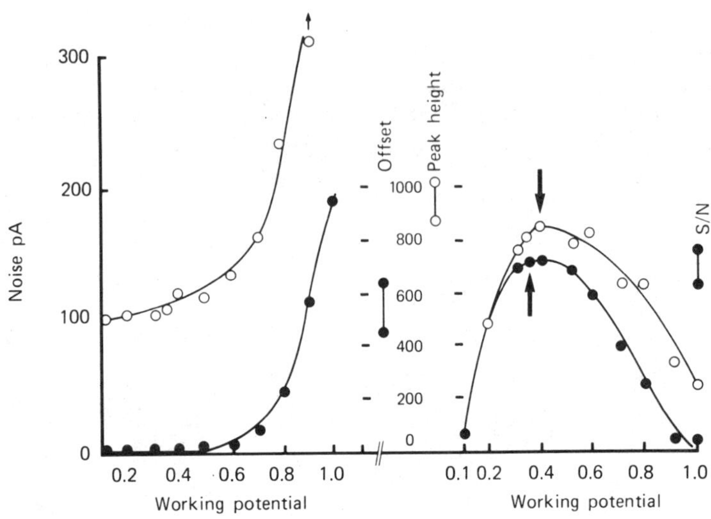

Fig. 1. Variation in electrode parameters for noradrenaline determined at a glassy-carbon electrode. *Left:* o, noise; •, offset. *Right:* o, peak height; •, signal/noise ratio. *Arrows* indicate maxima.

attached to the same BAS-4 detector with a TL-4 glassy carbon electrode. This was operated at the optimum potential for the detection of catecholamines, at which 500 fmol of noradrenaline should give full-scale deflection. In all cases noise levels increase with flow-rate but some pumps are more susceptible than others; the rapid-stroke ACS 300 pump gave the least noise. Clearly, with high back-pressure columns, most pumps should be operated at low flow-rates with some additional pulse damping if fmol amounts are to be measured. Hence cheap simple pumps with several damping devices could be quite satisfactory.

NOISE LEVELS OF THE ELECTROCHEMICAL DETECTORS

The noise generated by an electrode is very dependent on both operating potential and preparation of the electrode surface; any comparative evaluation must try to allow for such effects. Fig. 1 shows the various electrode parameters obtained for a BAS glassy-carbon electrode used to detect noradrenaline with HPLC conditions as in Table 1. Electrode noise levels are acceptably low at all working potentials below +0.8 V. Similarly the offset applied to

Table 3. Noise levels of 4 commercial electrochemical detectors.
Noise values represent the noise for a glassy-carbon electrode
measured peak-to-peak in pA at 0.6 V *vs.* Ag/AgCl with buffer
flowing at 1 ml/min using a damped Altex 110 pump.

Detector	Fast noise, pA	Slow noise, pA	Time Constant, sec
BAS 4 or 4A	16	40	2
Chromatix	95	400	2
Metrohm	180	10000	2
EDT	35	95	2

the cell is also acceptable in the same range. Above +0.8 V the
stability of the cell decreases, particularly if high-sensitivity
detection is to be performed. For noradrenaline, the maximum on the
voltammogram occurred at +0.4 V, which for work at moderate sensitivity - e.g. detection of 20 pmol - gives the maximum response; but
a lower maximum at 0.35 V is revealed by plotting signal-to-noise
ratio *vs.* voltage when noise is more pronounced, viz. detector set
at 1 nA full-scale deflection. Even such a reduction in working
potential to obtain maximum sensitivity may be too little to gain
improvements in selectivity. Even at low working potentials most
physiological fluids abound in electro-active species which can give
rise to interfering peaks and/or massive frontal peaks that mask the
peaks of interest. A reduction of 25% in the working potential was
found necessary to gain the necessary selectivity to measure thio-
malic acid in urine [7]. The maximum potentials quoted here may
appear low compared to those quoted in the literature. However, we
have observed that even the same electrode can change its surface
properties over a period of time; thereby its optimum working potential falls, its noise level increases, but its working sensitivity
apparently is unimpaired. One must therefore determine the necessary
operating parameters for one's system and check them at intervals.

Of the 5 detectors studied, the BAS 4 and 4A were very similar
in performance and can be considered the same. (This company has
since introduced two new models which have not been studied.) The
minimum noise levels achieved for the different detectors are shown
in Table 3.

Two detectors linked to identical HPLC systems were compared
for absolute sensitivities towards catecholamines (Table 4). The
two cells differed in design whilst both being of glassy-carbon
type; the BAS system employs a thin-layer cell while the EDT system
employs a wall-jet design with a working electrode of greater surface
area. The two electrodes gave comparable responses to catecholamines
and, provided that levels greater than 3-4 pmol were to be measured,

Table 4. Comparative performance of two commercial detectors
(gc denotes glassy carbon). The noise value is for 2 sec EC time
constant. The pmol value used for calculating the response is the
amount loaded onto the column.

| | | Response, pA/pmol | |
Cell type and surface area	Noise, pA	noradrenaline	adrenaline
BAS 4, thin-layer gc; 7 mm^2	40	286	306
BAS 4A, thin-layer 2 × gc; 14 mm^2	98	540	504
EDT, wall-jet gc; 19.6 mm^2	95	262	277

performed well. With a single electrode the noise level of the BAS
cell was less than half that of the EDT cell using identical time
constants, so improving the limit of detection 2-fold. With two
glassy-carbon electrode blocks maintained at the same potential (as
is standard practice in our laboratory), the noise level of the BAS
system more than doubled but the response increased by less than
2-fold. In such a configuration the surface area of the BAS cell
was still below that of the wall-jet cell but the response to catecholamines was double (Table 4). These findings complement those
reported by Bunyagidj & Girard [11], who compared the Hitachi 630
detector with an ESA Coulometric monitor: the latter gave a detection
limit similar to that of the BAS detector reported here, but the
Hitachi detector was some 40-fold less sensitive. The Hitachi detector is coulometric, as is the ESA, but the greater surface area
required to give 100% conversion entails a correspondingly greater
noise.

SELECTIVITY OF VARIOUS ELECTRODE MATERIALS

Using a flow-injection technique, the properties of various
electrode materials could be rapidly investigated - e.g. glassy-
carbon, carbon paste, gold and platinum. An ACS 300/01 pump was
coupled to a Rheodyne injection valve (20 µl) the outlet of which
was taken directly into a BAS TL-4 detector cell via 20 cm of 0.15 mm
i.d. PTFE tubing. The appropriate working electrode, suitably prepared, formed the lower half. Because of the pulse-free nature of
this pump it was still possible to operate the detector at up to
10 nA full-scale deflection with minimum baseline noise. With the
detector time constant set at 0.5 sec, the detector and recorder
together had an overall response time of ~0.8 sec; therefore peaks
of up to 70% full-scale deflection were reproducibly recorded.

All the electro-active compounds studied could be detected at
all electrodes but with differing sensitivities and selectivities;

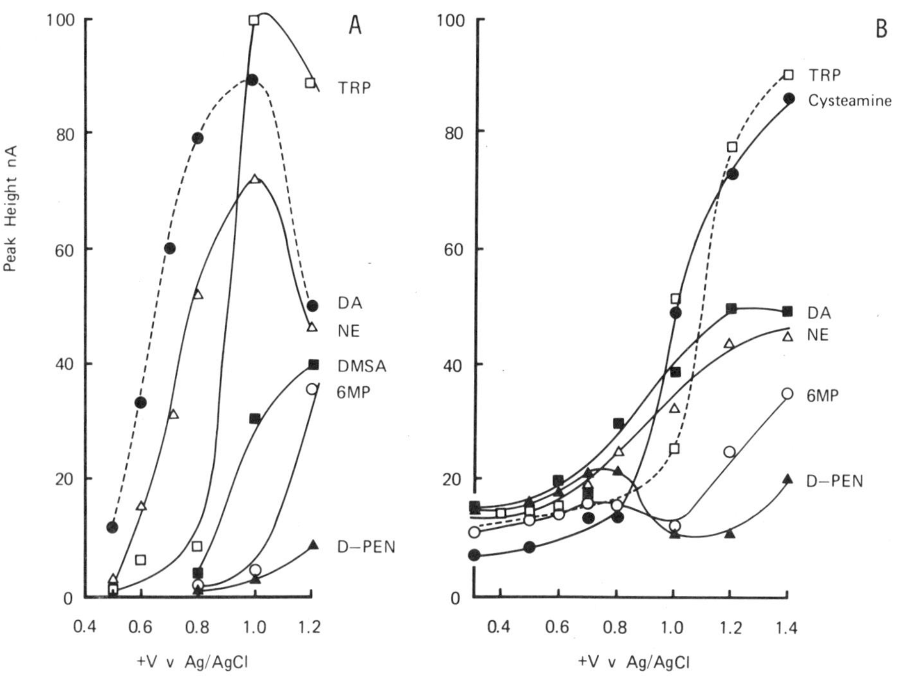

Fig. 2. Voltammograms for selected compounds determined by flow-injection analysis at A) glassy-carbon and B) gold working electrodes. TRP = tryptophan, DA = dopamine, NE = noradrenaline (norepinephrine), DMSA = dimercaptosuccinic acid, 6MP = 6-mercapto-purine, D-PEN = D-penicillamine.

this was particularly so for the thiols. Fig. 2 shows voltammograms obtained for selected compounds using glassy-carbon and gold elect-rodes. Fig. 3 shows the influence of the electrode material on the determination of a typical catecholamine, noradrenaline, and an anti-hypertensive thiol-containing drug, captopril.

Table 5 summarizes the optimum working potentials for all the compounds tested with the various electrodes. Silver was found to be a poor electrode since it rapidly lost sensitivity due to the formation of a black oxidized layer on the electrode surface. This was unfortunate since its initial response was always high. It is possible that with suitable buffer systems it could prove a good electrode material, particularly for thiols.

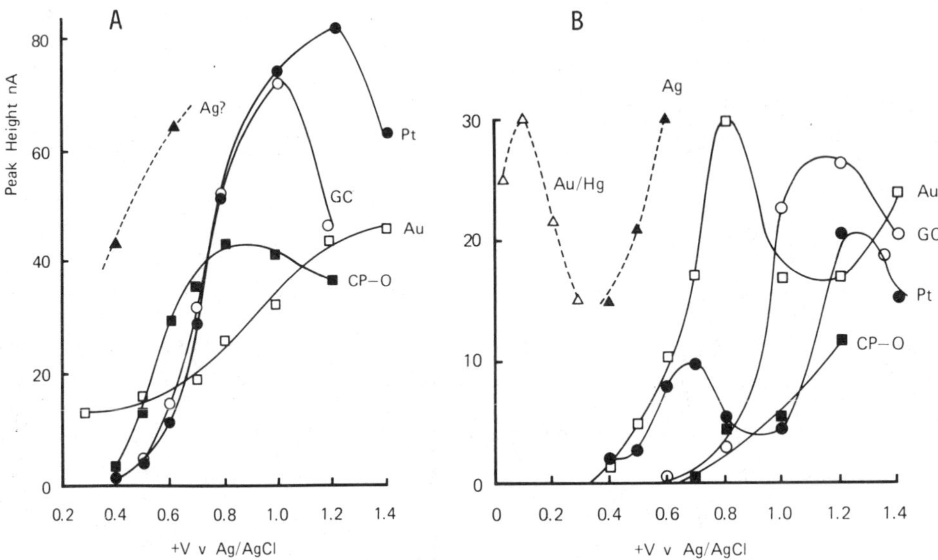

Fig. 3. Influence of working-electrode composition on the detection
of A) noradrenaline and B) captopril. GC = glassy-carbon,
CP-0 = carbon paste (BAS type 0).

The catecholamines were in general detected at potentials above
+0.7 V with all the electrodes except the mercury amalgam electrode.
The specific reaction of thiols with mercury gave lower maximum
potentials for all those thiols tested. Direct oxidation of the
thiol groups also occurred at potentials above +0.9 V with the other
electrode materials. At this potential oxidation to the disulphide
is probably the dominant reaction, and such high values have been
employed to measure D-penicillamine in physiological fluids [12]. But
for captopril with gold and platinum electrodes a second significant
maximum at 0.8 V and 0.7 V respectively was also obtained. Cysteine,
D-penicillamine, 6-mercaptopurine, α-mercaptopropylglycine (Thiola) and
N-acetyl-D-penicillamine also exhibited a maximum at these lower
potentials with the gold electrode. For cysteamine this lower peak
was even more specific since it occurred only at the platinum electrode and the response was effectively constant between 0.55 and 1 V.

The use of such a secondary (i.e. lower) voltage should allow
the development of both highly specific and sensitive HPLC–EC assays.
In particular the findings with cysteamine offer, assuming a suitable
chromatographic separation, a conceivable means of measuring this
compound, relevant to the use of cysteamine in treatment of cystin-
osis. The nature of the electrochemical reaction occurring at these
lower voltages is still unclear, and awaits further work.

Table 5. Voltammogram maxima for different compounds and electrode materials. Values were determined by flow injection. **Note:** sensitivity varied considerably. Two values for +V (*vs.* Ag/AgCl) are given where there was a secondary peak (see text).

Compound	Glassy carbon	Carbon paste	Gold	Platinum	Gold/ mercury
Noradrenaline	1.0	0.8	1.4	1.1	*
Adrenaline	1.0	0.8	1.4	1.1	*
Dihydroxybenzylamine	1.0	0.8	1.2	1.1	*
Dopamine	0.9	0.7	1.2	1.1	*
Metadrenaline	1.0	0.9	1.4	1.2	*
Dihydroxyphenylacetic acid (DOPAC)	0.9	0.7	1.2	1.2	*
Tryptophan	1.1	1.2	1.4	1.2	*
Uric acid	1.2	1.2	*	1.2	*
Cysteine	1.2	1.2	1.4, 0.7	1.4, 0.5	0.08
D-Penicillamine	1.2	1.2	1.4, 0.8	1.4	0.1
N-Acetyl-D-penicillamine	1.2	1.2	1.4, 0.7	1.4	–
Thiomalic acid	1.2	1.2	1.4, 0.8	1.4	0.12
2, 3-Dimercaptosuccinate	1.2	1.2	1.4, 0.8	1.4	–
Mercaptopropylglycine	1.2	1.2	1.4, 0.8	1.4	–
6-Mercaptopurine	1.2	1.2	1.4, 0.7	1.4	0.1
N-acetylcysteine	1.2	1.2	1.4, 0.7	1.4	–
Cysteamine	1.2	1.2	1.4	1.4, 0.7	–
Thioglucose	1.2	1.2	1.4, 0.8	1.4	–
Captopril	1.2	1.2	1.4, 0.8	1.2	0.07

* not detected (cf. – , signifying not determined)

CONCLUSIONS AND EXAMPLES OF ASSAYS ON PLASMA

Trace analysis requires not only good sensitivity but also good chromatographic techniques with efficient separation, as stressed throughout this series. C.R. Jones (cited at the start of this article) has especially considered the subtle problems in routinely working at trace levels, particularly with biological samples. Full benefit from the capability of EC detection coupled to HPLC to give high sensitivity requires care in the selection of equipment and in its daily operation with attendant difficulties. For electro-active compounds the sensitivity is comparable to that of current fluorescence detectors, which are likely to be more selective. EC detectors are prone to buffer and sample artifacts which can affect both sensitivity and selectivity. The latter can be improved by the correct choice of electrode and working potential, and then one can develop some uniquely sensitive and rapid assays.

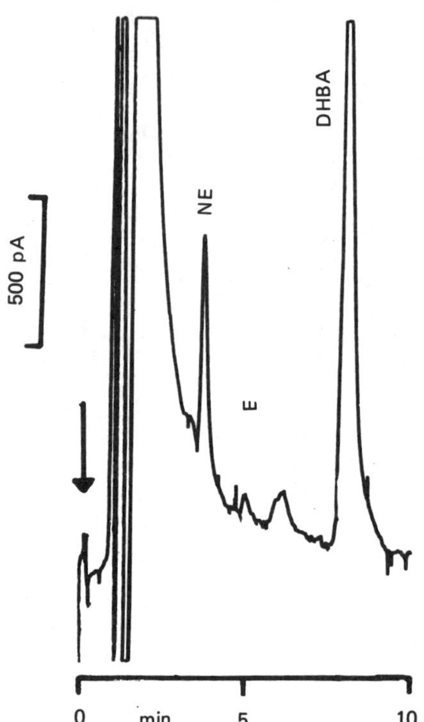

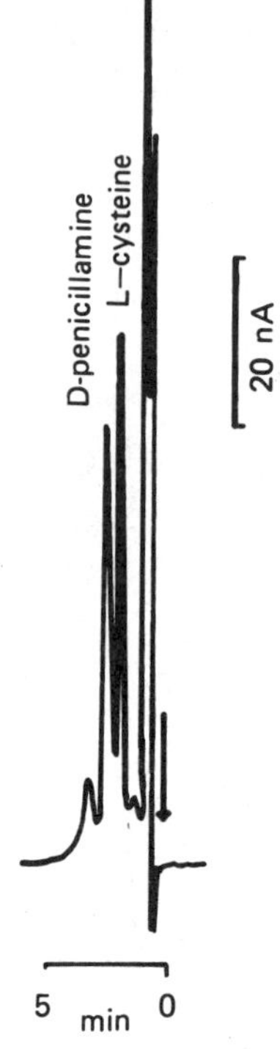

Fig. 4. Determination of catecholamines in normal basal plasma. A
2 ml sample was extracted onto
acid-alumina. Elution of analytes
into 125 µl of 0.5 M phosphoric
acid was followed by HPLC (100 µl
loaded).
Column: 3 µm ODS-Hypersil, 100 x
4.6 mm. Mobile phase: 50 mM
citrate-phosphate buffer pH 6.0
containing 1 mM octanesulphonic
acid and 10% (v/v) methanol.
BAS detector with dual glassy
carbon electrode at 0.5 V *vs.*
Ag/AgCl.
NE = noradrenaline, E = adrenaline;
calculated plasma levels 1.25 nM and
0.32 nM respectively.   DHBA, dihyd-
roxybenzylamine as internal std.

Fig. 5. Determination of
D-penicillamine and L-cys-
teine in plasma of a cystinuric patient receiving the
drug. Plasma (1 ml) precipitated with 100 µl salicyl-
sulphonic acid (4 M), and
20 µl of supernatant taken.
Column: 5 µm SCX silica,
100 x 4.6 mm. Mobile phase:
30 mM ammonium acetate buffer pH 2.3.
BAS detector with a gold
working electrode at 0.5 V
vs.   Ag/AgCl.

Through our understanding of the HPLC-EC systems outlined in this article, a number of assays for naturally occurring analytes and various drugs [3, 7] are routinely run in my laboratory. Fig. 4 illustrates the measurement of catecholamines in human plasma at pM levels using a glassy-carbon electrode sandwich. The catecholamines had been extracted from plasma onto acid-alumina, then desorbed for chromatography as indicated in the Fig. legend. Fig. 5 illustrates the simultaneous determination of cysteine and D-penicillamine in the plasma of a cystinuric patient receiving D-penicillamine for the dissolution of cystine calculi. The plasma proteins were acid-precipitated and the supernatant injected directly onto the column. The gold working electrode was maintained at +0.5 V in order to achieve the necessary specificity by reducing the influence of the frontal peak.

Acknowledgements

I am grateful to Dr. Pierre Bouloux of St. Bartholomew's Hospital, Dr. Sue Rudge of Nottingham City Hospital and Dr. Chris Smith of the Rayne Institute, University College Hospital for their help with various HPLC-EC systems. I am also grateful to ACS Ltd., BioTech Ltd. and Spectra-Physics for the loan of equipment.

References

1. Saetre, R. & Rabenstein, D.L. (1978) *Anal. Chem.* 50, 276-280.
2. Bergstrom, R.F., Kay, D.R. & Wagner, J.G. (1981) *J. Chromatog.* 222, 445-452.
3. Perrett, D. & Drury, P.L. (1982) *J. Liq. Chromatog.* 218, 97-110.
4. Helpern, J.A., Ewing, J.R. & Welch,K.M.A. (1982) *J. Chromatog.* 240, 491-492.
5. Kok, W. Th., Brinkmann, U.A. Th. & Frei, R.W. (1983) *J. Chromatog.* 256, 17-56.
6. Kreuzig, F. & Frank, J. (1981) *J. Chromatog.* 218, 615-620.
7. Rudge, S.R., Perrett, D., Drury, P.L. & Swannell, A.J. (1983) *J. Pharm. Biomed. Anal.* 1, 205-210.
8. Allemark, S. (1982) *J. Liq. Chromatog.* 5 *(Suppl. 1)*, 1-41.
9. Holly, J.M.P. & Makin, H.L.J. (1983) *Anal. Biochem.* 128, 257-274.
10. Wolf, T., Fritz, G.T. & Palmer, L.R. (1981) *J. Chromatog. Sci.* 19, 387-391.
11. Bunyagidj, C. & Girard, J.E. (1982) *Life Sci.* 31, 2627-2634.
12. Abounassif, M.A. & Jeffries, T.M. (1983) *J. Pharm. Biomed. Anal.* 1, 65-72.

#C-2

SPECIFIC ELEMENT CHROMATOGRAPHIC DETECTION BY PLASMA SPECTRAL EMISSION, AS APPLIED TO ORGANOMETALLICS AND GOLD-CONTAINING DRUGS

Peter C. Uden

Department of Chemistry, GRC Towers
University of Massachusetts
Amherst, MA 01003, U.S.A.

Inorganic gas and liquid chromatographic analysis is finding application in bioanalytical areas. Characterization methods for complex samples include specific element detection. Plasma emission spectroscopy exhibits notable advantages for such detection by virtue of versatility, sensitivity, selectivity and multi-element capability. Among the plasma systems which have been evaluated, the atmospheric pressure microwave plasma has shown wide suitability for analysis of both metallic and non-metallic elements; high-resolution capillary GC maximizes detection limits. Thus, trialkyl Pb compounds have been determined, directly and as organometallic derivatives, in aqueous samples down to ppb after suitable extraction and pre-concentration procedures. Other published examples include characterization of fluorine-containing metabolites in blood plasma and determination of boronate esters and steroidal carboranes.

For HPLC detection, the Inductively Coupled Plasma (ICP) and the Direct Current Plasma (DCP) are most suitable. Applications of the latter include Vitamin B_{12} and gold-containing anti-arthritic drugs.

The bioanalytical importance of effective separation has been stressed by R.W. Maickel [#A-1, this vol.]. In both high and low resolution chromatographic separation, identification of eluted peaks or developed bands is vital to assure the integrity of known species for quantitative purposes and to identify and characterize unknown members of multi-component mixtures. GC and HPLC are complementary high-resolution techniques offering high efficiency for multi-component systems at low analyte levels. While in GC practice the greatest

development in directly interfaced peak characterization has been
in the area of mass spectrometry (GC-MS), and recently vapour phase
IR spectroscopy (GC-IR) [1], other functional parameters of eluted
species, measurable in the vapour phase, have been employed, e.g.
thermal fragmentation [2] and mol. wt. measurement [3]. In conventional GC methodology, some standard detectors impart a degree of
specific information concurrent with detection, notably ECD* and (for
N and P) AFID (Vol. 7, this series); also, for S, flame photometry.

Increasing attention is being paid in GC to specific element
detection and monitoring since for many complex mixtures a single
run may be unable to achieve adequate resolution of the species of
interest, either among themselves or from matrix interferences as
commonly encountered in bioanalytical work. If the species share
some common property, the use of a detector responsive solely to
that particular parameter may serve to simplify qualitative and quantitative analysis, as exemplified by specific element detection. In
HPLC the eluate matrix is more complex than in GC since the liquid
mobile phase is present in the detector along with the eluates.
Eluate characterization is greatly facilitated by measurement of a
property not apparent in the mobile phase. Element-specific detection not only fulfils this function but gives many practical advantages.

Atomic spectroscopic methods are very appropriate for specific
metal monitoring in both GC and LC applications to trace-level bio-
analytes containing metal or non-metal hetero-atoms. While both
flame and furnace atomic absorption (AA) have been employed for
metal specific detection as has, to a limited extent, atomic fluorescence [4], these techniques suffer limitations in sensitivity, in
their inability to accomodate simultaneous multi-element measurements
and, in furnace AA, the necessity for discontinuous segmented pro-
filing [5].

In contrast with AA spectroscopy, atomic emission spectroscopy
has the notable merit of multi-element capabilities and may have a
linear dynamic range up to 5 or 6 orders of magnitude. The advent
of various accessible plasma sources, particularly in combination
with high-resolution monochromators to minimize spectral interferences, has led to a resurgence of analytical application of atomic
emission methods. Notably, microwave induced and sustained plasmas
(MIP) are used in GC for element-selective detection. Other plasma
systems also serve in both GC and HPLC, particularly the DC argon
plasma (DCP) and the inductively coupled plasma (ICP).

The major advantages of interfacing plasma emission spectroscopy with a column are the following.- (a) Ability to perform
speciation for many metals and non-metals, either before or within
the column; application may be either direct or by derivatization.

---

*electron-capture detection; cf. electrochemical connotation in HPLC.–*Ed.*

(b) Ability to tolerate non-ideal chromatographic conditions and elution conditions; the specificity of plasma emission enables the analyst to tolerate incomplete resolution, a factor of great importance in complex matrices; here the selectivity of the particular element is the primary concern. (c) Sensitivity of plasma emission detection (even sub-pg), surpassing any other GC detector except possibly ECD or MS. (d) Multi-element capacity of plasma emission. (e) Compatibility with existing chromatographic systems through the incorporation of simple interface devices.

THE ANALYTICAL PLASMA

An analytical plasma consists of a mass of ionized gas at a temperature of $4000-9000+^{\circ}K$ which is maintained by input of energy in various forms. In the DCP an electrical discharge is maintained through argon gas; in the ICP an electromagnetic field is generated at radio frequencies and in the MIP at microwave frequencies. The geometry of the DCP is shown in Fig. 1; it usually operates at 600-800 W. The ICP (Fig. 2) typically operates at powers of 1-5 kW, and the MIP (Fig. 3) at 50-250 W depending on the application.

A typical block diagram of a plasma emission system suitable for GC is shown in Fig. 4. An alternative parallel detector such as FID is often included. In liquid chromatography, effluent to be monitored may be carried with nebulization to an ICP or DCP; the MIP, however, is less suitable since it cannot tolerate typical liquid flow rates without extinction.

Emission from the plasma is focussed into a suitable optical analyzer which may be a monochromator, polychromator or echelle spectrometer. For metals and a number of metalloids and non-metals, detection limits, reproducibility and response linearity are comparable to or better than for flame AA and emission. The range of applications in both HPLC and GC has therefore increased rapidly, and in this article representative topics of general bioanalytical, toxicological and related interest will be reviewed and discussed. Chromatographic plasma interfacing has been reviewed elsewhere [6-8].

'SPECIATION' IN ELEMENTAL BIOANALYSIS

While the study of biological, toxicological and pharmacological responses to trace metals and organometallics has flourished, metals analysis is almost universally for total element with little capability for determining the chemical form of the element in the original sample. Thus, while total Pb may be readily determined in many samples, typically no knowledge is gained as to the nature of the element, whether metallic, ionic salt or organometallic Pb. Clearly the ability to 'speciate' Pb is cardinal to any study of biological responses for the metal. The combination of a separatory technique

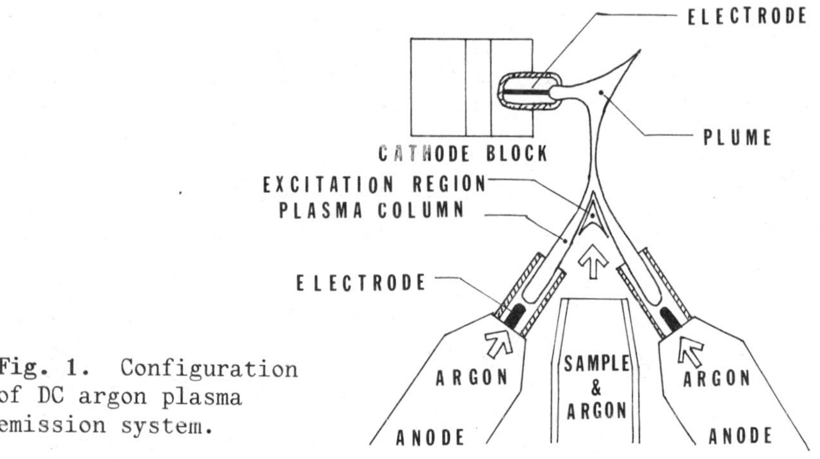

Fig. 1. Configuration of DC argon plasma emission system.

Fig. 2. Configuration of Inductively Coupled Argon plasma emission system (ICP).

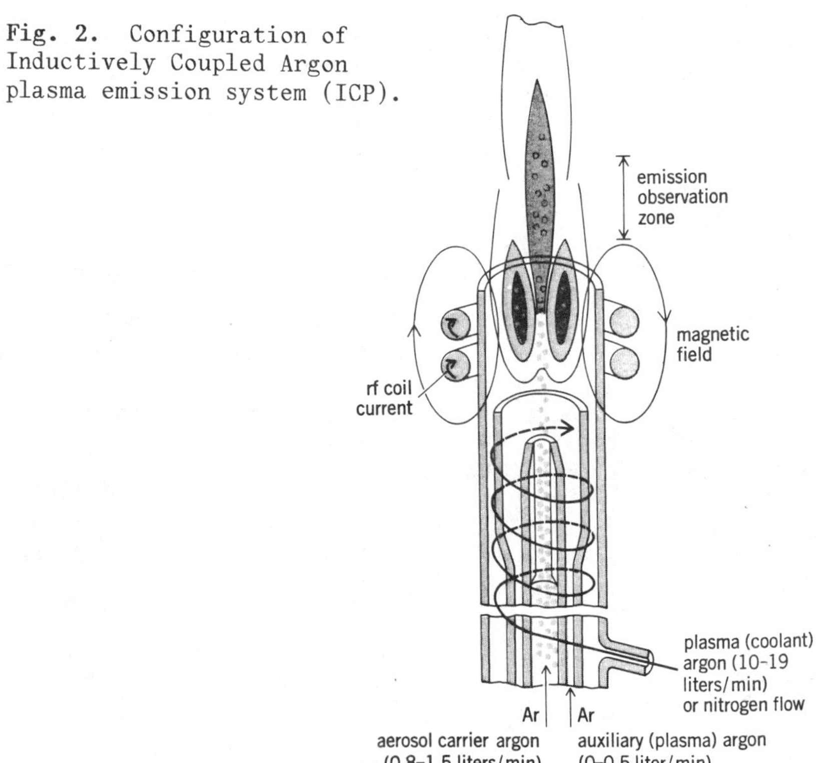

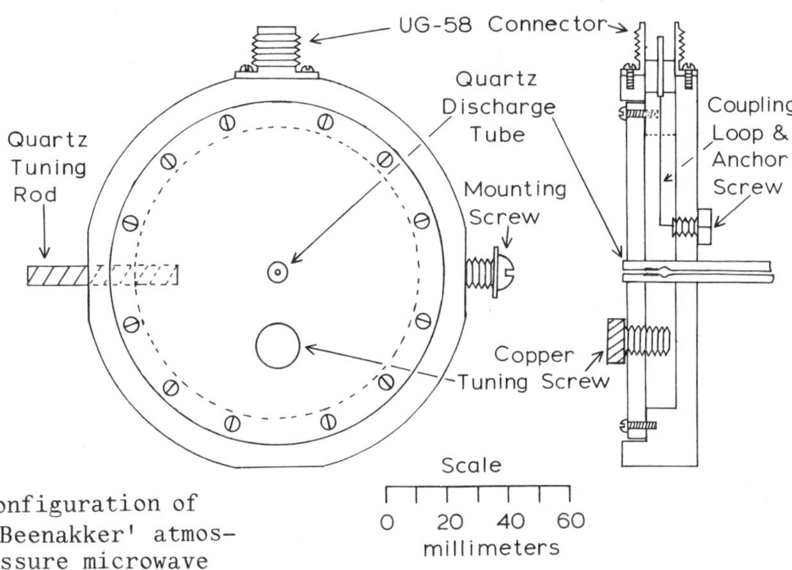

Fig. 3. Configuration of the TM_{010} 'Beenakker' atmospheric pressure microwave induced plasma cavity (MIP).

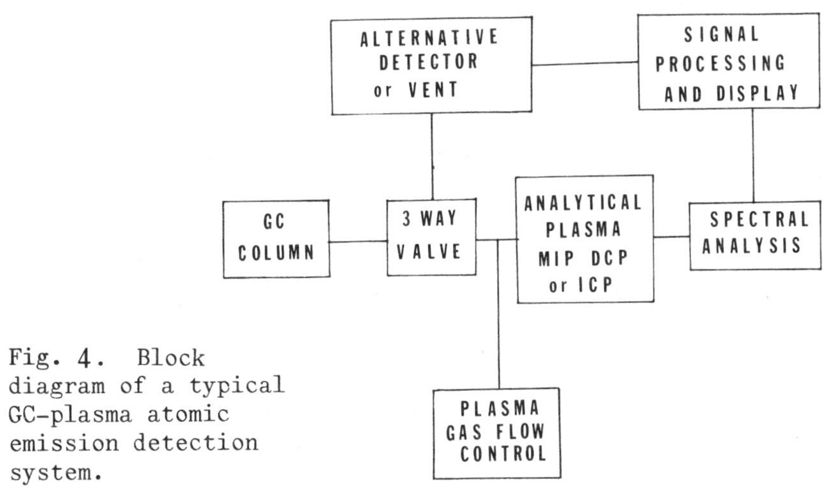

Fig. 4. Block diagram of a typical GC-plasma atomic emission detection system.

such as chromatography based on the overall chemical nature of the analyte followed by quantitative specific element detection goes far to address this need.   There is a notable abundance of speciation applications for GC and especially HPLC. A typical procedure for detailed and complete speciation would include specific element monitoring of the separatory process along with other interfaced detection modes such as MS, IR spectroscopy, fluorescence, etc., in series or in parallel.

GC DETECTION WITH THE MIP

A low-power, atmospheric pressure argon MIP was first employed extensively in GC in 1965 [9], being used to detect traces of I, Cl, Br, S and P in organic compounds. Since then, many developments in MIP interfacing have been reported and significant improvements have been made [6-8].

McLean et al. [10] employed a low-pressure helium plasma, sustained in a thick-walled quartz capillary. Pressure was adjustable from 0.25 torr, and microwave power from 100 to 200 W at 2.45 GHz was applied. Detection limits for C, H, D, F, Cl, Br and S ranged from 3×10^{-11} to 9×10^{-11} g/sec and for N and O were ~3 $\times 10^{-9}$ g/sec; selectivities *vs.* carbon were, however, usually below 1000. An important feature of the plasma system is that response is proportional to the number of atoms of a particular element present and is independent of structure. Thus it is possible, by comparing element response ratios of unknown compounds with those of a standard containing the same element, to determine empirical formulae. This system is now on the market.

The TM_{010} cylindrical resonance cavity described by Beenakker [11] (Fig. 3) gives increased efficiency of transfer of microwave power to the discharge, so that an atmospheric pressure helium (or argon) plasma can be sustained at the same low power levels as used with previous cavities. This offers a great advantage in GC applications. A further feature of this design is the ability to view light emitted from the plasma axially. For reduced-pressure helium plasma cavities it is viewed transversely through the walls of the quartz discharge tube. Deposition of materials on the discharge tube walls and devitrification of the quartz result in gradual attenuation of sample response with time. The addition of small amounts of oxygen or nitrogen to the helium to act as a scavenger gas reduces carbonaceous deposits, but deposition of metals and devitrification still present limitations for low-pressure cavities which are not experienced with the Beenakker design.

The TM_{010} cavity has been interfaced with packed (non-capillary) columns and with open-tubular columns. One limiting feature is its low tolerance for large eluting peaks, e.g. solvent. A practical limit is 50-100 µg of organic material entering the discharge. Greater amounts will extinguish the plasma and require cleaning or replacing the discharge tube. This drawback is avoided by either inserting a high-temperature low dead volume switching valve between the GC and MIP to divert large solvent peaks from the plasma [12], or by utilizing a fluid logic gas switching system [13]. When flexible high-resolution fused silica capillary columns are employed and if dual detection with a parallel FID is not required, an even simpler approach allows the column to be terminated within a few mm of the plasma. In this case sample injection splitting is employed to attain allowable plasma sample levels [14].

Besides specific element detection, a 'non-selective' mode from the GC point of view is carbon-specific detection for organics. Thus a series of alkanes (C_8 to C_{15}) was run on a glass capillary column with a 1:1 split to plasma (247.9 nm) and FID, using the fluid logic gas switching system. The absolute detection limit of 2-3 pg/sec for carbon rivalled or exceeded that of the FID.

ALKYL-LEAD CHLORIDE ANALYSIS

Organically bound Pb is a minor but important contributor to total Pb intake by humans and animals. It has been shown that alkyl Pb salts such as trialkyl Pb carbonates, nitrates and/or sulphates arising in tissues from rapid metabolic dealkylation of tetraalkyl Pb compounds are important in Pb toxicity [10]. The combination of high resolution and great inertness of fused silica columns now allows the elution of the trialkyl Pb chlorides at sub-ng levels [13].

Satisfactory chromatograms have been obtained for triethyl Pb chloride extracted from spiked tap water by an extraction/vacuum reduction procedure. The detection limit is 10-30 ppb.

Direct quantitative GC measurement of trimethyl Pb chloride or triethyl Pb chloride suffers, however, from two major difficulties. (i) Both compounds are thermally unstable and tend to decompose even at the lowest possible injection-port temperatures (~160-170°) required to give complete and rapid volatilization. (ii) Both compounds are very chemically reactive, giving some tailing of chromatographic peaks even with the most inert chromatographic columns available. Thus, consistent direct quantitative GC measurement of trialkyl Pb compounds is difficult.

If the compounds can be converted to tetraalkyl leads, quantitative determination is feasible. Use of n-butyl Mg bromide Grignard reaction for trimethyl- and triethyl-Pb chlorides is promising since tetra-n-butyl Pb can be used as an internal reference which will not interfere with speciation of methyl or ethyl tetra- or trialkyl Pb compounds. It also marks the termination of the GC-MIP Pb-specific analysis, since it is the last tetraalkyl Pb compound to elute [14].

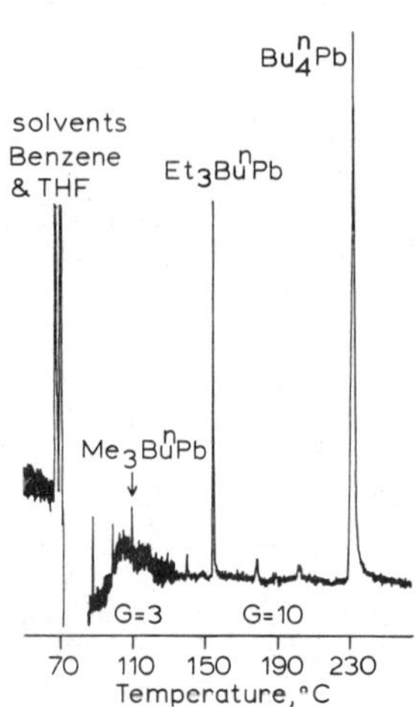

Fig. 5A. GC with Pb-specific detection, MIP at 283.3 nm, on effluent derivatized with butyl Grignard reagent. SP-2100 WCOT fused silica capillary column, 12.5 m. Gain = $G \times 10^{-8}$ A.

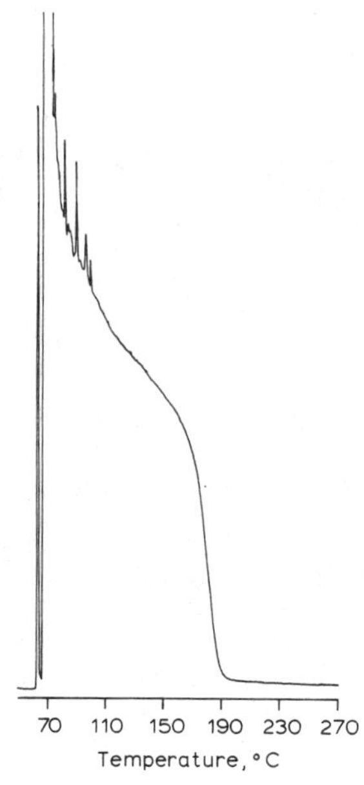

Fig. 5B. As for Fig. 5A, but carbon-specific detection: MIP at 247.9 nm. Gain = 1×10^{-5} A.

GC-MIP analysis for trimethyl- and triethyl-Pb ions in an industrial plant effluent showed the presence of the latter at a 19.0 ± 4.0 ppb level based on 4 replicates. Single 30 µl samples, however, contained insufficient trimethylbutyl Pb for detection, and the multiple sample pre-concentration technique was required. A sample chromatographed from a total volume of 300 µl subjected to pre-concentration gave the chromatographic peak seen in Fig. 5A, indicating an original trimethyl Pb concentration of ~5 ppb.

To check whether information could be obtained with a detector not of element-specific type, e.g. FID, the plant effluent analysis was repeated with the GC-MIP system monitoring at 247.9 nm (all other GC-MIP operational conditions were as for Pb). The background carbon response was vast (Fig. 5B), demonstrating the need for an element-specific detector such as the MIP, rather than merely a carbon-sensitive detector. To keep the analytical portion of the chromatogram on scale, the gain had to be so attenuated that the internal reference could not be detected even in the relatively flat portion of the chromatogram.

DETECTION OF FLUORINE-CONTAINING METABOLITES

In a contrasting application, a reduced-pressure helium microwave plasma was interfaced to a dual capillary column GC to provide

parallel FID, ECD and plasma-emission detection [15]. This system
was applied to the detection and quantitation of perfluorooctanoate
in human blood plasma resulting from industrial exposure. The selective MIP-fluorine channel greatly simplified the interpretation of
complicated FID and ECD capillary chromatograms and rendered quantitation of the analyte straightforward. The emission was monitored
by the 0.75-m spectrometer. Separate secondary slits and photomultiplier detectors were placed in appropriate spectral positions for
individual and simultaneous element monitoring (685.6 nm for fluorine, 495.7 nm for carbon). Each photomultiplier tube was linked to
a separate amplifier, to monitor and subtract a portion of the carbon
signal so that there was no 'ghosting' from the carbon continuum on
the other channels. The MIP system had 8 independent amplifiers,
each interfaced to an A/D converter, permitting the simultaneous
determination of up to 8 elements.

The selectivity of the microwave emission detector is apparent
when the MIP fluorine, FID and ECD chromatograms of human plasma
extracts are compared (Fig. 6). FID reveals that the capillary system has resolved >200 organic components. The ECD capillary chromatogram reduces the number of detected components to ~50 electron-
capturing species. In contrast, only one peak with a suspected isomer shoulder is observed on the MIP-fluorine channel. This compound
was identified as methyl perfluorooctanoate.

An important characteristic of the MIP is that it is an 'absolute' detector: a signal on a given channel represents a wt./unit
time of that element, virtually independent of the chemical nature
of the compound containing that element. Thus, in dealing with pre-
dominantly perfluorinated materials with little difference in
carbon-to-fluorine ratio, quantitation errors due to unresolved
isomers or other similar compounds are minimal.

The dual capillary system was also used to resolve the fluorine-
containing metabolites in the blood plasma of rats which had received
a single dose of 1H,1H,2H,2H-perfluorodecanol and had been sacrificed
at various times after dosing. The dosage was 400 mg/kg body wt.
administered by intubation as a 8 g/100 ml solution in corn oil.
Blood taken under anaesthesia from the descending aorta was immediately transferred to a heparinized tube and centrifuged to furnish
plasma. Diethyl ether was used for analyte extraction after acidifi-
cation with HCl solution (~2.5 M).

Fig.7 shows a pattern 2 h after dosage. The starting alcohol
(peak no. 3) is virtually lacking, and the perfluorooctanoate (peak
1) appears to be the final metabolic product. Fluorine NMR and
retention-time matching against known commercially available or synthesized compounds indicated that two of the intermediates after
diazomethane derivatization were 2H,2H-perfluorodecanoate (peak 4)

Fig. 6. Triple GC detection for methyl perfluorooctanoate in extracted human plasma.
Top, MIP fluorine at 685.6 nm (computer-normalized to give full-scale representation).
Centre, ECD.
Bottom, FID.
Column: DB 5 bonded silicone fused-silica capillary, 30 m.
Temp.: initially 70°, rising by 10°/min to 330°.

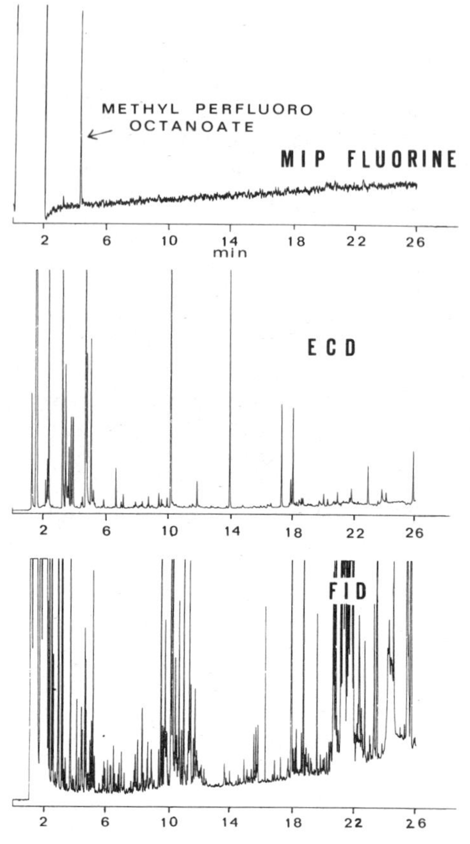

and its corresponding unsaturate, $C_7F_{15}CF=CHCOO^-$ (peak 2). One of the major metabolites (peak 5) appeared to contain esterifiable hydrogen. High inorganic fluoride levels in the rat plasma and the presence of perfluorooctanoate confirmed the metabolic defluorination of the starting alcohol.

BIOANALYTICAL APPLICATIONS OF THE BORON-SPECIFIC GC-MIP

Specific element analysis for boron is an attractive application for plasma emission detection since background is low and sensitivity is good. I.S. Krull et al. have reported some bioanalytical and toxicological applications for boron detection, and we have seen scope in the study of carborane-silicone pyrolysis [16].

Steroidal carboranes have been suggested for use in boron neutron capture cancer therapy [17]. Their high boron content, extreme chemical stability and extensive derivatization chemistry, and the high natural abundance of ^{10}B may make them good carriers for the introduction of boron to tumours, where localization may occur through binding to cytoplasmic receptor proteins. Krull et al. have

Fig. 7. Triple detection GC for fluorinated metabolites of 1H,1H,2H,2H-perfluorodecanol in rat plasma. *Top*, MIP fluorine at 685.6 nm (computer-normalized to give full-scale representation).
Centre, ECD.
Bottom, FID.
Column as in Fig. 6; same temp. programming.

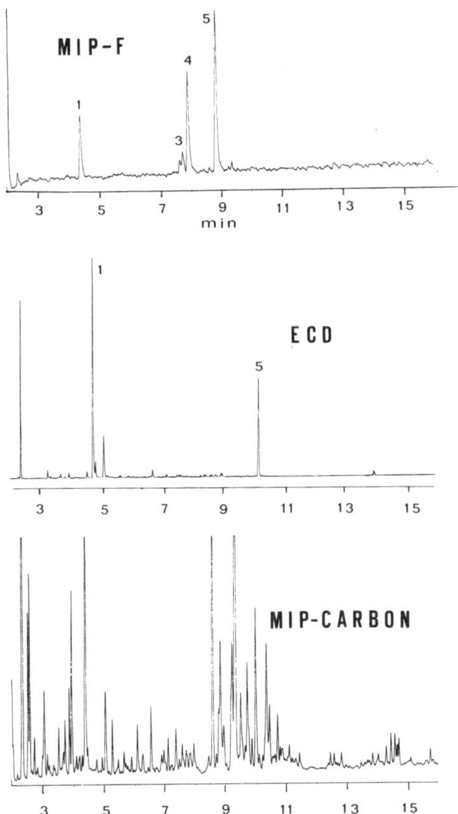

reported sensitive and selective boron detection with the atmospheric pressure MIP from 'Permabond[TM]' bonded-silicone packed columns [17]. Derivatized diol boronates have been run in our laboratory with this type of packing in a glass column (3 ft. x i.d. 2 mm; 250°). Thus, there was a good 249.77 nm MIP peak matching the FID peak (at 2.7 min) for a steroidal (acetate) carborane, 20-carboranyl ketone-3-β-acetate-22,23-bisnorcholenic acid. For such compounds the minimum detectable level was in the range 15-43 pg/sec.   In a parallel study [18], Krull et al. developed a GC-MIP method for n-butylboronate ester derivatives of catechols in human urine extracts, and for comparison ran silylated derivatives with FID detection.

ELEMENTS DETERMINED BY GC-MIP

To date more than 30 elements have been determined by GC-MIP. The most extensive survey was conducted with high-resolution fused silica columns [19]. Measurement features are shown in Table 1 for a selection of metals and metalloids. While in most cases applications have yet to be developed, detection limits are propitious.

Table 1. MIP detection limits, selectivities and linear dynamic ranges for some metals and non-metals [19]. The nm value refers to the emission wavelength. Values for pg/sec indicate peak width.

Element (& wavelength)	Detection limit		Selectivity vs. carbon[*]	Linear dynamic range
	pg (absolute)	pg/sec		
Co (240.7 nm)	18	6.2	1.9×10^5	10^3
Ni (231.6 nm)	5.9	2.6	6.5×10^3	10^3
Mn (257.6 nm)	7.7	1.6	1.1×10^5	10^3
Hg (253.7 nm)	60	0.60	7.69×10^4	10^3
B (249.8 nm)	27	3.6	9.25×10^3	5×10^2
Al (396.2 nm)	19	5.0	3.90×10^3	5×10^2
C (247.9 nm)	12	2.7	(1.00)	10^3
Si (251.6 nm)	18	9.3	1.58×10^3	5×10^2
Sn (284.0 nm)	6.1	1.6	3.58×10^5	10^3
Pb (283.3 nm)	0.71	0.17	2.46×10^5	10^3
P (253.6 nm)	56	3.3	1.06×10^4	5×10^2
As (228.8 nm)	155	6.5	4.70×10^4	5×10^2
Se (204.0 nm)	62	5.3	1.09×10^4	10^3

[*] Peak area/mol *vs.* that for carbon

THE ATMOSPHERIC PRESSURE DC ARGON PLASMA DETECTOR (DCP) FOR GC

For ready application of an interfaced plasma emission technique for chromatographic detection, the ability to use available spectroscopic instrumentation is important. This led to the development of a system based on commercially available atomospheric-pressure DC argon plasma [the Spectrascan System, SpectraMetrics (Smith Kline Beckman)] as an excitation source, together with a single-channel high-resolution echelle monochromator. This system was used for both GC detection (GC-DCP) [20] and HPLC monitoring (HPLC-DCP) [21]. For GC, interfaces have been designed for quantitative transfer of high-boiling metal complexes and organometallics without peak broadening or sample degradation.

An important criterion in interface design is the lack of hot or cold spots in the transfer line at which high-boiling components could be lost or labile species degraded. The transfer line must also have a small enough volume to avoid band spreading and loss of resolution, particularly for capillary column applications. Stainless steel, nickel and flexible fused-silica tubing have all been employed, the latter proving best [22].

Also of importance is the optimal positioning of the transfer-line exit orifice underneath the plasma excitation spot (Fig. 1). The orifice position is particularly critical when employing a fused-silica narrow-bore capillary column, which passes directly through the heated interface from the GC oven to the plasma. The usual experimental arrangement for GC with packed columns or glass capillary columns permits a 1:1 split (or other suitable ratio) at the end of the column between the transfer line to the plasma and a FID. In this way simultaneous detection may be obtained for all species and for the element of interest.

The DC argon plasma system has proved suitable for detecting metallic elements that can be efficiently excited, and also elements present in GC derivatizing groups, e.g. B and Si. Thus, TMS-anthranilic acid is detected well [22]. With silicon there is an advantageous absence of elemental spectral response from the quartz plasma tube used in the MIP system. Another considerable advantage, whilst due in part to the higher resolution of the echelle spectrometer, is the greater selectivity of silicon over carbon, the value for the MIP (Table 1) being only 1.58×10^3.

The following detection limits (pg/sec) and selectivities ($vs.$ C) have been established for GC-DCP, with a 3-electrode plasma jet [22]: Cr (267.7 nm): 4, 4×10^8; Si (251.6 nm): 25, 2×10^7; Sn (286.3 nm): 60, 2.5×10^6; Pb (368.3 nm): 100, 5×10^5; B (249.8 nm): 3, 3×10^5. Earlier use of a 2-electrode plasma jet [20] gave: Cu (324.7 nm): 5.6, $> 10^6$; Ni (341.4 nm): 320, $> 10^6$; Hg (253.6 nm): 65, 6×10^5. Compared with GC-MIP, the absolute detection limits for some elements such as lead and mercury are two orders of magnitude higher. Others such as boron are similar. For individual analyses the two systems vary in merit. The atmospheric-pressure MIP is well suited to high-resolution capillary column GC and its very low detection limits allow trace determinations to be made in small injected samples. It is limited, however, in its capacity to handle larger samples; packed column applications demand that solvent-venting systems be used. The DCP system is somewhat less convenient to interface, although fused silica transfer lines have helped to simplify this. As a **drawback,** for most elements the DCP has lower absolute sensitivity, but this is offset by its ability to handle large injected volumes without extinction of the plasma. Further, the very high selectivities shown $vs.$ carbon enable good detection in the presence of high backgrounds.

PLASMA ATOMIC EMISSION SPECTRAL DETECTION FOR HPLC

HPLC in its various modes has the notable advantage over GC that it is not restricted to analytes which are volatile or can be made so by derivatization or degradation. In the speciation of resolved analytes from complex matrices, specific element detection can be a great asset. For metals in liquid phase systems, atomic emission and absorption are the most attractive modes although atomic fluorescence and certain electrochemical detectors [featured elsewhere in this vol.-$Ed.$] are feasible.

Of the widely adopted plasmas for analytical spectroscopy, the ICP and the DCP have received most attention for HPLC interfacing. The lower powered MIP is not sustainable in the presence of solvents entering at typical chromatographic flow rates. A range of commercial plasma spectrometers are now available, some offering both DCP and ICP capabilities. Pending rigorous comparisons, it seems likely that the two systems are of comparable utility and have similar applicability. The somewhat greater sensitivity observed for some elements in the ICP may be offset by the more economical and solvent-independent operation of the DCP.

The outstanding need is for improved methods of sample delivery into the plasma. Most workers report minimum detection limits for elements studied that are usually 2-3 orders of magnitude worse than for the same elements subjected to direct analysis. Currently there is considerable interest in nebulizer design for both DCP and ICP to remedy this deficiency.

Our own effort has concentrated on the HPLC-DCP interface system. In principle, DCP instrumentation is capable of the same degree of multi-element analysis as any ICP. It is amenable to both aqueous and organic solvents entering the arc source, and is as free of serious source and/or spectral interferences as the ICP. The DCP may not be as matrix-independent for certain elements in the direct mode, but this is less of a problem in the HPLC-DCP mode. An advantage of HPLC-DCP interfacing is its stability and good performance with quite high flow rates, especially for normal phase-HPLC mobile phases, e.g. with hydrocarbon and halocarbon eluents. Linear response curves have been obtained for many metal species down to the low-ng-per-peak levels [21]. Amongst inorganic complexes studied, in normal- and reverse-phase modes, are mixed-ligand Co(III) and Cr(III) β-diketonates [23].

HPLC with DCP detection is an attractive technique for the analysis of organo-metallics in complex biological samples, and in general provides an added dimension of selectivity and sensitivity compared to most other HPLC detectors.

HPLC-DCP DETERMINATION OF VITAMIN B_{12}

Due to the complexity of the matrices in which vitamin B_{12} is encountered and its low levels, it has traditionally been analyzed by microbial techniques. These very reasons render the compound amenable to HPLC with specific element detection. Human blood plasma samples, with appropriate controls and spiked samples, were treated with 0.9 vol. of methanol and centrifuged. The clear supernatant was injected onto a 5 cm column with 5 μm C-1 bonded phase. The mobile phase was 1:1 methanol/water (0.8 ml/min). The analyte had a retention volume of 0.72 ml. A linear response (r = 0.9930) was

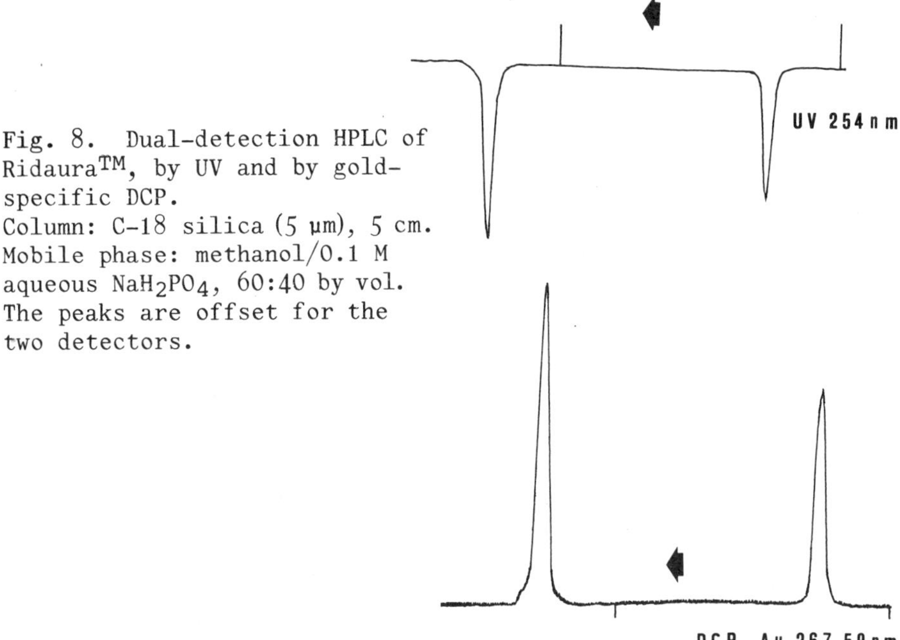

Fig. 8. Dual-detection HPLC of Ridaura™, by UV and by gold-specific DCP.
Column: C-18 silica (5 μm), 5 cm.
Mobile phase: methanol/0.1 M aqueous NaH_2PO_4, 60:40 by vol.
The peaks are offset for the two detectors.

obtained, with an analyte detection limit of 88 ng. The cobalt-specific response of the DCP detector greatly simplified the chromatogram and its interpretation, as interfering components were not manifest. Cobalt-specific detection was at 345.3 nm using the Spec-scan IV echelle spectrometer and 3-electrode plasma jet.

HPLC–DCP DETERMINATION OF RIDAURA

The gold-containing anti-rheumatic drug Ridaura, 2,3,4,6-tetra-O-acetyl-1-thio-D-glucopyranosato-5(triethylphosphine)gold (obtained from Smith, Kline & French Labs., Philadelphia, PA) in acetonitrile was chromatographed as in the legend to Fig. 8. At a flow rate of 2.2 ml/min the analyte retention volume was 8.25 ml; the column exhibited ~2500 theoretical plates. The detection limit corresponded to 56.4 ng of gold injected. Up to at least 600 ng of gold injected there was a linear calibration curve, with r = 0.9995. The HPLC–DCP determination of Ridaura and any possible gold-containing metabolites formed *in vivo* is under study.

Acknowledgements

The support of Spectrametrics Inc. (Smith, Kline Beckman) for the DC plasma study is acknowledged. The support and participation of numerous students and colleagues in this research area is gratefully recognized.

References

1. Penzias, G.J. (1973) *Anal. Chem. 45,* 890–895.
2. Groenendyk, H., Levy, E.J. & Sarner, S.F. (1970) *J. Chromatog. Sci. 8,* 599–605.
3. Lanser, A.C., Ernst, J.O., Kwolek, W.F. & Dutton, H.J. (1973) *Anal. Chem. 45,* 2344–2348.
4. Van Loon, J.C. (1981) *Am. Lab. 13* (5), 47–53.
5. Brinckman, F.E., Blair, W.R., Jewett, K.L. & Iverson, W.P. (1977) *J. Chromatog. Sci. 15,* 493–503.
6. Risby, T.H. & Talmi, Y. (1983) in *Critical Reviews in Analytical Chemistry,* Vol. 14 (Campbell, B., ed.), CRC Press, Boca Raton, FL, pp. 231–265.
7. Krull, I.S. & Jordan, S. (1980) *Am. Lab. 12* (10), 21–33.
8. Carnahan, J.W., Mulligan, K.J. & Caruso, J.A. (1982) *Anal. Chim. Acta 130,* 227–241.
9. McCormack, A.J., Tong, S.C. & Cooke, W.D. (1965) *Anal. Chem. 37,* 1470–1476.
10. McLean, W.R., Stanton, D.J. & Penketh, G.E. (1973) *Analyst 98,* 432–442.
11. Beenakker, C.I.M. (1976) *Spectrochim. Acta 31B,* 483–486.
12. Quimby, B.D., Uden, P.C. & Barnes, R.M. (1978) *Anal. Chem. 50,* 2112–2118.
13. Estes, S.A., Uden, P.C. & Barnes, R.M. (1981) *Anal. Chem. 53,* 1336–1340.
14. Estes, S.A., Uden, P.C. & Barnes, R.M. (1982) *Anal. Chem. 54,* 2402–2405.
15. Hagen, D.F., Belisle, J. & Marhevka, J.S. (1983) *Applied Spectroscopy 38B,* 377–385.
16. Sarto, L.G. jr., Estes, S.A., Uden, P.C. & Barnes, R.M. (1981) *Anal. Lett. 14,* 205–218.
17. Krull, I.S., Jordan, S.W., Kahl, S. & Smith, S.B. jr. (1982) *J. Chromatog. Sci. 20,* 489–498.
18. Jordan, S.W., Krull, I.S. & Smith, S.B. jr. (1982) *Anal. Lett. 15,* 1131–1148.
19. Estes, S.A., Uden, P.C. & Barnes, R.M. (1981) *Anal. Chem. 53,* 1824–1837.
20. Lloyd, R.J., Barnes, R.M., Uden, P.C. & Elliott, W.G. (1978) *Anal. Chem. 50,* 2025–2029.
21. Uden, P.C., Quimby, B.D., Barnes, R.M. & Elliott, W.G. (1978) *Anal. Chim. Acta 101,* 99–109.
22. Beyer, J.O. (1983) *PhD. Dissertation,* Univ. of Massachusetts, Amherst, MA.
23. Uden, P.C., Bigley, I.E. & Walters, F.H. (1978) *Anal. Chim. Acta 100,* 555–561.

#C-3

HPLC-ELECTROCHEMICAL DETERMINATION OF
CIS-PLATINUM ANTI-CANCER DRUGS

*I.S. Krull, X-D. Ding, C. Selavka and [†]F. Hochberg

Institute of Chemical
Analysis
Northeastern University
360 Huntingdon Avenue
Boston, MA 02115, U.S.A.

[†]Neurology–Oncology Division
Massachusetts General
Hospital
32 Fruit Street
Boston, MA 02114, U.S.A.

Liquid chromatography-electrochemical detection (LCEC) has been applied to the trace determination of drugs [chemical names below] termed cis-Pt, CBDCA and CHIP. cis-Pt, the parent compound, can be determined by either oxidative or reductive LCEC, with different detection limits. Both CBDCA and CHIP can be determined by direct LCEC, but the detection limits for CBDCA are too high for clinical or stability studies. With a new derivatization method for CBDCA and related Pt compounds CBDCA is quantitatively convertible to cis-Pt, determination of which by reductive LCEC allows CDDP assay in plasma at the 0.1 µg/ml level (100 ppb).

The LCEC methods finally adopted have been applied to drug-stability studies (water, plasma and saline infusion solutions) and also clinically, with single or dual electrodes. The latter, used in parallel and furnishing dual-detector response ratios, are beneficial to analyte identification and characterization and to overall detection limits.

The treatment of human neoplasms often involves the use of certain platinum-derived drugs, many having cis-oriented dichloro, diamino or substituted diamino ligands attached to the central Pt atom [1]. The three for which methods have now been developed are cis-dichloro-diammine platinum (cis-Pt)$^\theta$, cis-diammine–1,1-cyclobutane dicarboxylato–platinum (CBDCA) and cis-dichloro-trans-dihydroxy-di-isopropylamine-platinum (CHIP). In reported *in vivo* or stability studies, compound or analyte speciation methods as in trace metal

---

* addressee for any correspondence $^\theta$other terms: CDDP; cis-platin

analysis [2,3] have seldom been used. Clinically useful Pt-containing drugs mostly cannot be analyzed by conventional GC because of their ionic nature and low vapour pressures [cf. P.C. Uden, #C-2, this vol. -*Ed.*]. Though HPLC separations are possible, detection of the individual analytes is difficult with conventional HPLC detectors, e.g. UV [4], and total Pt has usually been determined. Metal analysis with speciation usually needs a detection step coupled on-line or off-line to a separation, GC or HPLC in most instances [2, 3].

The extensive HPLC studies of L.A. Sternson and colleagues [e.g. 5; see #C-6, this vol.] has involved detection by UV absorbance for appropriate Pt analytes, off-line continuous graphite furnace atomic absorption (GFAA), and recent trial of polarographic HPLC detection with a dropping Hg electrode [6]. Thus far, there have been no reports of successful LCEC using thin layer-type electrode cells (Bioanalytical Systems Inc.), applicable to use of both oxidative and reductive modes depending on the particular oxidation state of the Pt drug and metabolites or decomposition/hydrolysis products. We now report optimized conditions for, and features of, RP-HPLC assay of the above three Pt drugs, with single or dual detection using a glassy carbon or Au/Hg working electrode surface. Aqueous solutions, spiked blood/plasma and patient samples have been assayed.

SINGLE-ELECTRODE EC DETECTION OF Pt-DERIVED DRUGS

The customary cyclic voltammogram (CV) was determined initially for *cis*-Pt itself, with glassy carbon and, for reference, Ag/AgCl electrodes at Bioanalytical systems Inc. (by K. Bratin & R. Shoup) [7]. This study pointed to possible working potentials of ~+1.0 to 1.2 V oxidatively and −0.9 to −0.0 reductively. Fig. 1 shows a reductive LCEC run (at −0.1 V) on *cis*-Pt. Where the large peak due to oxygen in the mobile phase and sample (Fig. 1) interferes with the peak of interest [6, *inter alia*], oxygen can be removed from the HPLC solvent by de-gassing and from the injected sample by various techniques, and air excluded during the run. Detection limits for *cis*-Pt (~10 ng/ml for standard solutions) were better reductively than oxidatively, and might benefit from injecting 200 µl rather than 20 µl. Reductively there was linearity from ~10 ng/ml to at least 50 µg/ml.

Oxidatively (glassy carbon, +1.20 V; Fig. 2), *cis*-Pt showed a detection limit of ~160 ng/ml and good linearity. LCEC studies with aqueous samples [7] have included examination of infusion solutions for half-life, related to *cis*-Pt and NaCl concentrations. For plasma or blood, the dual-electrode approach (below) is better.

For CHIP, reductive LCEC conditions have been developed, as for *cis*-Pt but at 0.0-0.2 V; the detection limit seems to be ~100 ng/ml. Stability has been studied [7]. CBDCA was a difficult analyte; even oxidatively, it could be determined only at 5 µg/ml or above. An impractical potential (+1.40 V) was needed. An alternative approach

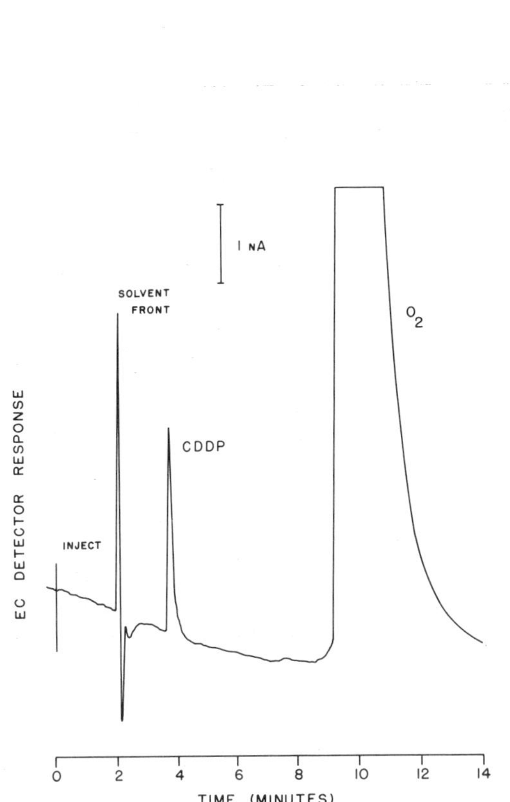

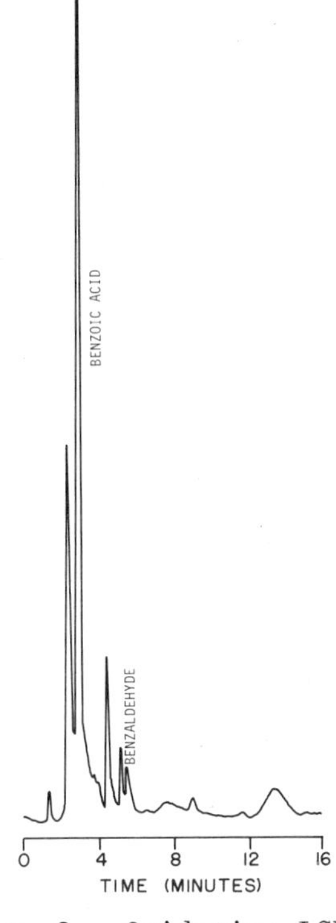

Fig. 1. Reductive LCEC of *cis*-Pt (CDDP), 630 ng/ml. C-18 10 µm column with mobile phase of 10 mM pH 4.6 acetate, 0.15 mM hexadecyl-trimethylammonium bromide (HTAB), 1 ml/min; Bioanalytical Systems LC-4B detector and TL-6A cell; Au/Hg electrode at -0.1 V. Cf. [7]. *By permission of Preston Pubns. Inc.*

Fig. 2. Oxidative LCEC comparable to Fig. 1; 160 ng/ml *cis*-Pt (much bigger peak if 1.25 µg/ml). Column run as in Fig. 1 except that the acetate buffer was 0.1 M and flow-rate 1.25 ml/min. LC-4A detector; glassy carbon electrode at +1.20 V. Cf. [7]. *As for Fig. 1.*

used acid-catalyzed solvolysis of CBDCA to remove the cyclic dilactone ring and replace this with two Cl atoms from HCl. Using dil. HCl as the solvolysis medium, quantitative conversion of CBDCA to *cis*-Pt took ~20 min at 50°. Oxidative and reductive LCEC verified the formation of *cis*-Pt concomitant with the oxidative disappearance of the CBDCA. LCEC conditions and detection limits for the derived *cis*-Pt were as in the *cis*-Pt assay. This trace assay method appears to be the first developed for CBDCA. (In a dual-electrode run shown in Fig. 3, CBDCA run as the untreated compound evidently gave a poor peak oxidatively, with +1.15 V; in another run, with +1.20 V, the peak for 40 µg/ml was bigger.)

DUAL-ELECTRODE EC DETECTION OF Pt-DERIVED DRUGS

With the recent interest in dual-electrode LCEC detection, we tried this for the Pt drugs [8]. Parallel (rather than series) dual-electrode operation was effective for all three, with benefit to specificity. This approach can also give calibration plots as a function of the working potential used, furnishing concentration/ response ratios characteristic of individual Pt compounds. Selecting suitable potentials in parallel operation alters the detectability of individual Pt analytes and the LCEC patterns. The overall analyte selectivity achievable by dual-electrode LCEC far surpasses that by LC-polarographic reduction or single-electrode LCEC operations. For some Pt analytes glassy carbon as well as Au/Hg-type electrodes can be used. Cancer-patient samples can be thus assayed. *(See below.)*

Working potentials and dual-detector ratios arc as follows with glassy carbon: CDDP: +1.05/+1.00 V, 3.0; CDDP: -0.50/+1.05 V, 2.6; CHIP: -0.50/-0.45 V, 2.0; CBDCA: +1.24/+1.18 V, 6.4; cf. Au/Hg:- CDDP: -0.01/+0.01 V; CHIP: -0.01/+0.01 V, 2.6. A number of dual-electrode ratios can be obtained quite quickly, and are characteristic of the drug of interest, so helping confirm the presence or absence of a Pt drug in a complex biological sample through getting several ratios - the more the better - for authentic drug and the peak in question.

Fig. 3 illustrates, for a mixture of *cis*-Pt, CHIP and CBDCA, the use in parallel of two glassy carbon electrodes, one operated oxidatively at +1.15 V and the other reductively at -0.40 V. With the chosen potentials, the oxidative chromatogram shows both *cis*-Pt and CBDCA and the reductive chromatogram shows only CHIP. Small changes in either of these working potentials drastically change the pattern and so allow selectivity to be varied easily from one injection to the next. Such speciation is most effective when dual-detector ratios can be obtained, which is not feasible under the conditions of Fig. 3 where the potentials preclude appearance of any of the drugs on both chromatograms simultaneously. Although both electrodes may be operated oxidatively, or both reductively, in the present case the oxidative/reductive mode gives more information and specificity. At +1.20 V and -0.46 V the reductive chromatogram shows *cis*-Pt but not CBDCA, enabling a dual-detector ratio to be obtained for the former. The dual mode favours specificity and also detectability.

Fig. 4 illustrates dual-electrode results for human plasma with *cis*-Pt spiked in. Both electrodes, Au/Hg, were operated near 0.00 V, one oxidatively at +0.01 V and the other reductively at -0.01 V relative to the Ag/AgCl reference electrode; but both chromatograms show oxygen peaks and thus seem to result from reductive processes despite the convention that one process be termed oxidative and the other reductive. The apparent detection limit for this drug in plasma is somewhat less than 0.5 µg/ml [8]. (Acetonitrile deproteinization can be done on whole blood.)

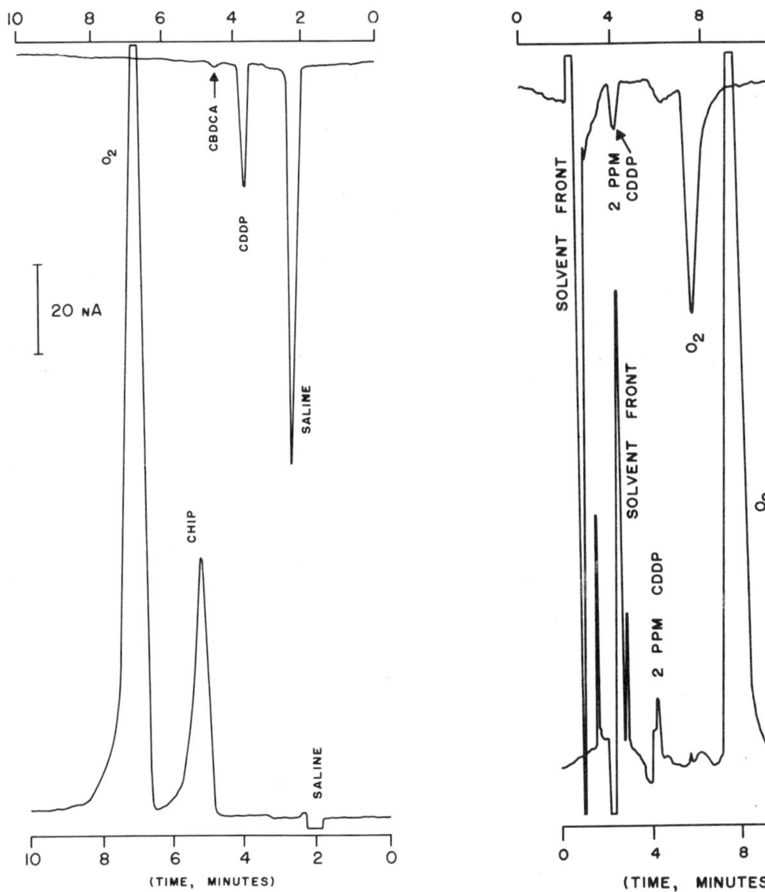

Fig. 3. LCEC run on drugs in saline with dual glassy carbon electrodes in parallel, oxidative at +1.15 V *(upper trace)* and reductive at -0.40 V *(lower trace)*. The injected sample contained *cis*-Pt (CDDP), CBDCA and CHIP (respectively 0.8, 1.6 and 1.6 µg). C-18 column; 10 mM pH 4.60 acetate/ 0.15 mM HTAP plus methanol (95:5 by vol.), 1.0 ml/min. Cf.[8](Fig. 4 also). *By permission of Marcel Dekker Inc.*

Fig. 4. LCEC run on plasma spiked with *cis*-Pt, 2 µg/ml (2 ppm), with dual Au/Hg electrodes in parallel, oxidative at +0.01 V *(upper trace)* and reductive at -0.01 V *(lower trace)*. C-18 column 250 X i.d. 4.6 mm; run as in Fig. 3 but without methanol; LC-4B EC detector.

We have shown that these newer dual-electrode approaches are applicable to samples from cancer patients as well as infusion solutions. For two of the latter, prepared to contain 171 and 75 µg/ml of *cis*-Pt, we obtained concordant values, viz. 153 ±2 and 72.8 ±5.3 µg/ml (mean ± S.D., n = 4). A reductive/reductive approach was used for these samples, and for plasma samples [8] from two patients given *cis*-Pt by i.v. infusion of 83 mg during 2 h.   The plasma levels attained were respectively 4.4 and 0.75 µg/ml;   for the 4 analyses performed on each sample the respective C.V.s were 20% and 12%. The Pt drug was undetectable in pre-infusion samples.   Evidently dual-electrode LCEC methods can be of help in cancer chemotherapy with *cis*-Pt and related Pt drugs.   Elsewhere [8] we give a fuller description; before loading, the plasma was deproteinized with TCA.

Acknowledgements

Authentic samples of all three Pt drugs were provided by Bristol Laboratories Inc. and Johnson-Matthey Inc.   A sample of Platin, the commercial formulation of *ci s*-Pt, was provided by Massachusetts General Hospital.   Samples of *cis*-Pt infusion solutions and of pre- and post-infusion blood or plasma were provided by Dr. Dave Henner of the Sidney Farber Cancer Institute, Children's Hospital, Boston.

This work was supported, in part, by a grant from the NIH Biomedical Sciences Research Support Grant No. RRO7143, Dept. of Health & Human Resources, to Northeastern University.   Mr. X-D. Ding held a Visiting Chinese Scholar Fellowship from the Government of China (P.R.C.).

References

1. Prestakyo, A.W., Crooke, S.T. & Carter, S.K., eds. (1980) *Cis-platin: Current Status and New Developments*, Acad. Press, New York.
2. Krull, I.S. (1984) in *Liquid Chromatography in Environmental Analysis* (Lawrence, J.F., ed.), Humana Press, Clifton, NJ, Chapter 5.
3. Schwedt, G. (1981) *Chromatographic Methods in Inorganic Analysis: Separation Methods……*, A. Hüthig, Heidelberg, 226 pp.
4. Vickrey, T.M., ed. (1983) *Liquid Chromatography Detectors* (Vol. 23, *Chromatographic Science Series*), Dekker, New York, 434 pp.
5. Riley, C.M., Sternson, L.A. & Repta, A.J. (1982) *Anal. Biochem. 124*, 167-179.
6. Bannister, S.T., Sternson, L.A. & Repta, A.J. (1983) *J. Chromatog. 273*, 301-318.
7. Krull, I.S., Ding, X-D., Braverman, S., Selavka, C., Hochberg, F. & Sternson, L.A. (1983) *J. Chromatog. Sci. 21*, 166-173.
8. Ding, X-D. & Krull, I.S. (1983) *J. Liq. Chromatog. 6*, 2173-2194.

#C-4

HPLC ANALYSIS OF CISPLATIN ANALOGUES IN BIOLOGICAL FLUIDS

D.R. Newell, Z.H. Siddik and K.R. Harrap

Department of Biochemical Pharmacology
Institute of Cancer Research
Belmont, Sutton, Surrey SM2 5PX, U.K.

The cisplatin analogues investigated relate to the search for a second generation Pt complex less toxic or more active than the parent compound. For cis-diammine(1,1-cyclobutane-dicarboxylato)platinum (CBDCA), normal-phase chromatography on silica with UV detection suffices down to levels of 10 µM in urine or plasma ultrafiltrates. Quantitation below 10 µM needs HPLC fraction collection followed by flameless atomic absorption spectrophotometry (FAAS; 0.5 µM detectable). These methods have been successfully applied to CBDCA in plasma and urine from patients and rats [1, 2]. For cis-dichloro- trans -dihydroxy bis (isopropylamine) platinum (CHIP), plasma ultrafiltrates were analyzed by HPLC on a µBondapak phenyl column with UV detection [3], or by water/methanol gradient elution from a µBondapak C-18 column followed by fraction collection and FAAS [4] (detection limits 1 µM and 0.04 µM respectively). Urines containing CHIP were analyzed by the latter method [3, 5]. These assays have been successfully applied to dog and human samples where besides CHIP, cis-dichloro-bis(isopropylamine)platinum has been found [3]. [Formulae are shown overleaf.-Ed.]

The RP analysis of cis -1,1-di(aminomethyl)cyclohexane-(sulphato)platinum (TNO-6) has been complicated by the poor recovery (<10%) of the complex from the HPLC column; yet the assay, entailing electochemical (EC) detection, has been invaluable in identifying the instability of TNO-6 in commonly used intravenous injection fluids [6]. Attempts to develop an HPLC assay for cis -1,2-diaminocyclohexane(4-carboxyphthalato)platinum (DACCP) have so far failed. However, an HPLC method for the trimellitic acid leaving group has been developed and used to demonstrate its urinary excretion [7]. HPLC techniques, maybe with EC or other improved on-line detection devices, will play an important role in the pharmacokinetic evaluation of these new drugs.

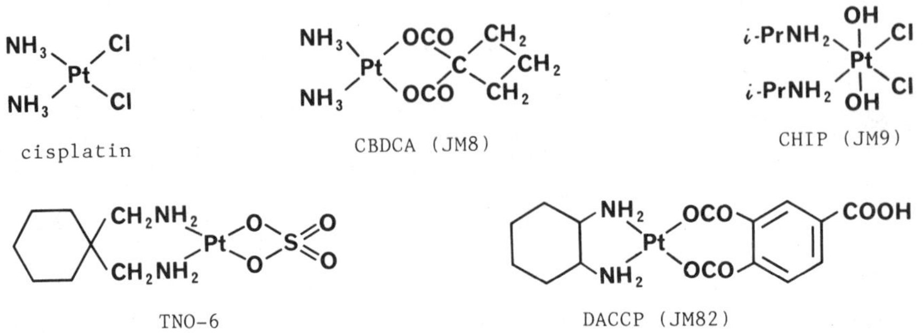

Fig. 1. Structures of cisplatin and four analogues (as named on the preceding page).

Cisplatin (*cis*-dichlorodiammineplatinum; Fig. 1) is undoubtedly one of the major success stories of cancer chemotherapy. When used alone for the treatment of ovarian cancer or in combination with other drugs against testicular teratoma, cisplatin produces good responses which are well maintained in many patients. Unfortunately, cisplatin therapy is associated with side effects including nephrotoxicity, neurotoxicity, ototoxicity, emesis and haematological toxicity. Of these the first three are particularly problematic as they can be cumulative and hence dose-limiting. Accordingly a number of groups have sought to develop a less toxic second-generation Pt complex which retains the beneficial activity of the parent compound. In these studies, recently reviewed by Harrap [8], particular attention has been paid to the selection of either non-nephrotoxic Pt complexes or alternatively compounds which are not cross-resistant with cisplatin. For four such complexes (Fig. 1), methodology has been sought for their analysis in biological fluids as an adjunct to current clinical evaluation.

Although flameless atomic absorption spectrophotometry (FAAS) remains the technique of choice for Pt analysis by virtue of its sensitivity and selectivity, FAAS is indiscriminate in terms of complex structure. Thus the majority of pharmacokinetic studies performed on cisplatin have been limited to a description of either total or free Pt levels in plasma and urine [reviewed in 1]. More recently HPLC methods have been developed which allow the specific determination of cisplatin in plasma and urine ([9, 10] and accompanying articles). Their application should allow a more rational appraisal of cisplatin pharmacokinetics and hence its toxicity and mechanism of action. During the pre-clinical and early clinical evaluation of the second-generation Pt complexes, attempts have been made to develop HPLC methods for use in conjunction with FAAS during bioanalytical

studies.   For CBDCA and CHIP these attempts have been successful.
However, satisfactory methods for the analysis of TNO-6 and DACCP
in biological matrices are as yet not available.

CBDCA (carboplatin)

A simple, rapid and fully automated method for the HPLC analysis
of CBDCA in plasma   ultrafiltrates and urine has been developed.
This method is suitable for the analysis of CBDCA in human urine and
plasma following therapeutic doses (300-500 mg/m^2) [1] and in rat
plasma and urine after non-toxic doses (20 mg/kg) [2].

In selecting an HPLC packing for the separation of CBDCA from
endogenous components two normal-phase media were examined, µBondapak
CN (Waters) and Spherisorb silica (PhaseSep).   Reverse phase (RP)
packings were not examined because of the insolubility of CBDCA in
organic solvents (C. Barnard, personal communication).   Ion-exchange
and ion-pair chromatography were also not examined because of the
possibility of chemical interaction between ions in the solvent and
the Pt complex with resultant changes in the structure of the complex.   Despite the use of solvents differing widely in polarity
(methanol, ethanol, isopropanol, hexane), µBondapak CN gave inadequate retention and excessive peak broadening.   Similarly, Spherisorb
silica, when used with water/ethanol or methanol/ethanol mixtures,
produced inadequate retention, again with poor peak shapes. However,
acetonitrile/water mixtures gave separations with good efficiency
and satisfactory retention.   A 9:1 mixture (by vol.) of acetonitrile
and water was chosen for routine use.

Methods for sample preparation prior to HPLC analysis should
ideally involve the minimum of sample handling consistent with
satisfactory compatibility between the HPLC system and the sample.
Initially the direct application of CBDCA-containing urine onto the
column was attempted.   Excellent separation of CBDCA from endogenous
urine components was observed (Figs. 2 & 3).   Thus for the analysis
of urine the only pre-treatment employed is the removal of any par-
ticulate matter (by allowing sedimentation prior to sampling).   When
plasma was mixed with the the HPLC solvent (specified above) precipitation was observed, indicating that removal of protein prior to
HPLC was required.   To facilitate comparison with studies performed
previously with cisplatin [9], centrifugal ultrafiltration was used.
An additional advantage of this technique is that it avoids possible
structural changes induced by chemical precipitants; thus, CBDCA is
unstable under acid conditions, hence acid precipitants are unsuitable [11].   However, for small plasma samples obtained from rodents
(20-100 µl) centrifugal ultracentrifugation is not practicable and
removal of protein in these instances was achieved by addition of
9 vol. of acetonitrile, then centrifugation (1,000 g), producing a
supernatant similar in composition to the mobile phase.   Using
larger volumes of control rat plasma spiked with CBDCA it was shown

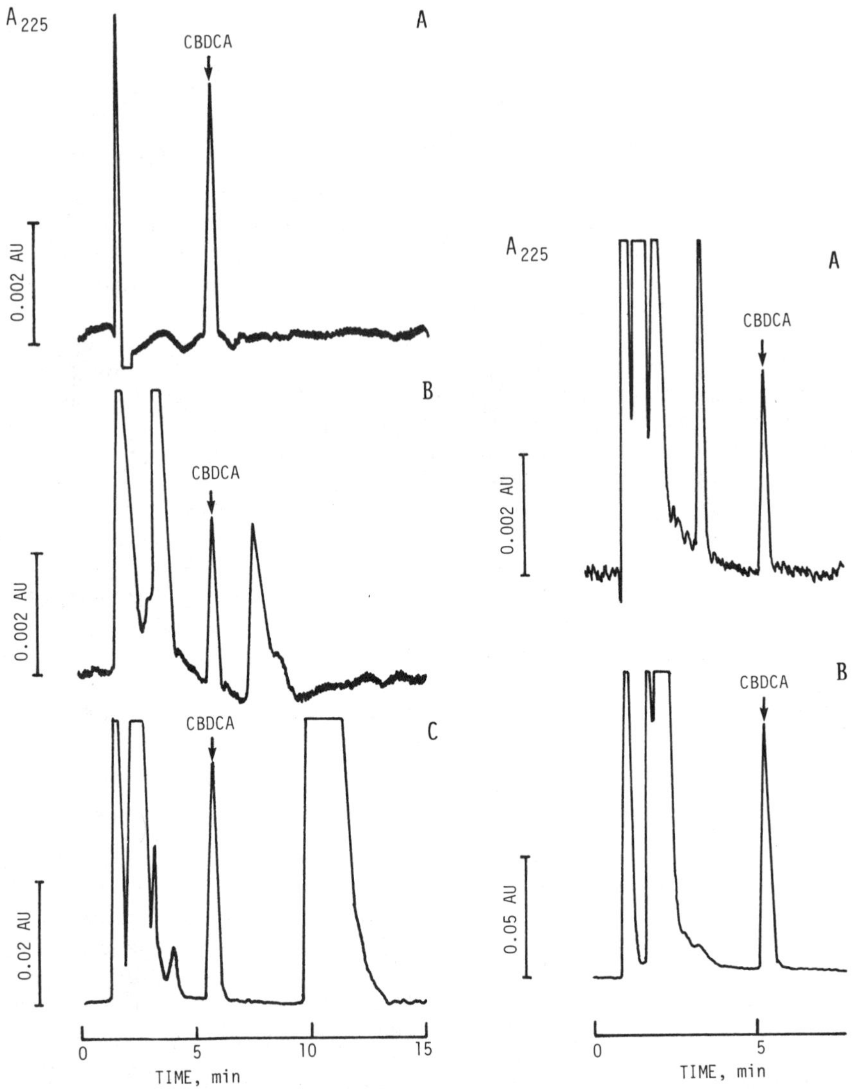

Fig. 2. HPLC chromatograms illustrating the analysis of CBDCA in human samples. All traces are from 10 µl injections; column conditions as in Table 2.
A. 100 µM standard dissolved in water. B. Plasma ultrafiltrate containing 64 µM CBDCA. C. Urine containing 940 µM CBDCA.
Reproduced from ref. [1], by permission of Cancer Research.

Fig. 3. HPLC chromatograms illustrating the analysis of CBDCA in rat samples. Column conditions as in Table 2.
A. 100 µl of supernatant following the addition of 180 µl of acetonitrile to 20 µl of rat plasma containing 80 µM CBDCA. B. 10 µl of rat urine containing 2.44 mM CBDCA.

that acetonitrile precipitation produces a similar recovery of CBDCA
to that achieved by centrifugal ultrafiltration (Table 1). For both
human plasma ultrafiltrates and rat plasma acetonitrile supernatants
the separation of CBDCA from plasma components was satisfactory, as
illustrated in Figs. 2 and 3.

    Having established suitable conditions for sample preparation
and separation, attention was focused on methods for detecting CBDCA
in the HPLC effluent. CBDCA absorbs weakly in the UV; hence in
the first instance UV detection at 225 nm was employed to establish
solvent composition and conditions for sample preparation. In calibration experiments UV detection was found to be satisfactory for
the analysis of CBDCA down to 10 μM (S/N ratio ~5 for a 10 μl aliquot
of plasma ultrafiltrate containing 10 μM CBDCA). For satisfactory
baseline stability the use of low-UV grade acetonitrile was found
necessary (Rathburn Chemicals Ltd., Walkerburn, Scotland). For
ultrafiltrates containing <10 μM CBDCA it was found necessary to
revert to performing FAAS on the HPLC effluent. To improve sensitivity, plasma ultrafiltrates (1 ml aliquots) were freeze-dried and
re-constituted in 200 μl of water prior to HPLC analysis of $\leq$30 μl
aliquots (application of >30 μl resulted in a split CBDCA peak).
The effluent with a retention volume corresponding to CBDCA (total
volume 3 ml) was collected, blown to dryness with nitrogen at room
temperature and reconstituted in 200 μl of 0.1 M hydrochloric acid,
100 μl of which was analyzed by FAAS. As shown in Table 1, this
method was adequate for the analysis of CBDCA in ultrafiltrates of
plasma containing 0.5-10 μM CBDCA. Although there was some loss of
Pt during this procedure (~35%), there was reassuring linearity.

Table 1. Assay approaches for CBDCA in plasma and urine.
Sample preparation was by ultrafiltration (ult.) or acetonitrile
precipitation (pptn.). Recoveries refer to CBDCA added to plasma
or urine; S.E.M. values are based on 6-10 observations.
All assays were linear over the range stated (r >0.99).

Species	Fluid	Sample preparation	Concentration range, μM	Recovery, %	Precision: C.V., %
Human	plasma	ult.	10-300	70-100	6
Human	plasma	ult.	0.5-10[*]	51 ±2	7
Rat	plasma	ult.	10-500	81 ±1	10
Rat	plasma	pptn.	10-500	80 ±2	7
Rat or human	urine	none	10-50,000	100	6

[*] FAAS after 5-fold concentration (225 nm detection in the other
approaches)

Table 1 also serves to summarize the methods available for the analysis of CBDCA in plasma and urine. Ultrafiltration, HPLC (cf. Table 2) and FAAS were performed as described elsewhere [1]. The stability of CBDCA in urine and plasma is such [1] that the HPLC analysis can be performed using an automatic sample processor. Thus for samples where CBDCA is detected by UV absorption the method is fully automated. The recovery values (as tabulated) are <100% for plasma samples due to protein binding which, whilst consistent in the strain and sex of rat used [2], was subject to some variability in man [1]. The mechanism of this binding, which is readily reversible at early times after drug administration (rat, 0–120 min; man, 0–300 min), is currently under study.

The HPLC methods described herein have been used in conjunction with FAAS during the pre-clinical [2] and clinical evaluation of CBDCA [1]. Thus Pt-containing species have been divided into three components, viz. protein-bound Pt, unbound Pt *not* in the form of CBDCA, and CBDCA itself. Comparison of the results with those for cisplatin has allowed us to speculate on the reasons for the low nephrotoxicity of CBDCA [1].

CHIP

CHIP (Fig. 1) has recently undergone phase I clinical evaluation at the Roswell Park Memorial Institute (Buffalo, NY). During pre-clinical studies in dogs and during the clinical trial, Pendyala et al. [3–5] developed and successfully applied HPLC assays for the analysis of CHIP in plasma and urine. The methods, summarized in Scheme 1, involve centrifugal ultrafiltration to remove protein from plasma and urine, followed by HPLC with either UV or FAAS detection. For HPLC with UV detection it is necessary to pass plasma ultrafiltrates through Dowex 2-X8 anion-exchange resin to remove interfering plasma components. HPLC is then performed as described in Table 2.

This method has been applied only to human plasma samples. For the analysis of CHIP in urine (or plasma samples if so desired) ultrafiltrates are applied to RP Sep-paks (Waters) and eluted with methanol/water mixtures of increasing eluotropic strength (30–100% methanol/water); recovery is again excellent. The eluates are pooled and freeze-dried, and the reconstituted residue analyzed by HPLC (Table 2), detection of CHIP being achieved by fraction collection and FAAS. Despite the greater time required the second method has the advantage of detecting Pt-containing species other than CHIP. Indeed, recent studies have shown that besides CHIP there are a number of other Pt-containing species in human plasma and urine. One of these was identified [3] as the Pt (II) species produced by the reduction of CHIP and consequent loss of the trans hydroxyl groups.

CHIP in plasma has also been analyzed by HPLC with EC detection [11], as described in accompanying articles by Krull and others.

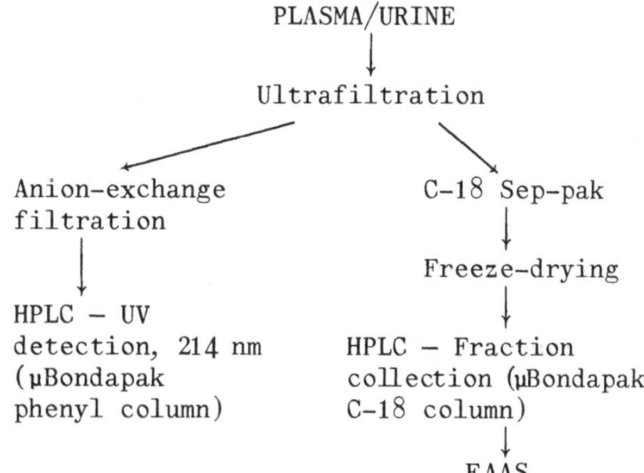

Scheme 1. HPLC analysis of CHIP in plasma and urine: procedures of Pendyala et al. [3-5].

TNO-6

TNO-6 (Fig. 1) was developed in Utrecht (Inst. of Applied Chemistry TNO) and has recently undergone clinical evaluation at a number of centres throughout Europe. In an attempt to develop suitable HPLC methodology, van der Vijgh and colleagues investigated HPLC with EC detection for the analysis of TNO-6 and the related dichloro complex TNO-1, *cis*-1,1-di(aminomethyl)cyclohexane(dichloro)platinum (II) [6]. Although this assay (summarized in Table 2) has yet to be applied to biological samples its use during stability studies has shown that in 150 mM NaCl, a common i.v. injection fluid, TNO-6 rapidly decomposes to TNO-1. Furthermore, TNO-6 also decomposes in glucose-containing injection fluids, albeit less rapidly [6]. That TNO-6 is a highly chemically-reactive complex is confirmed by the observation that <10% of the Pt in TNO-6 elutes from the HPLC column. This accounts for the poor sensitivity of the assay for TNO-6 when compared to TNO-1 [6]. A more suitable HPLC packing is currently being sought.

DACCP

DACCP has been the subject of clinical studies at the Memorial Hospital, New York [7]. In conjunction with these studies Leyland-Jones and co-workers have sought an HPLC assay for the parent complex (B. Leyland-Jones, personal communication). Unfortunately these studies have been unsuccessful and hence investigations have focused on the development and use of an assay for the trimellitic acid leaving group. This assay is a conventional ion-pair RP method which has demonstrated that trimellitic acid, whilst undetectable in plasma, is excreted in the urine (40% of the administered dose) [7]. In view of the doubtful clinical utility of DACCP and of difficulties encountered in its formulation [12] the value of further attempts to develop HPLC methods must be questioned.

Table 2. HPLC methods for the analysis of cisplatin analogues.

Complex	Stationary phase	Mobile phase*	Detection method	Limit of detection, μM	Ref.
CBDCA	Spherisorb silica	CH_3CN, 90; water, 10	UV, 225 nm	10	1
	Spherisorb silica	CH_3CN, 90; water, 10	FAAS	0.5	this art.
	C-18	MeOH, 25; buffer[†], 75	EC	13	11[Θ]
CHIP	μBondapak phenyl	MeOH, 10; water, 90	UV, 214 nm	1	3
	μBondapak C-18 (with gradient)	100% water → 100% MeOH	FAAS	0.04	4, 5
	C-18	MeOH, 25; buffer[†], 75	EC	0.2	11[Θ]
TNO-6	Nucleosil 10 C-18	MeOH, 10; 0.1 M $NaNO_3$, 90	EC	2	6

* The values signify the volumes mixed.
† Buffer = 0.01 M acetate pH 4.6, 0.15 mM hexadecyltrimethylammonium
Θ Relevant arts. in this vol. include #C-3.   [bromide.

CONCLUSIONS

Comprehensive pharmacokinetic studies call for information on
levels of both the parent compound and total drug-derived materials.
In the case of Pt complexes the latter can readily be achieved by
FAAS. The studies reviewed herein, and those concerning cisplatin
[9, 10], have shown that HPLC is a suitable technique for achieving
the former. The major limitation to the assays currently available
lies in the methods by which Pt complexes are detected in the HPLC
effluent. Although UV detection is suitable for higher concentrations, the more laborious procedure of fraction collection and FAAS
must be used for trace-level determinations. Routine on-line detection of Pt complexes after HPLC separation at all levels of therapeutic interest will, it is hoped, become feasible with alternative
detectors currently being studied (see the articles from the laboratories of P.C. Uden, I.S. Krull and L.A. Sternson - #C-2, C-3 & C-6).

Acknowledgements

The authors are indebted to Drs. Lakshmi Pendyala (Roswell Park Memorial Inst., Buffalo, NY), Wim van der Vijgh (Free University Hosp., Amsterdam) and Brian Leyland-Jones (National Cancer Inst., Bethesda, MD) for permission to quote information in press and for personal communications. We also thank Drs. P. Hydes and C. Barnard of the Johnson Matthey Research Centre (Sonning Common, Reading) for helpful advice during CBDCA assay development. The assistance of Drs. S.J. Harland and A.H. Calvert who performed the clinical studies at the Royal Marsden Hospital is gratefully acknowledged, as are the gifts of CBDCA from Bristol-Myers International Corp. (Brussels) and the National Cancer Institute (Bethesda, MD). We also thank Miss K. Balmanno for typing this manuscript. The work concerning CBDCA was supported by grants from the Medical Research Council and Cancer Research Campaign, U.K.

References

1. Harland, S.J., Newell, D.R., Siddik, Z.H., Chadwick, R., Calvert, A.H. & Harrap, K.R. (1983) *Cancer Res.* 44, 1693
2. Siddik, Z.H., Newell, D.R., Boxall, F.E., Jones, M., McGhee, K.G. & Harrap, K.R. (1984) in *Platinum Coordination Complexes in Cancer Chemotherapy [Proc. IV Int. Platinum Symp., 1983]* (Hacker, M., Double, E. & Krakoff, I. eds.), Martinus Nijhoff, Boston, pp. 90-102.
3. Pendyala, L., Cowens, J.W., Madajewicz, S. & Creaven, P.J. (1984) *as for* 2., pp. 114-125.
4. Pendyala, L., Greco, W., Cowens, J.W., Madajewicz, S. & Creaven, P.J. (1983) *Cancer Chemother. Pharmacol.* 11, 23-28.
5. Pendyala, L., Cowens, J.W. & Creaven, P.J. (1982) *Cancer Treatm. Rep.* 66, 509-516.
6. van der Vijgh, W.J.F., van der Lee, H.B.J., Postma, G.J. & Pinedo, H.M. (1983) *Chromatographia* 8, 333-336.
7. Kelsen, D.P., Scher, H., Alcock, N., Leyland-Jones, B.,Donner, A., Williams, L., Greene, G., Burchenal, J.H., Tan, C., Philips, F.S. & Young, C.W. (1982) *Cancer Res.* 42, 4831-4835.
8. Harrap, K.R. (1983) in *Cancer Chemotherapy I* (Muggia, F.M., ed.), Nijhoff, The Hague, pp. 171-217.
9. Himmelstein, K.J., Patton, T.F., Belt, R.J., Taylor, S., Repta, A.J. & Sternson, L.A. (1981) *Clin. Pharmacol. Ther.* 29, 658-664.
10. Riley, C.M., Sternson, L.A., Repta, A.J. & Siegler, R.W. (1982) *J. Chromatog.* 229, 373-386.
11. Krull, I.S., Ding, X-D., Braverman, S., Selavka, C., Hochberg, F. & Sternson, L.A. (1983) *J. Chromatog. Sci.* 21, 166-173.
12. Kelsen, D., Scher, H. & Burchenal, J. (1984) *as for* 2., pp. 310-320.

#C-5

THE THERAPEUTIC USE OF GOLD COMPOUNDS: ANALYTICAL ASPECTS

Roberta J. Ward

Division of Clinical Cell Biology
Medical Research Council Clinical Research Centre
Watford Road, Harrow, Middx. HA1 3UJ, U.K.

The therapeutic mode of action of gold is still unknown. Although the gold content of many biological materials has been determined by either atomic absorption spectrophotometry (AAS) or neutron activation analysis (NAA), no gold metabolites have as yet been identified. The monitoring of plasma gold levels during chrysotherapy correlates with the administered dose but not with either therapeutic efficacy or occurrence of toxic side-effects. In man after i.m. injection, plasma gold is mostly associated with albumin, while in the tissues of experimental animals the cytosolic subcellular fraction of the kidney has the highest levels of gold. The latter findings might relate to the detoxification and excretion of the metal.

During the past decade there has been renewed interest in the use of metal complexes as therapeutic agents. These include rhodium, ruthenium [1] and platinum [2] as anti-tumour agents, antimony (as an anti-parasitic agent) in leishmaniasis and schistosomiasis [3] and gold for the treatment of many rheumatic diseases including rheumatoid arthritis (RA) [4]. Gold [I] and gold [III] are the main oxidation states of gold and together with gold [0] have been investigated with respect to their effects on biological systems. The radioisotope ^{198}Au has been used for therapeutic scans; after administration it is selectively localized to the reticulo-endothelial system.

Gold [III] compounds have not been used therapeutically because of toxicity, but gold [I] compounds have found extensive therapeutic use. The commonly used drugs are gold [I] thiolates [e.g. gold sodium thiomalate and gold sodium thioglucose (Fig. 1)], which are administered i.m. in doses of up to 50 mg/week. Careful monitoring of the patients during therapy is necessary to minimize the occurrence of toxic side-effects such as albuminuria, leucopenia and skin rash.

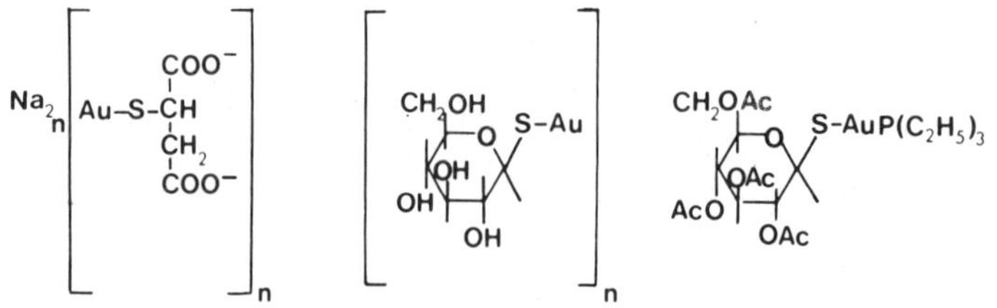

Gold sodium thiomalate (Myocrisin) — Gold thioglucose (Solganol) — (2,3,4,6-tetra-O-acetyl-1-thio-β-D-glucopyranosato-S)(triethylphosphine gold) (Ridaura)

Fig. 1. Chemical structures of commonly used gold [I] compounds.

Recently an orally administered gold [I] drug has been synthesized by complexing triethyl phosphine to tetra-acetyl thioglucose gold [1] (Fig. 1) and has been undergoing clinical trials to assess its efficacy in RA treatment [5]. Such changes in the structure of the gold [I] drug, e.g. the coordinate complex and the type and placement of the ligand, will dramatically alter the biological properties of the gold drug ranging from pharmacokinetic behaviour to interaction with macromolecules [6].

Quantitation of gold in biological fluids and tissues is done mainly by AAS and NAA, both of which are very specific with high sensitivity. Thereby gold has been determined in many biological fluids [7], and in the post-mortem tissues from one patient who had received high doses of sodium aurothiomalate [8], in attempts to correlate the local concentration of gold with either clinical efficacy or the occurrence of toxic side-effects. These investigations have been inconclusive apart from correlating the administered dose with the circulating level (Fig. 2). Hence there have been further studies of the distribution of gold between the cellular and protein components of the blood and amongst the subcellular organelles of the tissues.

INVESTIGATION OF GOLD COMPOUNDS IN BIOLOGICAL SAMPLES

The structure of the complexes formed after metabolism of the gold drugs has not been identified because of the limitations of the available analytical techniques. Sadler [9] studied the structure of many therapeutically used gold compounds with techniques such as extended X-ray absorption spectroscopy and Mossbauer spectrometry. The application of such methods to study the structure of gold complexes *in vivo* is severely limited because of the low concentrations of gold in biological samples and the interferences from biological matrices.

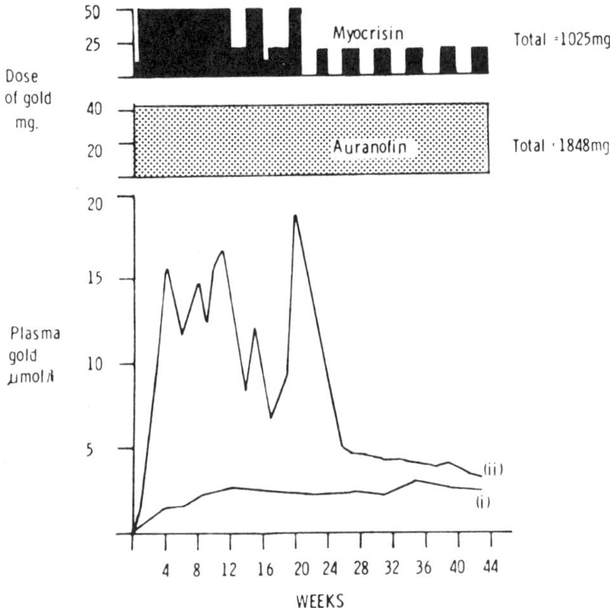

Fig. 2. Concentration of gold in the plasma of two patients with
rheumatoid arthritis after commencing treatment with either
(i) auranofin (Ridaura), 8 mg/day fixed dose, or (ii) myocrisin,
variable dose.

Erythrocyte gold levels, once thought to be informative (in
myocrisin-treated patients [10]), are high in all auranofin-treated
patients and thus may not be a useful parameter of clinical efficacy
or occurrence of toxic reactions.

The distribution of gold among the plasma proteins after i.m.
injection of myocrisin has been investigated by two techniques,
(i) electrophoresis, and (ii) gel filtration chromatography [7]. While
the former has the advantage that it can separate many plasma proteins from a relatively small amount of serum (~20 μl), there are
several disadvantages.-
(a) During the separation procedure gold may be redistributed to
other plasma proteins.
(b) Gold or its complexes may react with either the support media or
chemicals used.
(c) The cellulose acetate strip containing the particular gold complex may cause analytical problems during electrothermal atomic
absorption. The smoke emitted by the strip during the ash stage of
the atomization cycle may interfere with the atomization signal and
make interpretation of the answer difficult.
(d) Each protein band to be analyzed for gold needs manual insertion
into the atomization unit.

A preferred approach for separating the plasma proteins has therefore been gel-filtration (size-exclusion) chromatography. An elution buffer containing tris and 0.15 M NaCl was used to minimize any redistribution of gold between the proteins during Sephadex G-200 chromatography. The gold content of each of the eluted fractions was determined by both electrothermal AA and NAA. The latter technique was able to accurately quantitate the distribution of gold among albumin, globulins and low mol. wt. fractions, while electrothermal AA determined only the gold associated with albumin. The inability of electrothermal AA to determine the globulin and low mol. wt. associated gold was probably due to the presence of chloride in the elution buffer. This caused depression of the atomization signal by co-volatilizing with the gold during the ash stage of the atomization programme. Furthermore, the peak of the absorption spectrum of sodium chloride almost coincides with the major resonance line of gold (240 nm and 242.8 nm respectively), so that any sodium chloride not removed during ashing will contribute to the non-atomic signal during the final atomization stage.

Despite the disadvantages of both the electrophoretic and the gel-filtration techniques, both methods have shown most of the gold to be associated with albumin in the plasma of patients receiving sodium aurothiomalate with smaller concentrations associated with the globulin fraction and a low mol. wt. fraction. In one further study [12] the distribution of gold among the three plasma fractions was studied in 3 RA patients receiving long-term chrysotherapy by gel-filtration chromatography with NAA. At each time interval where the gold content of the three fractions was determined, the majority was associated with the albumin fraction. The authors suggested that the low mol. wt. gold may represent complexes of endogenous thiomalates, e.g. S-aurocysteine, as well as the free drug. There have been other studies on this low mol. wt. fraction as it may be the active moiety of the drug which is exchangeable with the tissues. Using ultrafiltrate techniques, either high [13] or low [14] percentages of the total gold have been associated with the ultrafiltrate. However, there is no evidence as yet that any of these gold-associated plasma fractions reflect a beneficial response to chrysotherapy.

Although gold has been determined in other easily available biological fluids, e.g. urine [15] and synovial fluid [15], there has been no characterization of these gold complexes or correlation of these results with clinical efficacy.

Investigation of tissues

Three techniques are available for determining gold in tissues: electron microscopy (e.m.), electron-pulse X-ray analysis, and subcellular fractionation of the cellular components before analysis of gold content by electrothermal AA or NAA [15].

(a) There has been wide use of e.m. autoradiography to observe the distribution of $[^{198}Au]$aurothiomalate after its injection into the synovial joint where it accumulates in the lysosomes of cells and macrophages of the synovium. There is a selective uptake of gold by the lysosomes which become progressively more dense and are then referred to as aurosomes. (Synovial lysosomal enzyme release causes much of the joint damage in RA patients; conceivably lysosomal gold enrichment may stabilize the lysosomal membrane [15].)

(b) Electron-probe X-ray analysis can detect as little as 10^{-18}-10^{-21} mol of gold, but if the gold is widely dispersed throughout the tissue the sensitivity will be reduced.

(c) Most studies of the distribution of gold in animal cells have used the differential centrifugation technique with all its limitations [as surveyed in companion vols., 'Biochemistry' subseries – *Ed.*]. The majority of the gold was associated initially with the cytosolic fraction of rat liver after i.p. injection of Na aurothiomalate, while 7 days later there was enrichment of the lysosomal (and mito-chondrial) fractions [16]. The content of gold in the kidney was 10-fold that of liver, being highest in the cytosolic fraction with 4 protein-associated gold components of unknown structural identity (experiments by Danpure & Ward). Concomitantly there was an increase in kidney copper concentration which could indicate that gold alters copper metabolism. Whether one of the therapeutic actions of gold is to alter the metabolism of copper in RA patients requires elucida-tion; the elevated concentrations of copper decreased towards the reference range in the plasma of RA patients responding clinically to gold [17].

CONCLUDING COMMENTS

Another area of investigation is the immune-system changes in RA. It is uncertain whether the presence of gold in immunological com-ponents is indicative of a direct site of action of gold [15]. Alto-gether the lack of a suitable animal model for RA is an investiga-tive handicap.

With the application of improved techniques in the next decade, notably the use of better subcellular fractions as obtainable with the Beaufay zonal ultracentrifuge and minimization of pressure dis-ruption of the organelles, identification of the structure of the gold complexes formed *in vivo* will be facilitated. This will help clarify the actual site of action of gold, and in turn lead to the development of other therapeutic gold compounds having maximum clinical efficacy with minimal toxicity.

References

1. Giraldi, T., Salva, G. & Bertoli, G. (1977) *Cancer Res. 37*, 2662-2666.
2. Prestayko, A.W., Crooke, S.T. & Carter, S.K., eds. (1980). *Cisplatin - Current Status and New Developments*, Academic Press, New York, 527 pp.
3. Black, C.D.V., Watson, G.J. & Ward, R.J. (1977) *Trans. Roy. Soc. Trop. Med. Hyg. 71*, 550-552.
4. Empire Rheumatism Council Research Subcommittee (1961) *Ann. Rheum. Dis. 20*, 315-333.
5. Prouse, P.J., Ward, R.J., Howard, A. & Gumpel, J.M. (1981) *Ann. Rheum. Dis. 40*, 524P.
6. Walz, D.T., DiMartino, M.J. & Griswold, D.G. (1982) *J. Rheum. Suppl. 8*, 54-60.
7. Ward, R.J., Fyfe, D. & Danpure, C.J. (1977) *Clin. Chim. Acta 81*, 87-97.
8. Gottlieb, N.L., Smith, P.M. & Smith, E.M. (1972) *Arth. Rheum. 15*, 16-22.
9. Sadler, P.J. (1982) *J. Rheum. 8*, 71-78.
10. Møller-Pedersen, S. & Møller-Graabeck, P. (1980) *Ann. Rheum. Dis. 39*, 576-579.
11. Brown, D.H. & Smith, W.G. (1980) *Chem. Soc. Rev. 9*, 217-240.
12. Danpure, C.J., Fyfe, D.A. & Gumpel, J.M. (1979) *Ann. Rheum. Dis. 38*, 64-370.
13. Campion, D.A., Olsen, R., Bohan, A. & Bluestone, R. (1974) *J. Rheum. 1, Suppl.*, 1-112.
14. Kamel, H., Brown, D.H., Ottaway, J.M. & Smith, W.E. (1976) *Analyst 101*, 790-797.
15. Shaw, C.F., III (1979) *Inorg. Perspect. Biol. Med. 2*, 287-355.
16. Lawson, K.J., Danpure, C.J. & Fyfe, D.A. (1977) *Biochem. Pharmacol. 26*, 2417-2426.
17. Ward, R.J., Doshi, P., Cashmore, C., Howard, A., Prouse, P. & Gumpel, J. (1983) in *Auranofin* (Capell, H., Cole, D., Marghari, K & Morris, R., eds.), Excerpta Medica, Amsterdam, pp. 389-397.

#C-6

CLINICAL ANALYSIS OF DIVALENT PLATINUM COMPLEXES WITH ANTI-NEOPLASTIC ACTIVITY

Larry A. Sternson

Pharmaceutical Chemistry Department
University of Kansas
Lawrence, KS 66045, U.S.A.

Analytical procedures for the clinical monitoring of Pt-containing anti-neoplastic drugs in biological fluids have been developed. Chromatographic separation of platinates has been achieved on solvent-generated (dynamic) ion-exchangers. The retention of platinum analytes can be increased or decreased by addition of organic modifiers and/or electrolytes to an aqueous mobile phase. Biological fluids can be directly applied to the chromatographic system using a column-switching approach in which the platinates are transferred from a silica pre-column to a second, RP, analytical column.

Sensitive on-line detection of platinates has been developed using either polarographic transducers or post-column reactors with spectrophotometric monitoring. The former involves reduction of analytes at a mercury cathode kept at 50°. The electrochemical process takes place at 0.00 V, offering a highly selective and sensitive method of detection. In the post-column reactor, platinates in the column effluent are mixed with dichromate solution and then sodium bisulphite to form strongly chromophoric products which are detected spectrophotometrically.

The unique structure and physicochemical properties of cis-platin have presented significant challenges in the development of analytical procedures sufficiently selective and sensitive for the

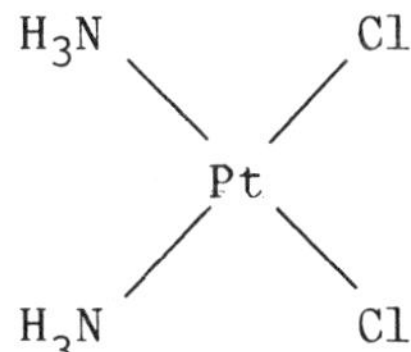

determination of the intact drug in biological fluids. Cisplatin undergoes facile nucleophilic substitution in aqueous solution. Such reactions hsve been implicated in its biotransformation, its inter- action with DNA, and expression of its toxicity [1].

Analytical techniques traditionally used to determine cisplatin in clinical samples include UV spectrophotometry (after derivatiza- tion) [2], X-ray fluorescence [3] and, most commonly, flameless atomic absorption (AA) spectrometry [4, 5] (cf. #C-2 by P.C. Uden, this vol.-*Ed*.]. These modes of detection are non-specific, respon- ding only to the total amount of Pt irrespective of the ligands coordinated with it.   Clinical evaluation of platinates requires specific methodology, capable of distinguishing platinates differing in ligand composition.

THE CHROMATOGRAPHY OF CISPLATIN

Due to the poor solubility of cisplatin in non-aqueous solution its chromatography is limited to reversed phase (RP) systems employ- ing aqueous mobile phases modified by polar organic solvents. The retention of cisplatin on various stationary phases was investigated using aqueous mobile phases modified with methanol (Fig. 1) [6]. Cisplatin was poorly retained on both silica and hydrophobic (μ- Bondapak C-18) RP columns, where its retention decreased slightly with increasing methanol concentration. In contrast, cisplatin was appreciably retained on stationary phases possessing cationic funct- ionalities by virtue of quaternary ammonium groups, either chemically bonded (SAX) or consisting of an adsorbed monolayer of hexadecyltri- methylammonium bromide (HTAB) on μ-Bondapak C-18 or ODS-Hypersil. With ostensibly aqueous mobile phases (methanol proportion ≤ 0.15), the stability of the modified stationary phase was maintained by the presence of a low concentration (~0.1 mM) of HTAB in the mobile phase.

Although cisplatin was retained on both chemically bonded and solvent-generated anion exchangers, the influence of methanol was different in the two systems (Fig. 1) [6]. With purely aqueous mobile phases, cisplatin was retained more strongly on the solvent- generated phase than on the chemically bonded anion exchanger. Increasing the methanol concentration decreased retention on the dynamic exchanger while increasing retention on the chemically bonded exchanger. With a methanol/water vol. ratio below 0.10, cis- platin retention on the modified RP column was independent of HTAB concentration in the mobile phase, suggesting that the contribution of mobile phase − solute −surfactant interactions to retention was negligible. At higher concentrations of methanol, cisplatin reten- tion depended on the concentration of HTAB in the mobile phase due to the adsorbed surfactant being more prone to become displaced. However, desorption could not account for the initial decrease in retention observed with increasing methanol concentration.

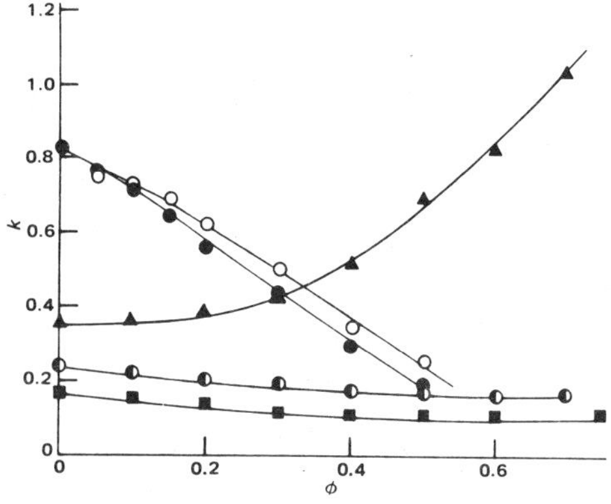

Fig. 1. Relationship between the capacity ratio (k) of cisplatin and the vol. fraction (ϕ) of methanol in a water-based mobile phase with various stationary phases at 30°:- ▲, strong anion-exchanger; ■, silica; ◑, RP (C-18); ○ and ●, RP onto which was adsorbed 0.99 µmol/m² of HTAB with, respectively, 0.1 mM and 0.01 mM HTAB in the mobile phase. *From [6], courtesy of Elsevier.*

Rationale for the retention

The retention of cisplatin on both types of cationic stationary phase can be rationalized in terms of ion-dipole interactions between the neutral Pt complex and the positively charged stationary phase. The difference in the chromatographic behaviour in the two systems can be explained in terms of the different environments in which these interactions occur and the influence of methanol on these environments. Cationic groups adsorbed onto a RP material exist in an apolar, hydrophobic environment, whereas the quaternary ammonium groups on the chemically bonded (SAX) column are in a polar environment proximal to the (hydrated) silica backbone.

In both systems, addition of methanol to the mobile phase results in its adsorption at the stationary phase-mobile phase interface (SMI), decreasing the polarity about the cationic groups of the bonded ion-exchanger while increasing the polarity experienced by the corresponding groups on the solvent-generated ion-exchanger. This increase in polarity at the SMI decreases the thermodynamic activity of the quaternary ammonium groups (as these functionalities become better solvated), reducing their ability to interact with the negative dipole of cisplatin and thereby reducing column retention. Conversely, partial replacement of the hydration layer on the silica-backed covalently bonded ion-exchanger by methanol reduces the polarity of the environment at the SMI, increasing the thermodynamic activity of quaternary ammonium groups manifested as enhanced retention of cisplatin with increasing methanol concentration.

Influence of electrolytes on the retention

The retention of cisplatin on dynamic anion-exchange columns can be further controlled by the addition of electrolytes to the mobile phase [7]. The addition of the monovalent electrolytes, Na

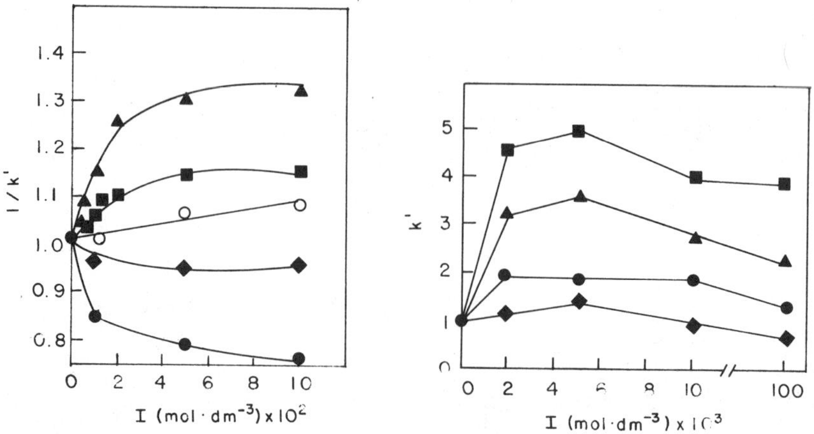

Fig. 2. Effect of salt concentration (I) on the capacity ratio (k')
of cisplatin, 1 mg/ml in the load sample; run at 30 ±0.1°.
Stationary phase μBondapak C-18 loaded with 1.31 μmol/m² HTAB; mobile
phase 100 μM HTAB in water plus salt as specified:-
Left: ■, Na bromide; ▲, Na nitrate; or Na acetate/acetic acid
buffer: ●, pH 7.0; ◆, pH 5.0; o, pH 3.5.
Right: citrate buffer: ■, pH 7.0; ▲, pH 5.5; ●, pH 4.0;
◆, pH 3.0.   *From [7], by permission.*

bromide or Na nitrate (both 0.1 M) to the mobile phase decreased the
retention of cisplatin, whereas polyvalent electrolytes (Na citrate
and sulphate produced a significant increase in retention (Fig. 2).
The presence of electrolyte in the mobile phase affects both the
electrical double layer and the surface tension of the eluent, thus
accounting for their influence on solute retention. These two factors act in opposition to one another, and the predominating process
determines the direction and magnitude of change in retention with
addition of salt.

Addition of bromide or nitrate to the mobile phase decreased
retention (Fig. 2), apparently by reducing the thermodynamic activity
of the cationic sites on the stationary phase due to creation of an
inner Helmholtz plane within the electrical double layer [8]. At
low concentrations, sulphate, citrate and acetate (at pH values
where the carboxylate is appreciably ionized) caused increased
retention of cisplatin (Fig. 2) owing to their effect on the surface
tension of the mobile phase as rationalized by solvophobic theory
[9]. At higher concentrations of polyvalent electrolyte, solute
retention decreased due to the secondary contribution of electro-
static effects [8]. According to solvophobic theory [9], retention
in HPLC systems employing aqueous mobile phase is described by:

$$\ln k = a + b + \gamma \Delta A / RT \qquad \text{(Eq. 1)}$$

which states that retention is governed by the sum of all possible contributions to solute-solvent-stationary phase interaction. The first term, a, is related to the properties of the mobile and stationary phases and is independent of the nature of the solute. In the present system, b may be taken as a measure of the ion-dipole interactions which contribute to retention. The third term is a measure of the hydrophobic interactions where γ is the mobile phase surface tension and ΔA is the decrease in hydrophobic surface area on binding of the solute to the stationary phase.

The addition of salt to the mobile phase results in an increase in surface tension according to

$$\gamma = \gamma_O + \tau m \qquad \text{(Eq. 2)}$$

where γ_O is the surface tension in the absence of salt, m is the molal salt concentration and τ is a constant related to the nature of the added salt [10]. Combining eqs. 1 and 2 gives:

$$\ln k' = a + b + \frac{(\gamma_O + m\tau)\, \Delta A}{RT} \qquad \text{(Eq. 3)}$$

which predicts a linear relationship between $\ln k'$ and τ at a fixed salt concentration. Table 1 shows the relationship between $k'_{cisplatin}$ at 0.1 M salt concentration and the reported values for τ (regression line: $\ln k' = 0.85\,\tau - 1.2$; $r = 0.998$; $n = 4$). These results support the hypothesis that the effects of salts on cisplatin retention arise from both the ionic contribution to the activity of the

Table 1. Relationship between τ values of electrolytes added to the mobile phase and the capacity ratios of cisplatin (k'). Mobile phase: 0.1 mM HTAB, 0.1 M electrolyte. Stationary phase: HTAB, 1.31 μmol/m^2 on an ODS-Hypersil column (100 × i.d. 4.6 mm). Temperature 30°. The τ values represent constants describing the effect of electrolyte on the surface tension of water (Eq. 2; [10]). $k' = (t_r - t_o)/t_o$.

Electrolyte	τ	k'
Sodium citrate (pH 7)	3.12	3.97
Sodium sulphate	2.73	3.22
Sodium nitrate	1.32	0.86
Sodium bromide	1.06	0.74

Fig. 3. Automated column switching system for the determination of cisplatin in urine utilizing silica and ODS-Hypersil RP columns. *From [11], by permission.*

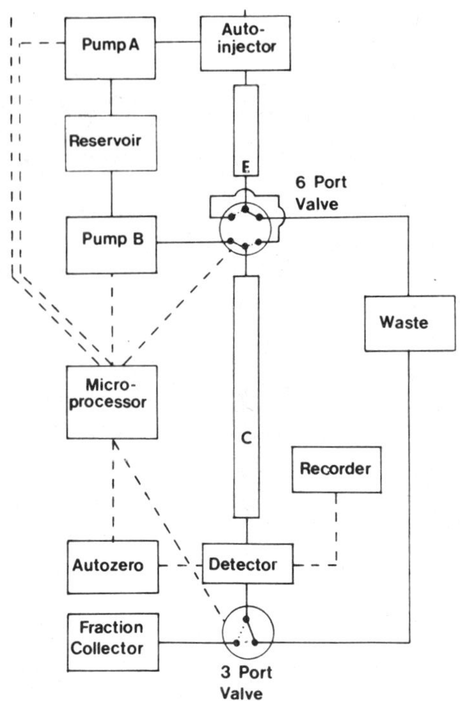

quaternary ammonium sites [leading to a decrease in a and b (Eq. 3)] as well as their influence on the hydrophobic term ($\gamma\Delta A/RT$). For salts with a large value of τ (e.g. citrate) cisplatin retention is governed by the hydrophobic term, while terms a and b dominate retention behaviour with salts having small τ values (e.g. nitrate and bromide).

Benefit from column-switching

In attempting to apply this chromatographic system to biological samples (plasma ultrafiltrate and urine), we found that even in the best situations, cisplatin co-elutes with some biological components. The problem is exacerbated by the long retention of several other endogenous components. This problem was overcome using a column-switching approach where the dynamic anion-exchanger was preceded by a silica pre-column (Fig. 3) [11]. The two columns, which exhibit different selectivities for Pt complexes, were linked in series through a 6-port valve under microprocesssor control. Initial sample clean-up was accomplished on the silica column (Column 1) with the analytical separation being carried out on a dynamic anion-exchange column. Adjustment of mobile phase composition in terms of organic modifier, pH and salt type and concentration allows significant control of retention behaviour of Pt complexes differing in ligand composition.

The use of column switching (1) provides rapid, inexpensive sample clean-up that can easily be automated, (2) offers powerful separating capacity since it combines column packings of widely different selectivities (such that on Column 1, biological material is strongly retained, whilst on Column 2, biological material elutes rapidly and cisplatin is strongly retained), (3) minimizes contamination of the analytical column since only the 'heart-cut' containing the compounds of interest is introduced onto the analytical column, and (4) shortens analysis time since late-eluting background peaks are not introduced onto the second column.

CHROMATOGRAPHING PRODUCTS FROM CISPLATIN REACTION WITH NUCLEOTIDES AND PROTEINS

Cisplatin is extensively biodegraded through non-enzymatic reaction with nucleophilic substances, notably amino acid residues and nucleotides [1]. The products are apparently responsible both for the chemotherapeutic activity and for the toxicity of cisplatin. Thus, methodology is needed to monitor these species in order to better evaluate changes in drug protocols involving cisplatin.

Nucleotides

Chromatographic systems have been developed to monitor the reaction of cisplatin with nucleotides (and their oligomers), a reaction that has been implicated in cytotoxicity [12]. Separations are achieved on columns packed with solvent-generated anion-exchangers using, as mobile phase, aqueous phosphate buffer in which pH and ionic strength are adjusted to control retention [12].

Proteins

Reactions of cisplatin with polypeptides and proteins (up to mol. wt. 60,000) have been followed by gradient elution RP (C-18) HPLC using aqueous phosphate buffer mobile phases modified with increasing amounts of acetonitrile [13]. Differentiation between Pt-containing and free protein was accomplished both chromatographically and through use of AA and spectrophotometric detectors placed in series. Chromatographic conditions were optimized to maximize resolution of nitrogenous components. In some cases, however, resolution of Pt-containing components and those devoid of metal was not possible. The two types were distinguishable by the serial detection approach. Reaction kinetics could be mapped using this technique, as exemplified by the reaction of cisplatin with metenkephalin (Fig. 4) where multiple products of varying stability were formed.

DETECTION OF PLATINUM IN COLUMN EFFLUENT

The physico-chemical properties of cisplatin do not lend themselves to detection at therapeutic levels (≤ 100 ng/ml) using most

commercially available detectors. Currently, Pt in column effluents
is monitored by off-line AA spectrophotometry [11-13]. This method
is sufficiently sensitive, but is laborious and time-consuming when
dealing with large numbers of samples. Accordingly, on-line detection is being developed. Two techniques have been investigated simultaneously: (1) polarographic detection, and (2) use of reaction
detectors.

Polarographic detection of column effluent.- A polarographic
detector specific for monitoring cisplatin following direct injection of biological fluid (e.g. plasma ultrafiltrate, urine) has been
developed [14]. A renewable Hg drop was selected for this application because of (a) requisite analyte adsorption on the Hg surface
prior to reduction, complicated by competing adsorption of mobile
phase surfactant; and (b) accumulation of the reduction product on
the electrode. Such cumulative surface effects degrade signal quality
and are difficult to overcome with solid electrodes. Both current-
sampled dropping Hg (DME) and hanging Hg drop (HMDE) electrodes provide atvantages over UV absorbance and off-line non-flame AA.

The detector was optimized with respect to cell configuration,
mode of measurement, drop size, drop time, cell temperature and flow
parameters. By raising the detector cell temperature to 60°, the
detector response to cisplatin (CDDP) was increased and shifted
anodically (from -10 mV at 25° to 0.0 V *vs.* Ag/Ag/Cl), thereby increasing detector selectivity as well as sensitivity for this compound. The minimum detectable quantities of cisplatin with DME and
HMDE are 1.8 ng and 70 pg respectively. Cisplatin can be determined
in untreated urine at levels below 100 ng/ml. The major disadvantages of this detector are (1) that not all Pt complexes respond,
and (2) the difficulty in using the detector for routine monitoring
due to its 'temperamental' nature.

Post-column reactors.- An alternative detection approach involves a post-column reactor that gives detectable derivatives from
cisplatin and its degradation products [15]. (Pre-column derivatization would call for a cisplatin-specific reaction.) We have found
that bisulphite [15,16] in the presence of a strong oxidizing agent (e.g.
dichromate) reacts notably rapidly with various divalent (but not
tetravalent) Pt derivatives to form strongly chromophoric product(s)
with UV absorbance at 290 nm. The reaction is a good candidate for
a detector because of its rapid kinetics, high yield, and strong
product absorptivity at a wavelength where most biological materials
do not interfere. This reaction has been utilized in a post-column
reactor (Fig. 5) where dichromate and bisulphite are mixed sequentially. Reaction delay times of more than 5 min could be achieved
without appreciable band broadening, using reaction coils of small
i.d. (0.3 mm) which were knit on a spool to give a cylindrically
braided coil, thus providing a tortuous path for the reagent streams
[17]. With this system [15], a linear response to cisplatin was

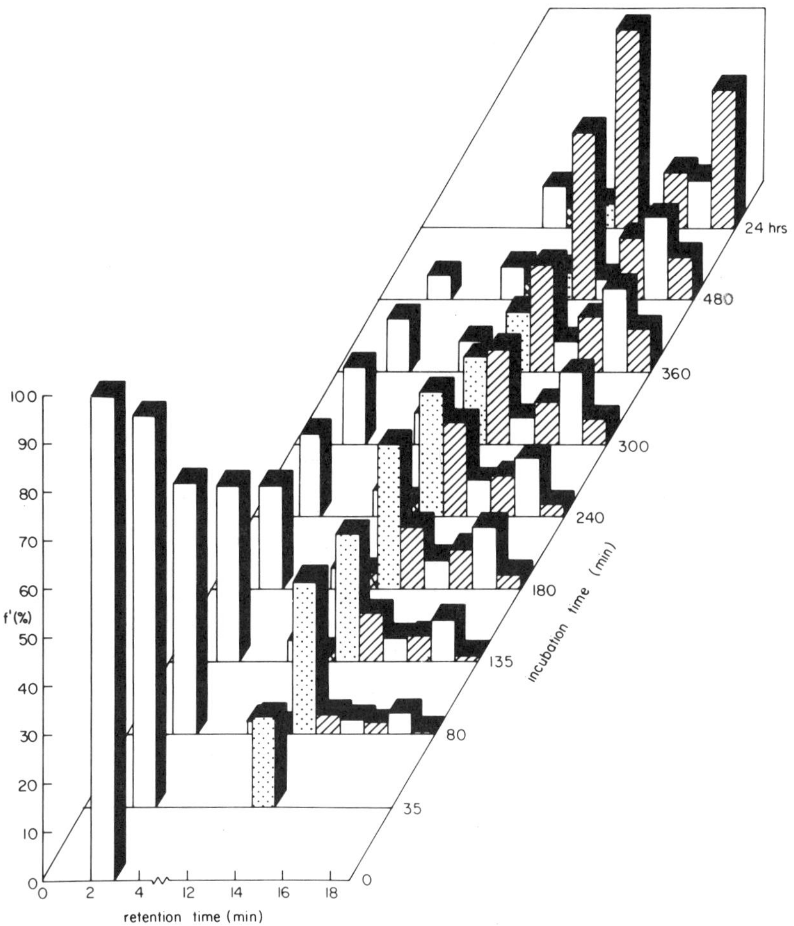

Fig. 4. Three-dimensional histogram showing the elution profile of Pt within samples of metenkephalin incubated with cisplatin at 37°. The Pt was determined by collecting 1.5 ml eluent fractions which were subsequently analyzed by flameless AA spectroscopy. The amount of Pt in each fraction is expressed in terms of the fraction of the total Pt injected onto the column (i.e. $f' = q_i/q_x$). *From [13]*.

obtained (and to other divalent Pt complexes) over the range 35–12,000 ng cisplatin/ml of biological fluid (Fig. 6). The reaction detector concept offers specificity provided both by the HPLC separation and a sensor responsive only to divalent Pt species.

Thus, it has become possible to specifically detect cisplatin and related compounds in the clinical context, e.g. to describe the pharmacokinetics in the management of testicular [18] and oesophageal [19] tumours.

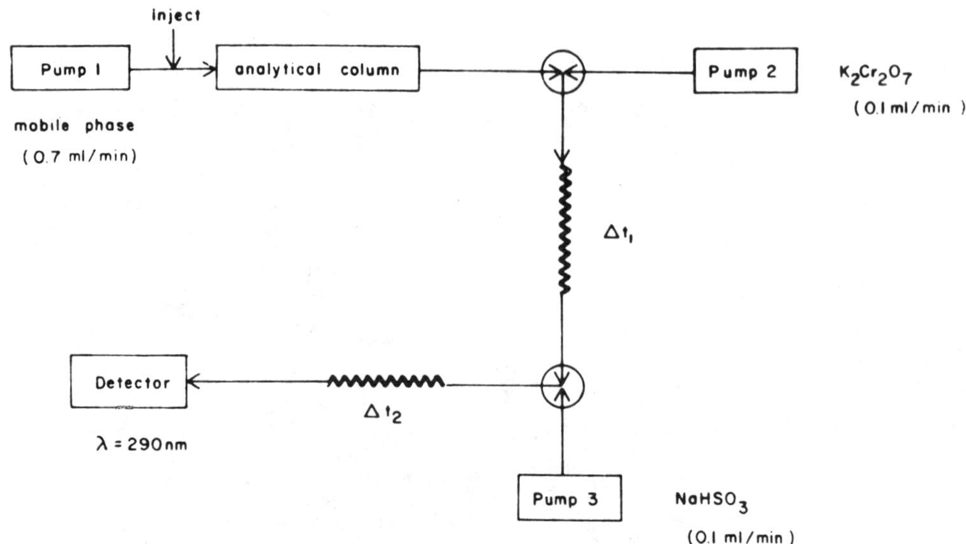

Fig. 5. Schematic representation of the Pt/bisulphite reaction detector in which the separated analytes are mixed sequentially with $K_2Cr_2O_7$ and $NaHSO_3$ and detected at 290 nm after the appropriate reaction delay time (Δt_1 and Δt_2).

References

1. Prestayko, A.W., Crooke, S.T. & Carter, S.K., eds. (1980) *Cisplatin, Current Status and New Developments*, Academic Press, New York.
2. Bannister, S.J., Sternson, L.A. & Repta, A.J. (1979) *J. Chromatog.* *173*, 333-342.
3. Bannister, S.J., Sternson, L.A., Repta, A.J. & James, G.W. (1977) *Clin. Chem. 23*, 2258-2262.
4. Bannister, S.J., Chang, Y., Sternson, L.A. & Repta, A.J. (1978) *Clin. Chem. 24*, 877-880.
5. Leroy, A.F., Wehling, M.L., Sponsellet, L., Solomon, W.S., Dedrick, R.L., Litterst, C.L., Gram, T.E., Guarino, A.M. & Becker, D.E. (1977) *Biochem. Med. 18*, 184-191.
6. Riley, C.M., Sternson, L.A. & Repta, A.J. (1981) *J. Chromatog. 217*, 405-420.
7. Riley, C.M., Sternson, L.A. & Repta, A.J. (1981) *J. Chromatog. 219*, 235-244.
8. Adamson, A.W. (1967) *Physical Chemistry of Surfaces*, 2nd edn., Wiley, New York.
9. Horvath, Cs., Melander, W..& Molnar, I. (1976) *J. Chromatog. 125*, 129-156.
10. Horvath, Cs. & Melander, W. (1977) *Arch. Biochem. Biophys. 183*, 200-215.
11. Riley, C.M., Sternson, L.A., Repta, A.J. & Siegler, R.W. (1982) *J. Chromatog. 229*, 373-386.

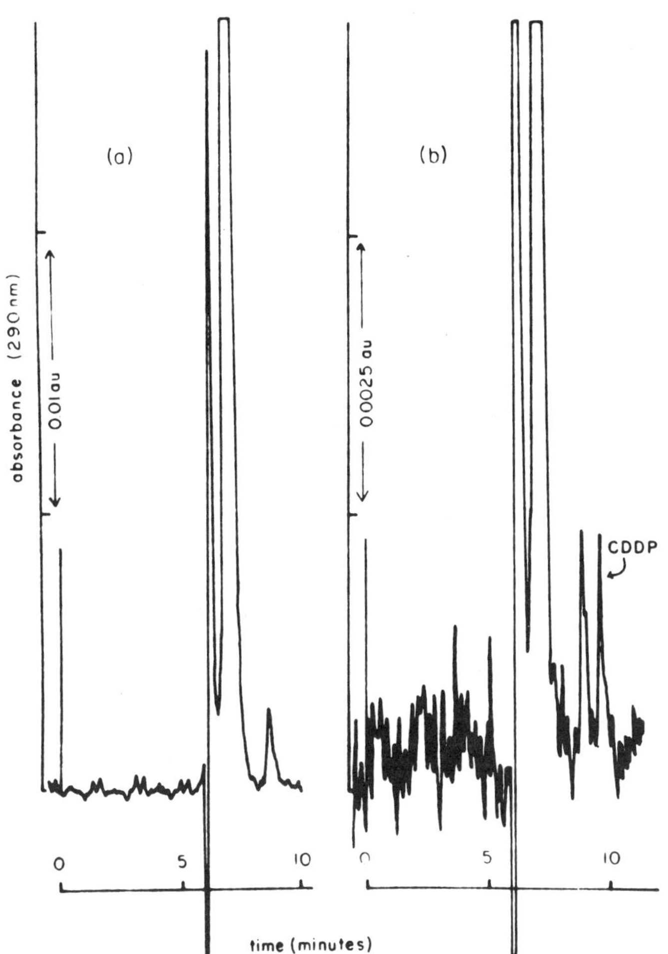

Fig. 6. Post-column reactor response to plasma ultrafiltrate
blank (a) and the same containing cisplatin (CDDP) (b). The CDDP
peak is the reactor response to 5.17 ng of Pt. Chromatography on
100 x i.d. 4.6 mm 5 μm Hypersil-ODS column coated with HTAB; mobile
phase 10 mM citrate (pH 4.5, 100 μM HTAB); detection at 290 nm.
The ultrafiltrate contained 25.2 μM CDDP, and 20 μl was loaded.
Reactor as in Fig. 5: 3.3 mM Na bisulphite, 26 μM K dichromate;
delay braids Δt_1 = 3.2 m x 0.3 mm i.d., and Δt_2 = 44.6 m x 0.3 mm
i.d.   *From [15], by permission.*

12. Riley, C.M., Sternson, L.A. & Repta, A.J. (1983) *Anal. Biochem.*
    130, 203-214.
13. Riley, C.M., Sternson, L.A. & Repta, A.J. (1982) *Anal.
    Biochem. 124*, 167-179.

14. Bannister, S.J., Sternson, L.A. & Repta, A.J. (1983) *J. Chromatog.* *273*, 301–318.

15. Marsh, K.C., Sternson, L.A. & Repta, A.J. (1984) *Anal. Chem.* *56*, 491–497.

16. Hussain, A.A., Haddadin, M. & Iga, K. (1980) *J. Pharm. Sci.* *69*, 364–366.

17. Engelhardt, H. & Neve, U.D. (1982) *Chromatographia 15*, 403–406.

18. Himmelstein, K.J., Patton, T.F., Belt, R.J., Taylor, S., Repta, A.J. & Sternson, L.A. (1981) *Clin. Pharmacol. Ther. 29*, 658–664.

19. Mattox, D.E., Sternson, L.A., von Hoff, D.D., Kuhn, J.G. & Repta, A.J. (1983) *Otolaryng.*, 271–275.

#C-7

SOLID PHASE DERIVATIZATION REACTIONS IN HPLC

I.S. Krull, S. Colgan, K-H. Xie, C. Santasania,
[†]U. Neue, [†]R. King, [†]A. Newhard and [†]B. Bidlingmeyer

Institute of Chemical
 Analysis & Department
 of Chemistry
Northeastern University
360 Huntingdon Avenue
Boston, MA 02115, U.S.A.

[†]Research and Development
 Departments
Waters Associates Inc.
34 Maple Street
Milford, MA 01757, U.S.A.

Heterogeneous as compared with homogeneous type derivatizations, either off-line or on-line, pre- or post-column, offer more advantages than disadvantages. This is especially true when solid phase reactions are fully compatible with HPLC mobile phases, and can be performed quantitatively in 'real time', on-line, at ambient temperatures under HPLC pressure conditions. Some polymeric or solid-supported reactions/ reagents have now been developed and applied to both on-line and off-line HPLC separations, usually with UV detection. Such approaches permit the use of difference chromatography, first a dummy run without reagent and then a repeat with the solid phase reactor (SPR) on line. The comparison aids analyte identification and specificity, especially when the nature of the reaction has been well defined for particular classes of analytes.

Particular reactors, polymeric or silica-gel supported, efficiently and selectively reduce or oxidize various classes of aldehydes, ketones and alcohols, possibly post-column. The polymeric reagents are fully compatible with RP-HPLC conditions and solvents including methanol and acetonitrile. With borohydride on silica gel there is full compatibility with normal-phase HPLC. The polymeric permanganate SPR suits either HPLC mode, and gives the expected oxidation products from many alcohols and aldehydes. Applicability of the different SPRs to 'real samples' has been shown. Lifetimes etc. have been investigated.

Homogeneous derivatization reactions to aid photometric, electrometric or other HPLC detection modes have been reviewed in this series (Vol. 7; R.W. Frei) and elsewhere [1-6]. Pre-injection reactions, many of which are applicable to GC or TLC also, are seldom used post-column because they would be tedious, non-continuous, subject to artifacts or contamination, and difficult to automate on-line as distinct from off-line. Yet amongst on-line reactions, many are fully compatible with the post-column mode, possibly automated, if not the pre-column mode [1, 2, 6]. Indeed, various HPLC manufacturers have recently introduced automated post-column reactors, wherein the homogeneous derivatization solution is mixed in a low-dead-volume mixing chamber with the HPLC effluent. The resulting solution is then passed into another low-dead-volume heated reaction chamber where the desired derivatization takes place; products and unreacted derivatizing reagent(s) then enter the optical cell. Automation of the HPLC and the post-column reactor instrumentation is fully achievable; but in general, homogeneous derivatization in HPLC has very serious disadvantages that we have detailed [7-11].

Application of solid phase reactions (SPRs) in the HPLC field [2, 7-13] has so far been slight, although such reagents and reactions, notably polymeric, have long been used in synthetic organic chemistry. Seemingly most HPLC practitioners are not oriented towards synthetic organic chemistry literature, and *vice versa*. Yet immobilized enzymes have been used for some years with much success in HPLC, especially pre-column on-line [12], and we became aware that solid phase reagents might be advantageous in HPLC for various applications [7]. To some extent these expectations have been borne out, although much work remains to be undertaken on HPLC-SPR, both with solid support (silica gel, alumina, Florisil, etc.) and with polymer support or attachment, ionic or covalent.

We have looked especially into the applicability of silica-supported or polymer-attached (ionically) reagents for very specific reductions or oxidations of various chemical classes. Amongst such solid phase reagents already prepared and evaluated are: (1) a polymeric borohydride reducing reagent for off-line and reverse phase (RP) HPLC interfacing; (2) a polymeric permanganate oxidizing reagent for off-line and on-line RP-HPLC and normal phase (NP) HPLC interfacing; and (3) a silica-gel supported borohydride reducing reagent for off-line and on-line NP-HPLC interfacing. Carbonyl-group reduction is effected by (1) and (3), and oxidation of various alcohols or aldehydes by (2), always with the same products as formed in solution under otherwise identical conditions. Mostly these SPRs have been used pre-column, on-line, but at least one has also served in the post-column on-line mode.

Improved analyte identification has been possible in HPLC-SPR by the use of difference chromatography, wherein the same sample is injected twice. With a dummy pre-column on-line SPR, the retention time or capacity factor is determined for the analyte that is to be derivatized; then a run is performed with the SPR in place. One of

the two resulting chromatograms represents only the starting analyte, and the second shows its disappearance to varying degrees and
usually the appearance of one or more SPR products. Such chromatograms, or rather the differences between them, typify individual
analytes and can be used to confirm or deny the presence of a particular analyte in a complex sample matrix. However, this difference
approach does not yet provide improved sensitivity or detection
limits, although this could eventually be feasible.

SOLID PHASE POLYMERIC REDUCTIONS FOR CARBONYL COMPOUNDS

One of the uses of polymeric reagents in the past decade has
been to remove aldehyde traces from alcohol process streams [13],
the reagent having the borohydride anion ionically linked to cationic
sites on a polystyrene-divinylbenzene polymer. Various borohydride
or other inorganic reducing agents have been attached to polymeric
backbones, and widely used in synthetic areas.

We have now used for aldehydes a polymeric borohydride reagent,
prepared in-house and coupled with RP-HPLC in a 'real-time' fashion,
pre-column [8]. Unreactive compound types included ketones, amides,
acid chlorides, aryl halides and *N*-nitroso drivatives. Only the
aldehydes tested, e.g. benzaldehyde, cinnamaldehyde, *p*-nitrobenzal-
dehyde and 2-naphthaldehyde, underwent 100% reduction to their
expected alcohols under the particular conditions of separation,
derivatization and detection. Each aldehyde was shown to completely
disappear with the SPR in place, and a single alcoholic product was
evident with a similar UV absorption pattern, this being attributable
to the molecule's aromatic portion which is not involved in the
reduction. With a non-aromatic aldehyde the reduction to a non-UV
absorbing alcohol would entail loss of UV absorbance if the carbonyl
posessed any. Difference chromatography is of value mainly when a
starting analyte peak is clearly detectable with the dummy column
on-line and the final derivative peak is at least as great.

Polymeric reductions of aldehydes have been accomplished with
diverse mobile phases, including methanol/water, ethanol/water and
acetonitrile/water in varying ratios. Conventional HPLC columns
and flow-rates are applicable. Each SPR has been characterized in
respect of active borohydride incorporation by standard titrimetric
determinations or elemental analysis for boron, assumed to be in
borohydride form. With a reasonably high loading of active borohydride (column 5 cm and i.d. 4.6 mm, containing ~1.6 g of reagent)
the lifetime may exceed 6 months, and no reactivity between the SPR
and conventional RP solvents becomes apparent.

One application has been to a commercial sample of cinnamon
spice, where the dummy run showed a methanol extract to contain cinnamyl alcohol and cinnamaldehyde (Fig. 1A). The latter vanished with
the SPR on-line, and the former was augmented (Fig. 1B) as expected.

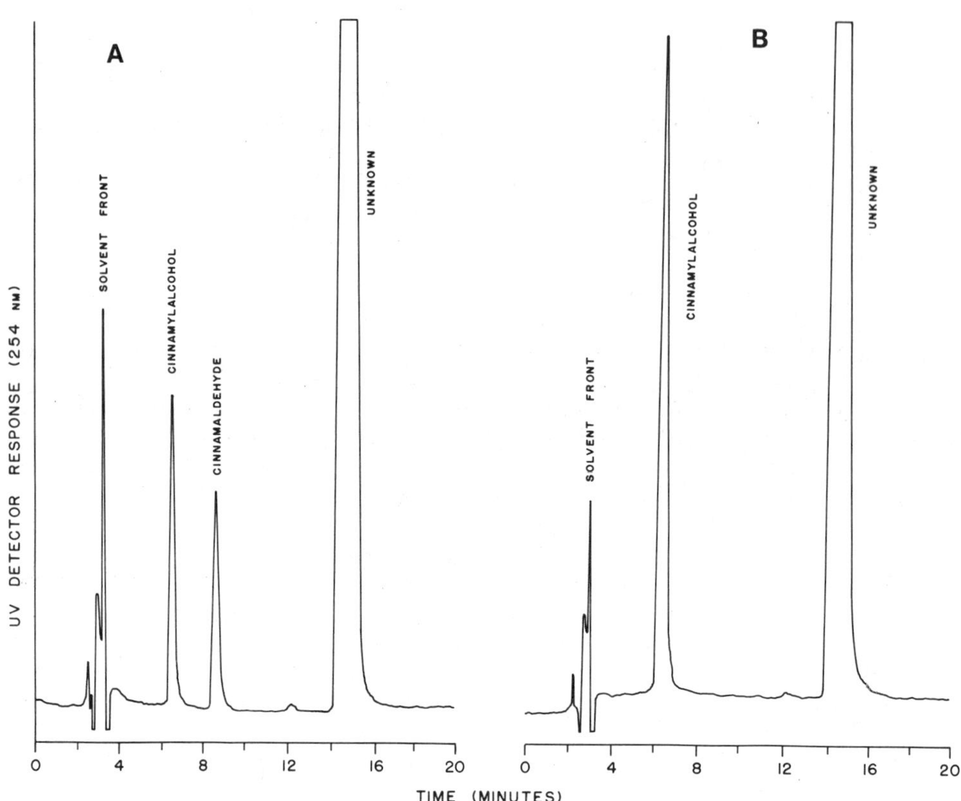

Fig. 1. Use of a polymeric borohydride pre-column SPR to determine cinnamaldehyde in a cinnamon specimen (see text) [8]. (A), dummy run; (B), with SPR on-line. C-18 analytical column, with aqueous 50% (v/v) acetonitrile at 1 ml/min. *By permission - Marcel Dekker.*

Similarly [9] a mouthwash, Lavoris, was analyzed for cinnamaldehyde.

BOROHYDRIDE ON SILICA FOR REDUCING CARBONYL COMPOUNDS

Borohydride/silica gel SPRs as long used in organic syntheses have now been prepared in-house and applied on-line to HPLC, pre- or post-column [9].   The carbonyl compounds tested again included aldehydes, ketones and acid chlorides.   Particular aldehydes could be distinguished with an appropriate mobile phase and SPR temperature, with conventional silica-gel HPLC packings, organic mobile phases, and commercially available HPLC equipment. With this particular SPR the reduction temperature influences the rate sufficiently to permit class and possibly compound identification, and quantitation even where an analyte co-elutes with a non-carbonyl compound. The SPRs can be characterized as above (iodine titration; elemental boron by inductively coupled plasma emission).   Such NP-HPLC-SPR approaches have been applied to certain vitamins and to cinnamon, and thus complement RP-HPLC with polymeric borohydride SPRs. However borohydride/silica SPR lifetimes are shorter (though at least 50-100 runs can be done with NP solvents) unless they are carefully protected from water and reducible impurities in the HPLC mobile phases. Relatively high loadings of the borohydride reagent can be accomplished, with benefit to lifetime, by repeated application of fresh borohydride solution to the silica gel; but the loading level does affect SPR reactivity, and % reductions for certain analytes may be governed by the history and state of the SPR.

Borohydride/silica SPRs could presumably be used off-line, as for permanganate or polymeric borohydride SPRs. Unlike the latter SPR, the borohydride/silica SPR can effectively reduce ketones and some acid chlorides as well as aldehydes; but it is likewise inert towards esters, nitro-aromatics, *N*-nitroso derivatives, alkyl/aryl halides and amides [9]. Usually the HPLC mobile phase has consisted of various isopropanol/n-hexane admixtures, at 0.4-1.0 ml/min, with a 214 nm fixed-wavelength detector for analyte and its product; but detection was at 254 nm in an application to a hexane extract of cinnamon [9]. Here, using a μ-Porasil column with 0.3% (v/v) isopropanol in hexane, we obtained the cinnamyl alcohol as a broad peak as late as 32 min although the flow-rate was increased from 1 to 4.3 ml/min at 13 min (cinnamaldehyde ran at 10 min). The elution time could, if desired, be reduced by recourse to a gradient, raising the isopropanol content of the mobile phase to give a more polar ratio.

For Vitamin K_3 (menadione), with 1% isopropanol in hexane, the dummy-run peak position was at 5 min, but the SPR (pre-column) reduction products as known from analogous borohydride reductions in solution were too polar to be eluted, notwithstanding acceleration of flow-rate at 6 min, although the K_3 peak disappeared.

POLYMERIC PERMANGANATE OXIDATIONS FOR ALCOHOLS AND ALDEHYDES

The many polymeric inorganic oxidizing reagents mostly incorporporate chromate or chromic acid in the organic synthesis context [13]. With polymeric permanganate, a somewhat newer reagent now prepared for pre-column (off- and on-line) HPLC-SPR, there is effective oxidation of both primary and secondary alcohols, most aldehydes and various ketones [11]. The SPR can be used with both RP- and NP-HPLC conditions and mobile phases, and derivatization extent is governed by the mobile phase, SPR permanganate activity and loading, residence time, and SPR temperature. Difference chromatography has again been utilized, and emphasis placed on UV detection, often enhanced after SPR oxidations; but any other detection method should suit too, if able to detect the starting material or the oxidation product. With a combination of detectors, enhanced specificity should be feasible.

In test runs with C-18 μBondapak columns and water/acetonitrile (1:1 by vol.) at 0.8 ml/min, values for % oxidation by pre-column

(on-line) SPR at 46° have included: o-aminobenzyl alcohol, 100%; sali-
cylaldehyde, 100; *p*-methoxyphenol, 100; *p*-methoxynaphthol, 100; hyd-
roquinone, 42; benzyl alcohol, 11; benzaldehyde, 15;   *p*-nitrobenzyl
alcohol, 52; *p*-nitrobenzaldehyde, 53.   Oxidations as thus shown by
difference chromatography are substantially greater in RP solvents
than in NP (hydrocarbon) solvents, as expected from the corresponding
solution oxidations.   This SPR used off-line with a longer reaction
time also gives more efficient oxidation, likewise as expected; the
improvement is 2-3 fold for secondary alcohols, where acetonitrile
solutions are used off-line or acetonitrile/water mixtures off-line
and on-line.

With a hair shampoo containing benzyl alcohol, its peak was seen
in a complex 'native' pattern, whereas a simpler pattern was generated
with the polymeric permanganate SPR off-line now showed nothing in
its position but two new peaks, one at the correct time for benzoic
acid and the other for benzaldehyde, which has indeed been shown to
undergo oxidation to benzoic acid under these HPLC-SPR conditions.

With a standard riboflavin sample, run similarly but with 15%
acetonitrile, the dummy run showed a peak at 9 min.   In its place
there was a much more complex chromatogram with the oxidation SPR
off-line, in part because riboflavin has a number of alcohol functions
that can undergo oxidation under these SPR conditions, such that a
number of possible products could be formed.   Yet this can help in
analyte identification, more than if a single oxidation product arose.
More complex derivatization chromatograms are therefore not necessarily detrimental to analyte identification; indeed they can provide
improved specificity in the overall HPLC-SPR analysis.

Acknowledgements

We acknowledge the interest and encouragement of colleagues
within both Northeastern University (NU) and Waters Associates,
especially B.L. Karger, B.C. Giessen, R. McNeil, R. Shansky and Wm.
LaCourse.   This work would not have been possible without the assistance and continued interest provided by K. Weiss at NU and C. Rausch
at Waters. Financial assistance and technical support for this joint
R & D program was provided to NU by Waters Assocs., Millipore Corpn.

K-H. Xie is a Visiting Chinese Scientist at NU, from the Analytical Institute of the Chinese Academy of Sciences, Beijing. The
Government of the People's Republic of China, and this Academy, are
thanked for letting him undertake studies and research at NU.

Part-support also came from a NIH Biomedical Research Support
Grant, No. RR07143, Department of Health & Human Resources, to NU.

References

1. Lawrence, J.F. (1981) *Organic Trace Analysis* (see Chap. 7), Academic Press, New York, 288 pp.
2. Frei, R.W. & Lawrence, J.F., eds. (1981) *Chemical Derivatization in Analytical Chemistry, Vol. 1: Chromatography* (see Chaps. 1, 3 & 4), Plenum, New York, 344 pp.
3. Knapp, D.R. (1979) *Handbook of Analytical Derivatization Reagents*, Wiley, New York, 741 pp.
4. Blau, K. & King, G., eds. (1977), *Handbook of Derivatives for Chromatography*, Heyden, London, 576 pp.
5. Snyder, L.R. & Kirkland, J.J. (1979) *Introduction to Modern Liquid Chromatography*, 2nd edn. (see Chap. 17), Wiley, New York, 863 pp.
6. Stewart, J.T. (1982) *Trends Anal. Chem. 1*, 170–174.
7. Krull, I.S. & Lankmayr, E.P. (1982) *Am. Lab. (May)*, 18–32.
8. Krull, I.S., Xie, K-H., Colgan, S., Neue, U., Izod, T., King, R. & Bidlingmeyer, B. (1983) *J. Liq. Chromat.* 6, 605–626.
9. Krull, I.S., Colgan, S., Xie, K-H., Neue, U., King, R. & Bidlingmeyer, B. (1983) *J. Liq. Chromat.* 6, 1015–1035.
10. Xie, K-H., Colgan, S. & Krull, I.S. (1983) *J. Liq. Chromat. Revs. 6 (Suppl. 2)*, 125–155.
11. Xie, K-H., Santasania, C., Krull, I.S., Neue, U., Bidlingmeyer, B. & Newhart, A. (1983) *J. Liq. Chromat.*, 6, 2109–2127.
12. Bowers, L.D. & Bostock, W.D. (1982) in *Chemical Derivatization in Analytical Chemistry, Vol. 2: Separation and Continuous Flow Techniques* (Frei, R.W. & Lawrence, J.F., eds.), Plenum, New York, pp. 97–138.
13. Mathur, N.K., Narang, C.K. & Williams, R.E. (1980) *Polymers as Aids in Organic Chemistry*, Academic Press, New York, 258 pp.

#NC(C)

NOTES and COMMENTS relating to

HPLC detection, and determination of metal-complex drugs

Comments related to particular contributions:

#C-1 & #NC(C)-3, p. 196
#C-2 to #C-5, p. 197
#C-7 & #NC(C)-2, p. 195

#NC(C)-1

A Note on

POST-COLUMN REACTION DETECTORS FOR TRACE ANALYSIS IN HPLC

U.A. Th. Brinkman and R.W. Frei

Department of Analytical Chemistry
Free University
de Boelelaan 1083
1081 HV Amsterdam, The Netherlands

Detection is still a relatively weak point in HPLC, which is a major drawback since trace-analytical work requires the use of highly sensitive and selective detection principles. One such method, and a promising one, is post-column derivatization, which in our laboratory has usually been aimed at producing a fluorescent derivative. [Early studies were outlined by R.W. Frei in Vol. 7, this series.-*Ed.*] We generally derivatize in coiled glass or PTFE capillaries, or stainless steel. For slow reactions that call for segmentation as a means of suppressing band broadening, both air and solvent segmentation are used. In the latter case the organic solvent – often a chlorinated aliphatic hydrocarbon – simultaneously acts as segmentation liquid and as extractant for the product formed from the analyte of interest during the post-column reaction.

The feasibility of ion-pair formation with a fluorigenic counter-ion such as 9,10-dimethoxyanthracene sulphonate (DAS) has been demonstrated [1-3] with the pharmaceutically important amine drugs chloro- and bromopheniramine, and secoverine, and with the herbicide metabolite hydroxyatrazine. After their separation on a CN-bonded phase with an acidic, highly aqueous mobile phase, the amines are converted into their amineH$^+$ DAS$^-$ ion pairs and extracted into tetrachloroethane or chloroform in an on-line post-column extraction detector. After phase separation the organic phase is continuously monitored in a fluorimeter. The aqueous phase contains the excess DAS (Na salt) which is highly fluorescent and thus interferes, and goes to waste. Applications to blood and serum analysis have been reported.

True chemical derivatization has mainly been carried out with o-phthalaldehyde (OPA) as reagent. Catecholamines, primary amino acids and anilines are readily converted (~20 sec; no segmentation) into fluorescent derivatives. This system has been selected in our laboratory for model studies with miniaturized HPLC equipment (1 mm i.d. columns), and also in studies [4] on N-methylcarbamates. Compounds of the latter type are, in a first step, hydrolyzed in a post-column reactor packed with an anion-exchange resin and heated to 80-100°. The methylamine formed is then reacted with OPA in a second post-column module and the reaction product monitored in a fluorimeter.

Post-column conversion of non–fluorescent analytes into fluorescent reaction products by means of UV irradiation has been used [5] to determine drugs such as clobazam, desmethylclobazam, demoxepam and - with enhancement of their native fluorescence - phenothiazines in serum and urine. Other successful applications [6, 7] involve vitamin K_1, diethylstilboestrol and several chlorophenols. Optimum irradiation times often are only 20–30 sec, and detection limits are frequently ~0.1 ng. (Applications of this approach to other analytes such as tamoxifen are described elsewhere in this vol.-*Ed.*)

Ligand exchange represents an approach that we have applied [8] to organosulphur compounds. The method is based on an exchange reaction between the non–fluorescent (quenching!) Pd(II)-calcein complex and the analytes which displace an equivalent amount of highly fluorescent calcein from the complex. There is no need to extract from the water-rich mobile phase, but air segmentation is required because the reaction is rather slow (10–15 min at 50–60° often being used). The widely used penicillamine, several further S-containing amino acids and low-mol. wt. thiols, thioethers, etc. have been determined in relevant matrices.

References

1. Brinkman, U.A.Th., Lawrence, J.F., van Buuren, C. & Frei, R.W., (1982) *Rec. Devel. Chromatog. Electrophor. 10*, 247–259.
2. Reddingius, R.J., de Jong, G.J., Brinkman, U.A.Th. & Frei, R.W. (1981) *J. Chromatog. 205*, 77–84.
3. van Buuren, C., Lawrence, J.F., Brinkman, U.A.Th. Honigberg, I.L. & Frei, R.W. (1980) *Anal. Chem. 52*, 700–704.
4. Nondek, L., Frei, R.W. & Brinkman, U.A.Th. (1983) *J. Chromatog. 282*, 141–150.
5. Brinkman, U.A.Th., Welling, P.L.M., de Vries, G., Scholten, A.H.M.T. & Frei, R.W. (1982) *J. Chromatog. 217*, 463–471.
6. Lefevere, M.F., Frei, R.W., Scholten, A.H.M.T. & Brinkman, U.A.Th. (1982) *Chromatographia 15*, 459–467.
7. Werkhoven-Goewie, C.E., Boon, W.M., Praat, A.J.J., Frei, R.W., Brinkman, U.A.Th. & Little, C.J. (1982) *Chromatographia 16*, 53–60.
8. Werkhoven-Goewie, C.E., Niessen, W.M.A., Brinkman, U.A.Th. & Frei, R.W. (1981) *J. Chromatog. 203*, 165–172.

#NC(C)-2

A Note on

POST-COLUMN REACTION WITH FERRIC CHLORIDE
FOR THE DETECTION OF D-PENICILLAMINE

[1]D. Witts and [2]I.D. Wilson

[1]Laboratory of Toxicology
and Pharmacokinetics
University College Medical
School, Rayne Institute
London WC1E 6JJ, U.K.

[2]Department of Drug
Metabolism
Hoechst UK Ltd.
Walton Manor, Walton
Milton Keynes MK7 7AJ, U.K.

D-Penicillamine (β,β-dimethylcysteine) is used in the treatment of a wide variety of diseases, including rheumatoid arthritis (RA), cystinuria, Wilson's disease and heavy-metal poisoning. Many toxic side-effects have been observed during the treatment of RA with penicillamine, and there is a wide variability in therapeutic effect. A simple, rapid and specific assay is required to aid in the investigation of the toxic side-effects, to provide pharmacokinetic data and to monitor patient compliance. In the past many approaches have been tried in an effort to develop suitable assays, including GC, colorimetry and amino-acid analyzer methods [1]. More recently, HPLC methods have been developed. These have used electrochemical (EC) detection [2], pre-HPLC derivative formation [3] or post-column reaction in order to detect the drug [1] (which lacks a suitable UV chromophore and does not fluoresce). However, all of these methods lack specificity in that they are general methods detecting either thiols or sulphur-containing compounds. We have attempted to develop an assay with a higher degree of specificity for penicillamine based on the intense blue complex formed by the reaction of penicillamine with ferric chloride. In this method ferric chloride is mixed with the eluent from the HPLC column in a simple packed-bed post-column reactor. The method has been compared with our pre-existing assay which uses post-column reaction with Ellman's reagent to detect thiols [1].

The column was 150X i.d. 4.6 mm, slurry-packed with Whatman SCX cation-exchanger. Mobile phase (phosphate-citrate buffer, pH 5.0, 0.015 ionic strength) and reagent (usually 50 mM ferric chloride) were each pumped at 1 ml/min using LDC Constametric III pumps (LDC

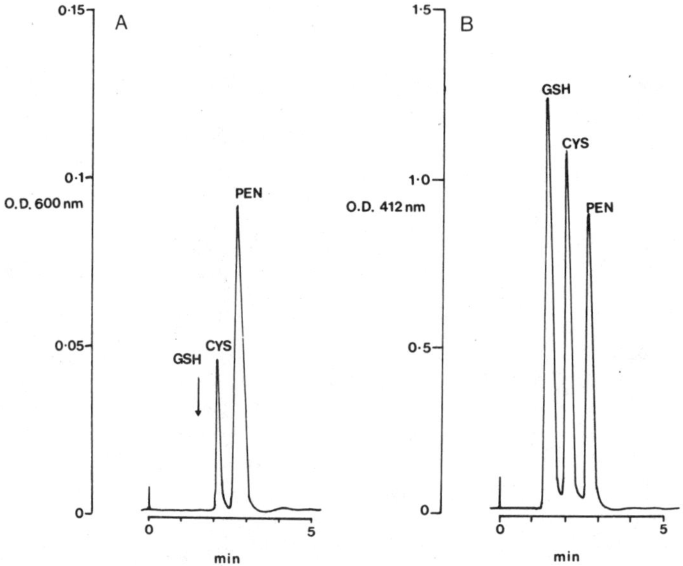

Fig. 1. HPLC of a mixture of glutathione (GSH), cysteine (CYS) and penicillamine (PEN), with post-column detection by use of: **A**, ferric chloride reagent; **B**, Ellman's reagent [5,5'-dithiobis(2-nitrobenzoic acid)].

Ltd., Stone, Staffs., U.K.). The effluent was monitored at 600 nm using a Vari-chrom variable wavelength UV-VIS detector (Varian Associates, Walton-on-Thames). The post-column reactor was constructed as described by Little et al. [4], consisting of a stainless steel tube 150 x i.d. 2 mm, dry-packed with 40 µm glass beads. Samples were injected onto the column via a Rheodyne 7120 loop injector equipped with a 20 µl loop.

Initially we tried the same ion-pair RP-HPLC system as used in our earlier work with detection by Ellman's reagent [1]; but the substitution of ferric chloride resulted in the formation of a flocculent brown precipitate (presumably ferric hydroxide) when mobile phase and reagent were mixed. The conditions used for ion-exchange HPLC did not cause precipitation, and were adopted for these studies. Firstly the effects of ferric chloride concentration on peak height and reproducibility were investigated. These were unaffected by concentration within the range 25-100 mM but were both impaired with below 25 mM. With the chosen concentration of 50 mM, the assay was linear for penicillamine with on-column loads of 30 ng to 15 µg (as high as tested). With a 20 µl injection this corresponds to concentrations of 1.5-750 µg/ml. Fig. 1 shows results with a standard

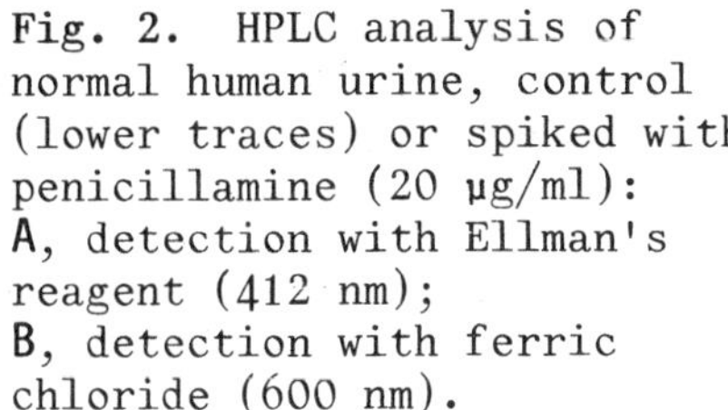

Fig. 2. HPLC analysis of normal human urine, control (lower traces) or spiked with penicillamine (20 µg/ml): **A**, detection with Ellman's reagent (412 nm); **B**, detection with ferric chloride (600 nm).

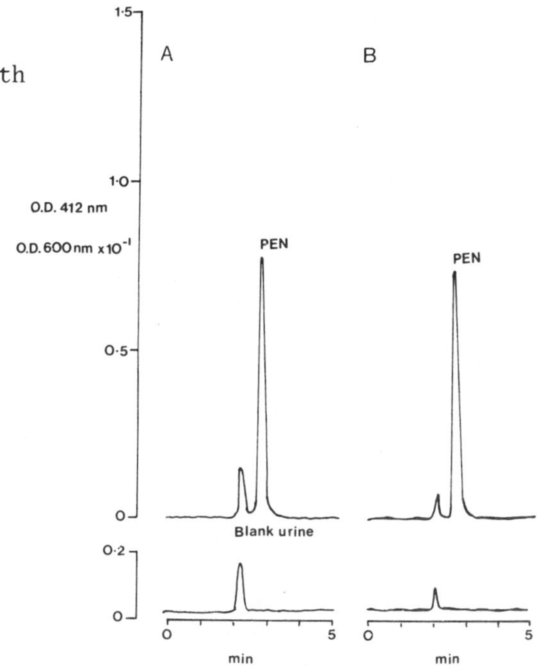

mixture containing equimolar amounts of three thiols – GSH, cysteine and penicillamine. With Ellman's reagent the three are equally well detected. With ferric chloride the peak corresponding to GSH is lost, and that for cysteine reduced by ~70% compared to the peak for penicillamine.

In Fig. 2 (upper part) a sample of human urine spiked with penicillamine and analyzed under the same conditions is shown. With Ellman's reagent a peak for penicillamine is observed together with an endogenous peak corresponding in retention time to cysteine. On substituting ferric chloride for Ellman's reagent the endogenous peak is much reduced compared with the penicillamine peak, and thus probably does represent cysteine. A practical limit of detection for penicillamine in urine is probably ~100 ng (on column). With a 100 µl loop this would represent 1 µg/ml. Concentrations of penicillamine that we have encountered in urine from RA patients range from 11 to 60 µg/ml (following an oral dose of 1 g). The method is therefore quite adequate for the detection and determination of penicillamine in urine. Plasma concentrations of penicillamine encountered in our laboratory have not been higher than 3 µg/ml. Clearly without some concentration step (or the injection of much larger amounts of sample) the ferric chloride assay would be barely sensitive enough for all but peak drug levels. Elsewhere in this vol. (#C-1) HPLC-EC is stated by D. Perrett to be applicable to penicillamine determination in plasma during treatment to remove cystine calculi.

Our intention was to obtain an assay with greatly improved specificity for penicillamine compared to the existing methods, and in this we have been partially successful. The use of ferric chloride does provide an increase in specificity insofar as glutathione (and a number of other thiols, e.g. homocysteine) are not detected. However, the assay is not completely specific as cysteine is still detected, albeit with much reduced response. Besides being useful for detecting and assaying penicillamine in urine, the approach can aid identification of GSH and cysteine in liver and kidney extracts through the differential response of these compounds to Ellman's reagent and ferric chloride.

Acknowledgements

We thank Prof. A.E.M. McLean for the provision of laboratory facilities.

References

1. Beales, D., Finch, R., McLean, A.E.M., Smith, M. & Wilson, I.D. (1981) *J. Chromatog. 226*, 498-503.
2. Saetre, R. & Rabenstein, D.L. (1978) *Anal. Chem. 50*, 276-280.
3. Lankmayr, E.P., Budna, K.W., Muller, K., Nachtmann, F. & Rainer, F. (1981) *J. Chromatog. 222*, 249-255.
4. Little, C.J., Whatly, J.A. & Dale, A.D. (1979) *J. Chromatog. 171*, 63-71.
5. Shaw, I.C., McLean, A.E.M. & Boult, C.H. (1983) *J. Chromatog. 275*, 206-210.

#NC(C)-3

A Note on

AN EVALUATION OF A TWO-ELECTRODE COULOMETRIC DETECTOR

Robin Whelpton

Department of Pharmacology and Therapeutics
The London Hospital Medical College
Turner Street, London E1 2AD, U.K.

We find the ESA Coulochem (Model 5100A) electrochemical detector to be versatile, stable and quite easy to use. The unusual arrangement that the two electrodes (in series) are porous graphite through which the eluent flows provides a large surface area, and efficiencies approaching 100% are claimed; but the risk of 'clogging' must always be borne in mind and steps taken to minimize it. There is a graphite pre-filter, and filtration of eluent through 0.2 μm filters is recommended. When clean, the back-pressure developed by cell and pre-filter is ~250-500 psi at 1 ml/min. Contamination can increase the pressure markedly; hence the cell must be used with caution if it is 'downstream' from another detector. We have used the detector in our current development of an assay for physostigmine [1]. The eluent (methanol + 10% v/v 0.1 M pH 8.9 NH_4NO_3) was pumped (ACS Series 300/02) through a 150 × i.d. 4.6 mm column of 3 μm Spherisorb.

Temperature and its control.- The rate of electrochemical reactions is temp.-dependent. If the rate is low and the residence time in the cell is insufficient to give 100% reaction, increasing temp. will increase the rate and hence the detector signal. To this end the cell is fitted with an integral heater to maintain a temp. of ~40°. Maintaining the cell above ambient temp. also reduces baseline drift due to fluctuations in laboratory temp. Unfortunately, use of the heater adversely affected baseline noise. Whether the effect was due to electrical interference or to changes in flow with temp. we cannot say; but the fluctuations corresponded to the heater cutting in and out. To investigate the temp. effect on the response to physostigmine we housed the detector and column in a GC oven. Peak heights and areas increased with temp., by 50% between 21° and 50° if the flow were 0.5 ml/min (Table 1). Increasing the temp. also helped reduce the back-pressure, although the background signal also rose.

Flow rate and buffer concentration.- Reducing the flow rate and buffer ionic strength reduced the background signal and noise.

Table 1. Influence of flow rate and temp. on detector response. Both heights (cm) and, *in italics,* areas (arbitrary units) are given. Each value is the mean of 3 determinations.

Temp.	1.0 ml/min		0.5 ml/min		0.3 ml/min	
21°	7.4	*4.14*	8.0	*5.94*	6.7	*6.46*
36°	11.0	*10.5*	10.6	*13.5*	9.7	*16.09*
51°	11.3	*9.0*	12.5	*14.7*	10.8	*17.1*

The noise reduction was not attributable to a fall in pumping frequency as this was constant at 23 cycles/sec. The benefit from weaker buffer is in contrast to our observations with an amperometric detector with glassy carbon electrodes. The latter system was more stable with higher electrolyte concentrations. Changing detectors therefore calls for re-optimizing the conditions.

MODES OF USE   *[Cf. #F-1, this vol.- Ed.]*

1. Screen mode: the first electrode, E_1, is set at a lower potential to oxidize unwanted materials.   The compound of interest is determined at the second electrode, E_2.   The use of this mode also produces a lower background current and a more stable baseline from E_1. More easily oxidizable metabolites, e.g. phenols, can be determined at E_1 and the drug at E_2 If one compound is fully oxidized at E_1 the compounds can be resolved electrochemically rather than chromatographically.   Further, in this mode the E_1/E_2 signal ratio serves to check peak purity.   This has been very useful for physostigmine, as some plasma samples have contained a co-eluting impurity.

2. Redox mode: the compound is oxidized and then reduced, if it can produce a stable oxidation product.   A third recorder output gives the difference $E_1 - E_2$; hence the signal is doubled.   This mode has not been advantageous for physostigmine and, as noise is also increased, it may give little benefit to sensitivity; but redox can aid peak identification since the reaction is irreversible in some cases.

3. Difference mode: E_1 and E_2 at same potential (such that at E_1 most of the drug is oxidized), and output difference monitored. Compounds that undergo little oxidation give very similar peaks at E_1 and E_2 and thus a very small difference.   This mode helps smooth baseline drift and get a quick result after the instrument has been turned on.

Some general features.- A pre-valve 'guard' cell pre-oxidizes the eluent, thus reducing noise. Current-voltage curves are obtainable easily by a scan. The auto-zero helps in high-sensitivity work.   With a second cell the controller can furnish two single-electrode detectors.

Reference

1. Whelpton, R. (1983) *J. Chromatog. 272,* 216-220.

#NC(C)-4

A Note on

A COMPARISON OF THE MOVING BELT AND DIRECT LIQUID INTRODUCTION INTERFACES FOR HPLC–MS OF RANITIDINE AND ITS METABOLITES

[1]L.E. Martin, [1]Janet Oxford, [2]D. Dixon and [2]R. Schuster

[1]Glaxo Group Research Ltd.   [2]Hewlett-Packard Gmbh
Ware, Herts. SG12 ODJ, U.K.   Waldbronn, W. Germany

Combined HPLC–MS is very useful in the study of non–volatile compounds, including many drugs and their metabolites. [Pertinent arts. in earlier vols. include #C–6 in Vol. 7 by D.E. Games et al.– *Ed.*] We have employed HPLC–MS in studies on ranitidine and its metabolites, ranitidine-*N*-oxide, ranitidine-*S*-oxide and *N*-desmethyl-ranitidine. Destruction of these compounds was found at the high temperatures required to evaporate the water in reversed phase (RP) systems from a moving belt transfer system [1]. When 1 µg of compound was injected on-column and a normal phase (NP) system containing no water was used, it was possible to obtain chemical ionization (CI) mass spectra with ammonia as the reagent gas. The mass spectra showed extensive fragmentation of the compounds, and $[M+H]^+$ ions in low abundance were present only in the ranitidine and desmethyl-ranitidine spectra.

The use of an NP system restricted the volume of aqueous biological fluids that could be injected on the column without deterioration of the chromatographic performance. The maximum amount of urine which could be injected was 10 µl. The limit of detection for ranitidine using deuterated (d–3) ranitidine as internal standard and the selected ion recording (SIR) technique was 1.05 µg/ml urine. We therefore investigated the use of the direct liquid introduction probe system (DLI) for RP–HPLC–MS of ranitidine and its metabolites. The system has been used to interface the Hewlett-Packard HPLC 1084B system to a HP 5985B quadrupole MS fitted with a cryogenic pumping system. The system has been described by Melera [3].

Ranitidine and its metabolites were analyzed by HPLC–MS using a gradient-elution RP system (Fig. 1). The HPLC flow rate was 0.3–0.4 ml/min, and the maximum amount of eluent which could be admitted

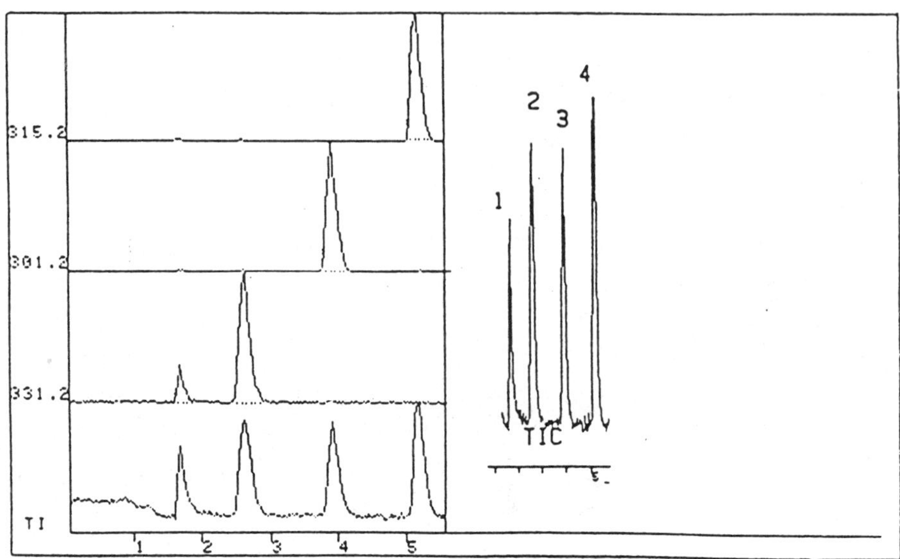

Fig. 1. HPLC–MS of ranitidine (257 ng; peak 4), its *N*-oxide (270 ng; peak 1) and *S*-oxide (320 ng; peak 2), and its desmethyl metabolite (280 ng; peak 3): total ion current chromatogram (TIC) and *(left)* MS at the stated m/z values.
Instrumentation: HP 5985B LCMS equipped with an HP 1084B HPLC system. Column 200 x i.d. 2 mm with 5 µm C–18 packing. Eluent: 0.4 ml/min, acetonitrile/water/triethylamine (pH 8 to 9) with a gradient of 15% to 80% (v/v) acetonitrile.

to the MS source was 30–40 µl/min. The total ion current chromato-gram (Fig. 1) shows that ranitidine and its metabolites were well resolved. Characteristic CI mass spectra were obtained for each com-pound when 200–300 ng of compound were injected on–column (Fig. 2). The $[M+H]^+$ ion was present in high intensity. Up to 50 µl of urine could be injected on–column without detriment to the chromatography.

The $[M+H]^+$ ion was the base peak in the mass spectrum of ranitidine, ranitidine *S*-oxide and desmethylranitidine. Some frag-mentation of the ranitidine-*N*-oxide molecule had occurred, but the $[M+H]^+$ ion was present in high intensity. When the SIR stable iso-tope technique was used, the limit of detection at a signal-to-noise ratio of 3:1 was 80 ng ranitidine/ml urine. The ratio of the peak area of ranitidine to d–3 ranitidine *vs.* the concentration of rani-tidine was linear.

Fig. 2 *(opposite)*. Chemical ionization (CI) mass spectra of ranitidine and its metabolites.

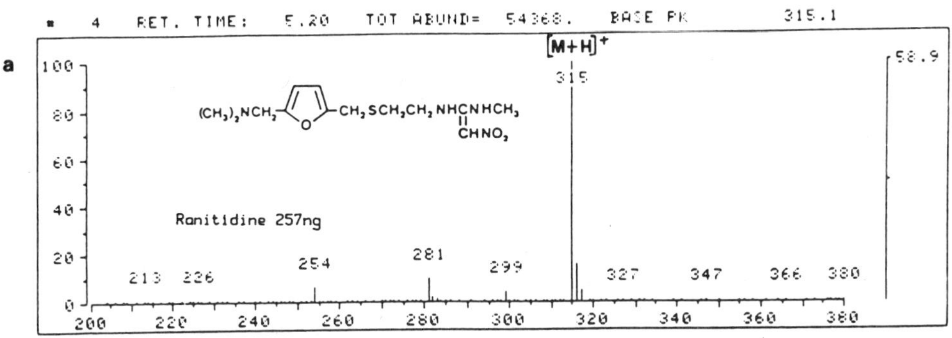
* 4   RET. TIME: 5.20   TOT ABUND= 54368.   BASE PK   315.1
a   [M+H]+
315
58.9
(CH$_3$)$_2$NCH$_2$ ... CH$_2$SCH$_2$CH$_2$ NHCNHCH$_3$
CHNO$_2$
Ranitidine 257ng
213 226   254   281   299   327   347   366   380
200   220   240   260   280   300   320   340   360   380

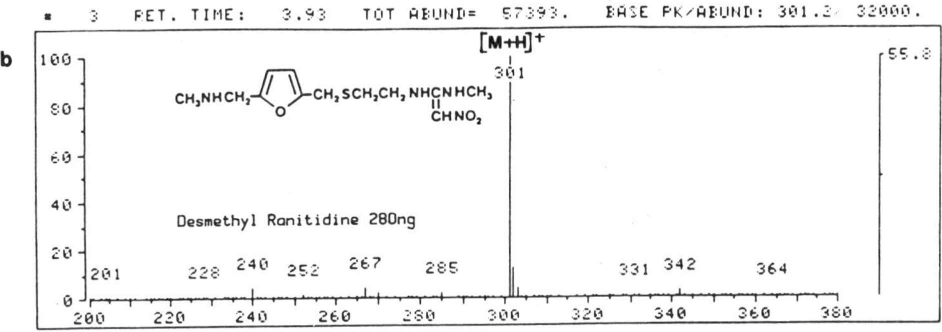
* 3   RET. TIME: 3.93   TOT ABUND= 57393.   BASE PK/ABUND: 301.3 32000.
b   [M+H]+
301
55.8
CH$_3$NHCH$_2$ ... CH$_2$SCH$_2$CH$_2$ NHCNHCH$_3$
CHNO$_2$
Desmethyl Ranitidine 280ng
201   228 240 252   267   285   331 342   364
200   220   240   260   280   300   320   340   360   380

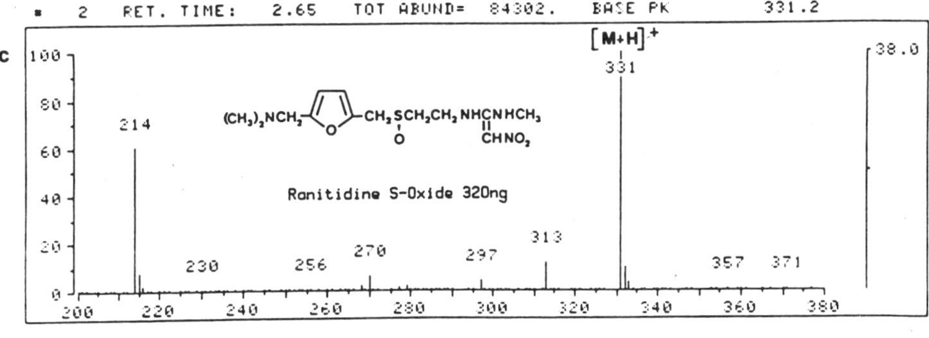
* 2   RET. TIME: 2.65   TOT ABUND= 84302.   BASE PK   331.2
c   [M+H]+
331
38.0
214
(CH$_3$)$_2$NCH$_2$ ... CH$_2$SCH$_2$CH$_2$ NHCNHCH$_3$
O   CHNO$_2$
Ranitidine S-Oxide 320ng
230   256   270   297   313   357   371
200   220   240   260   280   300   320   340   360   380

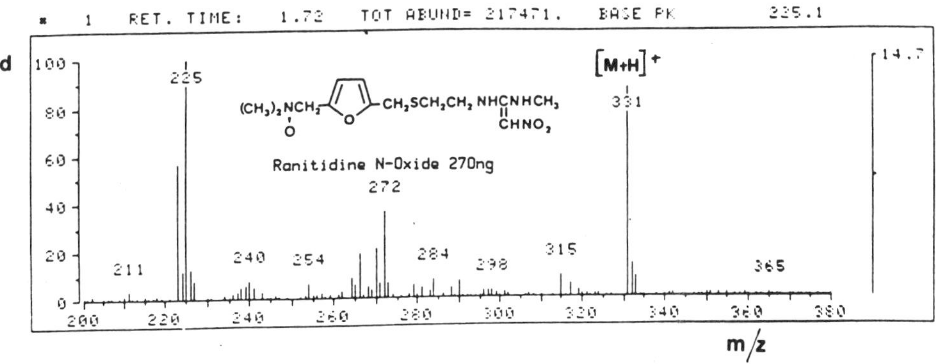
* 1   RET. TIME: 1.72   TOT ABUND= 217471.   BASE PK   225.1
d   [M+H]+
225
14.7
(CH$_3$)$_2$NCH$_2$ ... CH$_2$SCH$_2$CH$_2$ NHCNHCH$_3$
O   CHNO$_2$
Ranitidine N-Oxide 270ng
331
272
211   240 254   284 298   315   365
200   220   240   260   280   300   320   340   360   380
m/z

This study has shown that the DLI interface system is more suitable than the moving belt interface system for qualitative and quantitative RP–HPLC–MS analysis of thermolabile molecules.

References

1. Martin, L.E., Oxford, J. & Tanner, R.J.N. (1981) *Xenobiotica* *11*, 831–840.
2. Martin, L.E., Oxford, J. & Tanner, R.J.N. (1982) *J. Chromatog.* *251*, 215–224.
3. Melera, A. (1980) in *Advances in Mass Spectrometry*, Vol. 8B (Quayle, A., ed.), Heyden, London, pp. 1597–1615.

Comments on material in #C
(not adhering to the foregoing sequence)

Comments on #NC(C)-4, L.E. Martin et al. - MS linked to HPLC

Janet Oxford, *in reply to* L.A. Sternson.- For thermally stable compounds, the belt interface offers greater sensitivity than direct liquid introduction (DLI), but the conditions used to remove solvent destroy many organic molecules. Such is the case with ranitidine, so that DLI gives 200-fold greater sensitivity even though 90% of the analyte is lost (while ~50% is transferred from the moving belt). In general *(answer to* U.A. Th. Brinkman), taking account of the limitations of the belt system, DLI is preferable at least with thermo-labile compounds.

Question by D. Perrett.- As your comparison of DLI with belt was done with columns of different dimensions, was sensitivity comparison valid, since the 2 mm column used with DLI would have aided sensitivity anyway? Do you intend to use a true micro-bore system? *Reply by* L.E. Martin.- The 2 mm columns have higher capacity than micro-columns, are less affected by endogenous material in biological samples, are commercially available, and give reproducible chromatography. When this work was performed, 1 mm columns were not commercially available, and even today their reproducibility can vary.

Thermal spray detector (ref. cited by Editor).- This device [1] enables up to 2 ml/min of HPLC effluent to be introduced directly into the ion source. For MS in the CI mode, no ion source is needed if ions are present in the effluent.

1. Blakeley, C.R. & Vestal, M.L. (1983) *Anal. Chem.* 55, 750-754.

Comments on #C-7, I.S. Krull et al.- POST-COLUMN REACTORS
        also #NC(C)-2, D. Witts & I.D. Wilson

I.S. Krull, *answering* J.A.F. de Silva.- We don't know whether a matrix such as plasma is affected by post-column reactions; the only biological matrix we have so far dealt with is milk (for penicillamine). *Question by* E.P. Lankmayr, concerning quantitation in relation to the decreasing concentration of reactant (e.g. reducing agent, on the solid phase) as the run proceeds.- The reaction rate will be falling, and affect quantitation. Have you observed such phenomena and, if so, compensated for the effect? *Reply.-* The inorganic reagent loadings on our polymeric or silica-supported reactors

are typically ~50 mg/g.   Since we are derivatizing usually only µg
or ng amounts of analyte in each run, we have not observed serious
depletion of the original reagent on the support.   Hence reaction
rates have not changed with time,   even after using one particular
borohydride reactor for months in organic reduction reactions, off-
line and on-line.   However, even if continued use of a reactor were
to deplete active reagent,   quantitation would still be possible
through injecting analyte standards before and after the sample.

Question by Roberta J. Ward.- Has the ferric chloride technique
for penicillamine been applied to biological fluids such as plasma
where other thiols, e.g. albumin and cysteine, are present in high
concentration? *Reply by* D. Witts: the low drug concentration has
precluded study of plasma.

Comments on #C-1, D. Perrett, & #NC(C)-3, R. Whelpton – LC-EC

D. Perrett, *answering a question on* PFBC derivatization of
amines "under aqueous conditions":   the derivatizations took place
in the presence of an organic solvent such as diethyl ether, i.e. in
a 2-phase system.   R.D. McDowall, *addressing both contributors
(also* C.A. Marsden*).-* The cell heater incorporated in the ESA detector is indeed useless at low sensitivities.   Views on the relative
merits of amperometric and coulometric detectors would be of interest. (R. Whelpton: having had problems with the BAS detectors, I
prefer the coulometric ESA detector, which is notably, robust. C.A.
Marsden: whilst amperometric detectors are preferable for catechols
and the ESA detector has shown some advantages for neuropeptides,
precautions must be taken: e.g. each of our detectors is plugged
into a separate ring, and each electrode is enclosed in a Faraday
cage [cf. #F-1].) In our laboratory [Smith, Kline & French, U.K.] we
have found no differences between the BAS and ESA detectors in the
signal-to-noise ratio. However, agreeing with C.A. Marsden, precautions must be taken in operating EC detectors. Our BAS detectors
are plugged into an uninterruptable power supply (UAS). The cells and
columns are mounted in a large Perkin-Elmer column oven to act as a
Faraday cage and thermostatted at 40°. If there is a break in the
power supply, the detector can take up to 24-48 h to reach an acceptable baseline. Regarding the ESA detector, we just connect it to
a voltage stabilizer and use it without any other precautions.

Assay of blood for captopril (ref. cited by Editor).- Derivatization of this thiol drug to render it EC-responsive (+0.9 V) is
performed by treating the whole-blood sample with *N*-(4-dimethylamino-
phenylmaleimide and, after extracting with diethyl ether, adding
acetone to the aqueous layer and concentrating the latter. An adsor-
ption/elution step (C-18 cartridge; acetonitrile) is followed by
drying down and re-dissolving the residue in methanol for HPLC [1].

1. Shimada, K., Tanaka, M., Nambara, T., Imai, Y., Abe, K. & Yoshinaga, J.
   (1982) *J. Chromatog. 227*, 445-451.

Comments on #C-2 to -5, P.C. Uden; I.S. Krull; D.Newell; R.J. Ward
    - METAL COMPLEXES/ANTI-CANCER PLATINUM DRUGS

    P.C. Uden's *response to remark on negative-intercept calibration curves* (L.A. Sternson).- This feature (for gold thioglucose & vitamin B_{12}) is unexplained. HPLC losses are prone to occur with organometallics. *Reply to a question:* up to 5 ml/min of HPLC eluate can be fed into the spectral emission spectrophotometer.

    I. S. Krull, *answering* A.G. Fogg.- Well-behaved hydrodynamic voltammograms (HVs) indeed have plateaus, but it is on the sloping fronts that the two potentials are chosen for obtaining response ratios in the twin-electrode voltammetric detector.

    D. Newell, *replying to* Roberta Ward: since no CBDCA metabolites have been shown to exist, there has been no correlation of the clinical efficacy of Pt drugs with any metabolite.

    Z.H. Siddik, *question to* R.J. Ward *who replied thus:* the study of distribution in the rat [see (c), late in art. #C-5] did not include auranofin in respect of the liver, but in kidney this drug too showed a high concentration, with a parallel increase in Cu concentration.

    Refs. noted by Editor, including a post-column reactor.- A *cis*-Pt (DPP) study from a New Zealand laboratory [1] has furnished time-course distribution data for rat liver and kidney. Amongst particulate subcellular fractions, there was preferential localization in the microsomal fraction, and in mitochondrial/lysosomal fractions (liver) or the 'nuclear' fraction (kidney). However, localization was mainly in the cytosol, partly bound to a metallothionein-like protein; implications for the drug's therapeutic and toxic effects are discussed. A publication from L.A. Sternson's laboratory, not cited in #C-6, deals with urinary *cis*-platin as estimated by HPLC with a solvent-generated anion-exchanger and column switching; AA or (more sensitive) on-line polarography was the final step [2]. With clobazam/fenbendazole analytes (including metabolites), Uihlein & Schwab [3] have described a novel design of post-column reactor for UV generation of fluorescence [cf. tamoxifen studies that follow].

1. Sharma, R.P. & Edwards, I.R. (1983) *Biochem. Pharmacol. 32,* 2665-2669.
2. Riley, C.M., Sternson, L.A., Repta, A.J. & Bannister, S.J. (1982) *J. Pharm. Pharmacol. 34,* 826.
3. Uihlein, M. & Schwab, E. (1982) *Chromatographia 15,* 140-146.
Also (ref. noted by Co-Editor):-
    Reece, P.A., McCall, J.T., Powis, G. & Richardson, R.L. (1984)
    J. Chromatog. 306, 417-423.- CDDP in a plasma ultrafiltrate can be determined (after chloroform extraction and drying down) by HPLC on a Cyano Spheri-5 column at 40° as its diethylthiocarbamate adduct, made at the outset; detection at 254 nm.

Section #D

TAMOXIFEN AND OTHER ANTI-CANCER DRUGS

#D-1

DETERMINATION OF THE ANTI-CANCER DRUG AMSACRINE
IN BIOLOGICAL FLUIDS BY HPLC

J.W. Paxton

Department of Pharmacology and Clinical Pharmacology
University of Auckland School of Medicine
Private Bag, Auckland, New Zealand

Amsacrine (AMSA) is a 9-anilino-acridine derivative which is undergoing extensive clinical evaluation.

Requirement	*A sensitive and specific assay for AMSA in plasma (and urine) over the range 0.1 to 10 μM^* with 0.5 ml plasma samples.*
End-step	*RP-HPLC on a radial compression column with an acetonitrile/water mobile phase containing 10 mM 'TEAP'. Detection at 254 nm.*
Sample preparation	*After internal standard addition and hexane extraction at pH 3-4, the plasma is ether-extracted at pH 9 with borate present in high concentration. The extract (6 ml) is dried down, and the residue taken up in 0.1 ml of methanol for HPLC.*
Comments	*Accurate and precise results (C.V. < 4%) obtainable down to 50 nM, with 5.5 min run time (the hexane eliminates a late-running component); pharmacokinetic use is thus feasible. No interferences found.*

Amsacrine, viz. 4'-(9-acridinylamino)methanesulphon-*m*-anisidide, is a 9-anilino derivative with a sulphonamide group (formula overleaf) which is undergoing extensive evaluation for various solid tumours and haematological malignancies. Significant activity has already been demonstrated against acute leukaemia in adults [1]. For investigation of AMSA pharmacokinetics in leukaemic patients, the primary requisite was a sensitive and specific assay for AMSA in plasma. In addition a urinary assay was required to examine renal

---

* The authors' usage has been altered, throughout the article, from 'μmol/l' to 'μM'; 1 μM = 394 ng/ml (the preferred basis).- *Eds.*

elimination.   As pharmacokinetic studies were also contemplated in
laboratory animals, the plasma aliquot could not exceed 1 ml.   The
following assay was developed [2].

ASSAY PROCEDURE FOR PLASMA

(1)   To duplicate 0.5 ml plasma aliquots, add 2 nmol of internal
      standard (i.s.; see below).
(2)   Adjust plasma pH to 3-4 with 0.12 ml 0.5 M HCl and wash with 5 ml
      of n-hexane for 20 min; centrifuge at 1720 g for 10 min.
(3)   Re-adjust plasma pH to 8-9 with 0.5 ml satd. $Na_2B_4O_7$ and   extract
      with 6 ml diethyl ether (peroxide-free); centrifuge.
(4)   Transfer ether layer to tapered glas tube and evaporate ($35°$, N_2).
(5)   Re-constitute in 0.1 ml methanol and run 20-40µl by HPLC.
(6)   Run with a radially compressed C-18 column (we used a 100 × 8 mm
      Waters Radial-Pak column); acetonitrile/water (40:60 by vol.)
      with 10 mM triethylamine phosphate (TEAP) as mobile phase, at
      7 ml/min.
(7)   Detection is at 254 nm, and quantitation by peak-area ratios.

Choice of internal standard (i.s.)

Use of o-AMSA as a suitable HPLC i.s. (Fig. 1) for plasma was
mentioned in a terse description [3] of separation on 5 µm Spherisorb-
ODS with an isocratic mobile phase of methanol/0.01% Na_2HPO_4 (85:15
by vol.).   Using the same mobile phase and a Waters µBondapak C-18
column, we were unable to obtain baseline separation of o-AMSA and
AMSA.   Hence a more suitable analogue of AMSA was sought as i.s.

Various mono-, di- and tri-substituted methyl and methoxy derivatives of the parent compound (**R,** Fig. 1) were examined in our chromatographic system, with the following results.-
a) Parent compound R and AMSA co-eluted.
b) Insufficient resolution was obtained between AMSA and $3'-CH_3-R$ and
$2'-OCH_3-R$.
c) Baseline separation was achieved between AMSA and a number of analogues including $3-CH_3-R$; $3-OCH_3-R$; $4-CH_3-R$; $3-CH_3,3'-CH_3-R$ and (tri-
substituted) $3'-OCH_3,4-CH_3,5-CH_3-R$.

An i.s. which eluted after AMSA was chosen to lessen the possibility of interference by any metabolites and hydrolysis products
which would be expected to be more polar than the parent drug and so
elute from the column before AMSA.   This is evident from Fig. 2A,
where the reported *in vitro* thiolytic cleavage products of AMSA,
4-amino-3-methoxy-methanesulphonanilide *(not shown)*, 9(10*H*)-acri-
done and the 9-aminoacridine [4] had t_r 0.60, 1.47 and 2.16 min respectively.   9-Aminoacridine has previously been identified as a
metabolite in the plasma and urine of patients receiving AMSA [5].
From compounds which gave baseline separation and eluted after AMSA,

Fig. 1. Structure of parent compound **R**,
viz. 4'-(9-acridinylamino)methanesulphon-
anilide [cf. footnote overleaf – *Eds*.].

AMSA = 3'-OCH$_3$-**R**

o-AMSA = 2'-OCH$_3$-**R**

Plasma i.s. = 3-CH$_3$-**R** (procured from
Synthesis Division, Cancer Chemotherapy
Laboratory, School of Medicine, Auckland).

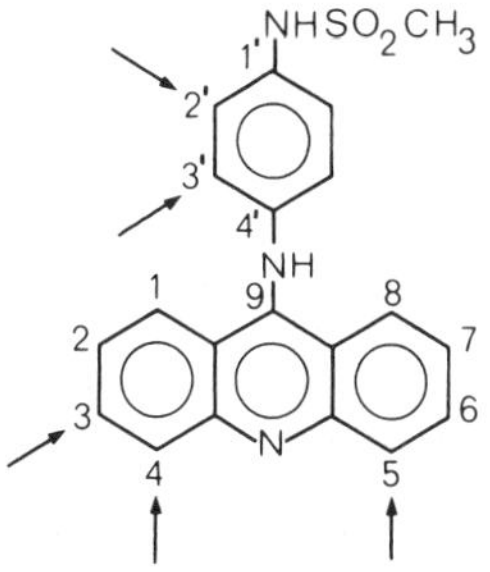

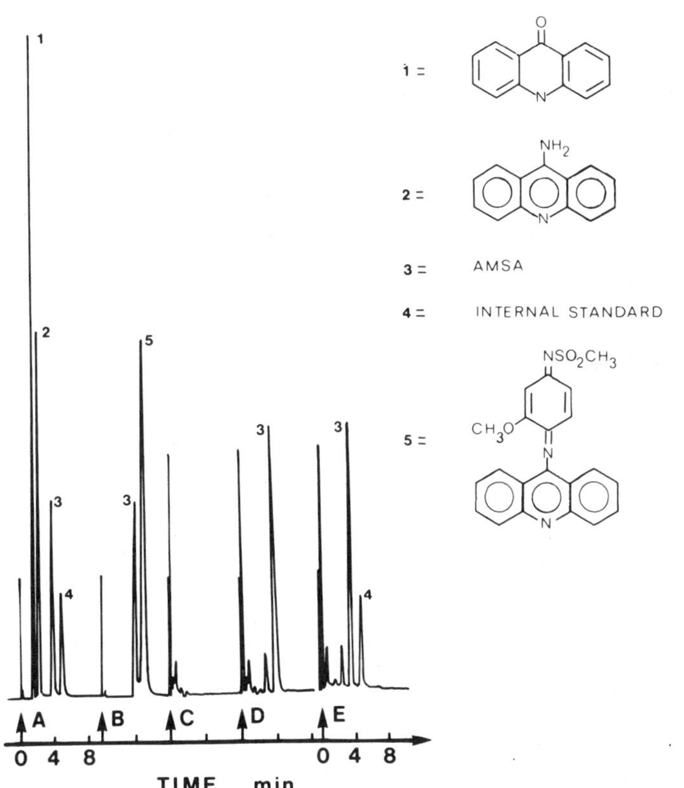

Fig. 2. HPLC of AMSA and degradation products.
A: standard methanolic solution of 9(10*H*)acridone (1), 9-amino-
acridine (2), AMSA (3) and i.s. (4).
B: standard solution of AMSA (3) and primary oxidation product,
the quinone diimine (5).
C: pre-infusion patient's plasma extract.
D & E: post-infusion plasma extract, with and without added i.s.

3-CH$_3$-R (Fig. 1)* was chosen as it eluted just after AMSA, allowing
a run-time of 5.5 min for each sample. The di- and tri-substituted
derivatives with their additional lipophilic groups tended to have
long retention times, resulting in longer analysis time.

One possible compound which might interfere with our chosen
i.s. is a primary oxidation product of AMSA, the quinone diimine
(5 in Fig. 2). However, this compound was not observed in any ether
extractions of post-AMSA infusion plasma without added i.s. (trace D
in Fig. 2). Similarly, patients' pre-infusion plasma samples (trace
C) contained no peaks which might interfere with AMSA or the i.s.

The i.s. had a pKa (7.49) very similar to that of AMSA (7.43)
and exhibited similar solubility properties in organic solvents [6].
No significant difference in peak-area ratio was observed between a
standard solution of AMSA/i.s. and plasma spiked with the same solution and taken through the assay procedure. Moreover, equimolar
concentrations of the i.s. and AMSA in the HPLC mobile phase showed
very similar UV absorption spectra [2]. Although the maxima for
both were at 270±2 nm, detection at 254 nm with the available fixed
wavelength detector entailed only 20% loss of sensitivity.

EXTRACTION PROCEDURE

Quantitative extraction of radiolabelled AMSA has been reported
with ethyl acetate from rat plasma at pH 7.2 [7]. However, as AMSA
is a weak base with pKa 7.43, extraction of alkalinized plasma was
considered preferable; but patients' plasma yielded significant
quantities of late-running endogenous compounds which interfered
with subsequent injections. These compounds were removed by a single
hexane wash of the plasma at pH ~3, as suggested by Malspeis et al.[3].

Diethyl ether in place of ethyl acetate allowed a marked reduction in overall assay time without loss of extraction efficiency.
The greater volatility of the diethyl ether allowed significant time-
saving in the evaporation step which offset the increased time due
to the hexane wash.

Addition of ^{14}C-AMSA to plasma samples which were taken through
the entire method gave an overall recovery of AMSA in the range
65-70%, indicating that the use of a suitable i.s. in this assay is
mandatory to correct for AMSA losses.

CHOICE OF CHROMATOGRAPHIC CONDITIONS

The use of the RP mode and mobile phase modifiers for analyzing
compounds containing amine groups is well documented [8]. The com-

---

* 4'-(3-methyl-9-acridinylamino)methanesulphon-anilide, if named by
analogy with AMSA (see start of text); authors' naming altered.-*Eds.*

bination of a radial compression column and a mobile-phase modifier
allows the rapid separation of many amines with excellent peak
shape. With a conventional column resolution was rather inadequate.

An acetonitrile/water mixture (40:60 by vol.) containing 10 mM
TEAP as modifier was found to be the optimum mobile-phase composition
for separation of AMSA and the i.s. and for a minimal run time.
Reduced concentrations of TEAP resulted in longer retention times.
The use of the commercially available but relatively expensive modifier, dibutylamine phosphate (D-4 reagent, Waters Assoc.) at 10 mM
in the same solvent system resulted in ~50% reduction in retention
times of AMSA and i.s. If the acetonitrile content of the mobile
phase were increased by 10%, similar retention times could be
achieved in our system using the considerably cheaper TEAP.

CHARACTERISTICS OF PLASMA AMSA ASSAY

The AMSA/i.s. peak-area ratio was linearly related to AMSA concentration from 0 to 20 µM. The standard curve was set up over the
range 0.1-10 µM (Fig. 3) and was represented by the equation which
accompanies Fig. 3 along with r and P values. Using 0.5 ml plasma,
the lower limit of the assay was 50 nM. This could be reduced to
~20 nM by increasing the volume of plasma used and by decreasing the
amount of i.s.

The accuracy and the intra-assay precision of the assay was
determined by 8 replicate analyses of 3 quality-control plasma pools
containing 10, 5 and 1 µM AMSA. Mean recoveries ranged from 104 to
115% and mean C.V.'s from 1.78 to 2.66%. Quality-control pools were
included in each subsequent assay to check for assay drift and
inter-assay variation, and the results for the subsequent 15 assays
are shown in Fig. 4. The overall inter-assay precision for these
measurements for each pool was less than 4% (C.V.). These determinations were undertaken over a 4-month period using aliquots from
the plasma pools stored in capped glass tubes at -20°, indicating
that AMSA is stable when thus stored. Methanolic solutions of the
i.s. similarly stored for the same time period were also stable.

One possible instability problem concerns the slow oxidation of
the free-base form of AMSA that has been reported to occur with bulk
material during storage [4]. The primary product of this oxidation
(compound 5, Fig. 2) elutes directly after AMSA and would interfere
with quantitation of the i.s. In our assay procedure, the free-base
form may occur when the plasma pH is re-adjusted to 8-9 before extraction with ether. However, extraction of several patients' post-
infusion plasma samples adjusted to pH 9 with and without i.s. addition indicated no post-AMSA peak that might interfere with the i.s.
(Fig. 2).

Further specificity studies indicated that a number of other

Fig. 3. Standard curve for AMSA in plasma.
In the equation, y is the AMSA/i.s. peak-area ratio and x the AMSA concentration (µM).

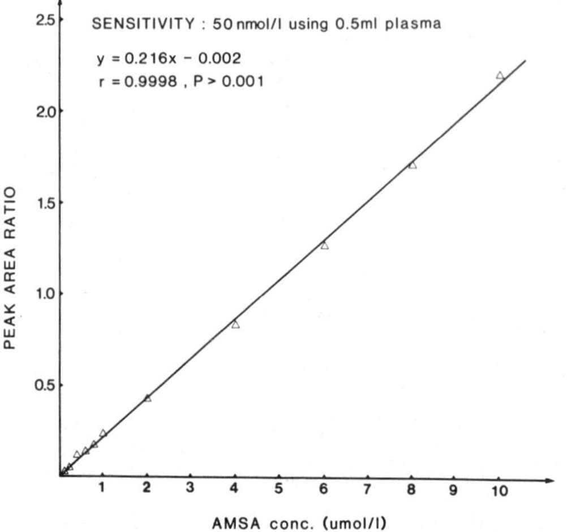

Fig. 4. Quality-control chart for 15 plasma assays undertaken over a 4-month period, with 3 plasma pools (see text; QH, QM & QL). The initial point is the mean ± S.D. of 8 replicate analyses of each pool.

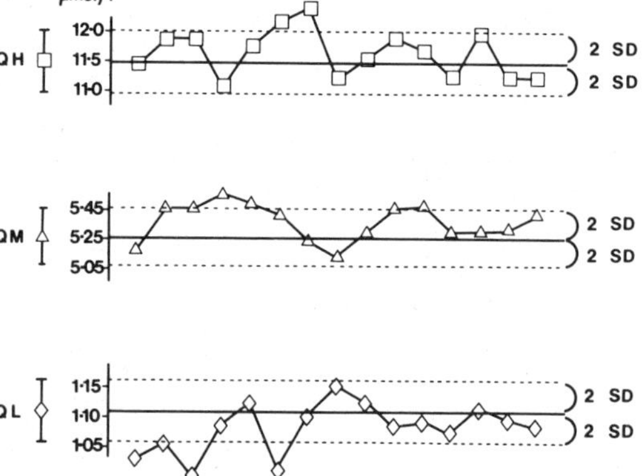

anti-cancer drugs including adriamycin, chlorambucil, cytosine arabinoside, 5-fluorouracil, lomustine, melphalan, methotrexate, prednisolone, 6-thioguanine, vincristine and vinblastine did not interfere with the AMSA or i.s. peaks under the conditions used.

CONCLUSIONS ON PLASMA ASSAY

We have developed a relatively robust HPLC assay for the determination of AMSA in 0.5 ml of plasma. It appears to be specific for AMSA and of good accuracy, with acceptable sensitivity and excellent precision, and has been successfully applied to the study of AMSA pharmacokinetics. In leukaemic patients, with plasma AMSA ~9 µM just after infusions (200 mg/m^2; daily for 3 days), levels were typically ~5 µM at 4 h post-infusion, 1 µM at 12 h, and under 0.5 µM at 24 h. The values tended to be higher after the third infusion than after the first. In rabbits given an infusion of 5 mg/kg, the mean levels fell from 11 µM initially to 2 µM at 4 h and 0.5 µM at 8 h.

With use of a C-18 guard column, >1000 injections of biological samples were chromatographed before a noticeable deterioration of column occurred. Reduction of the acetonitrile proportion to 38:62 restored the original resolution, and to date a further 900 injections have been run on the same column.

One problem encountered with this plasma assay has been apparent reduction of AMSA concentration in haemolysed blood samples. When plasma showing gross haemolysis was added to normal plasma containing AMSA and i.s., their peaks were both significantly reduced. The reason for this loss of AMSA and i.s. with haemolysis is not known. Possibly AMSA and i.s. become degraded through displacement of the anilino side-chain by glutathione released from the red cells. Such non-enzymatic nucleophilic attack by glutathione or protein thiol groups, with consequent displacement of the side-chain of AMSA and formation of the 9-thioacridine derivative, has been reported to occur in whole mouse blood [9]. However, the problem may be overcome by careful collection of blood samples to avoid haemolysis.

URINARY AMSA ASSAY

The plasma assay was applied to urine with certain alterations. The hexane wash was not required, as ether extracts of patients' urines adjusted to pH 9 did not show the late-running peaks seen with plasma. Hence sample-preparation time was much shortened.   An occasional problem was the presence in some urinary samples of a peak (X in Fig. 5) similar in retention to the plasma i.s. The 4-methyl derivative of AMSA was chosen as an alternative (IS' in Fig. 5); it had a rather longer t_r than AMSA (such that the run-time per sample was increased to 7.5 min), had a similar pKa (7.39) [6], and was similar in its ether-extractability from urine.

Characteristics of the urinary AMSA assay

Due to the greater possibility of interference by metabolites in urinary assays, AMSA detection and quantitation was carried out simultaneously at 254 nm and 405 nm. At the latter wavelength quantitation by peak-height measurements rather than by peak-area integration of peak area was superior, as a decreased signal/noise ratio was observed. Use of 405 nm detection entailed a 5-fold loss in sensitivity.

The urinary assay showed similar features to the plasma assay, with a linear standard curve over the same concentration range at both wavelengths (Fig. 6). Recoveries of AMSA spiked into 3 urine pools ranged from 94% to 106% measured at 254 nm, and from 94% to 111% at 405 nm. Mean C.V.s for 8 determinations ranged from 2.2% to 3.8% and from 2.1% to 7.4% respectively. Inclusion of these pools in 4 consecutive assays gave C.V.s for inter-assay precision of 1.4-3.7% and 3.2-4.6% respectively. AMSA values for urines from

patients receiving AMSA therapy were also not significantly different
when quantitated at 254 or 405 nm.

These data offer evidence of the homogeneity of the peak that
elutes at the retention time of AMSA, and suggest that AMSA is being
determined in urines of patients without interference by metabolites
or endogenous compounds.

Acknowledgements

Thanks are extended to Dr. B. Baguley and colleagues at the
Cancer Chemotherapy Research Laboratory, University of Auckland
School of Medicine, for the synthesis of pure AMSA and analogues, and
to Dr. R. Varcoe for providing samples from her patients.

This work was supported by a grant from the Auckland Medical
Research Foundation. J.W.P. is a Senior Fellow of the Medical
Research Council of New Zealand.

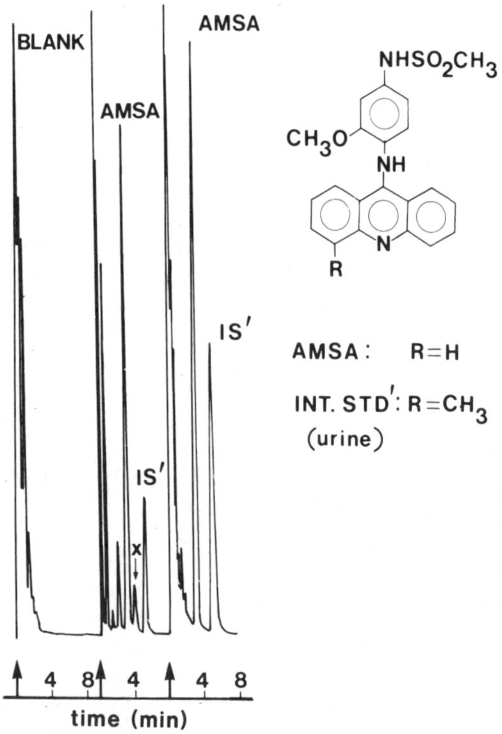

Fig. 5.   HPLC of urine extracts.
From left to right:
- a pre-infusion patient's urine;
- a post-infusion sample with the new i.s. added (see text; denoted IS'), manifesting interfering compound (X);
- a similar sample, lacking interfering compound.
Detection was at 254 nm.

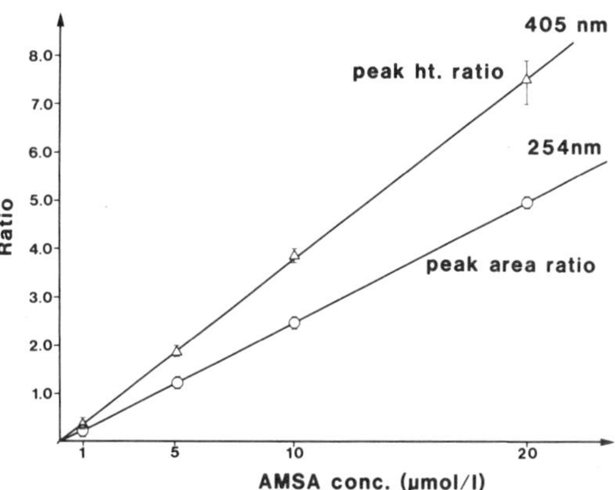

Fig. 6. AMSA standard curves for urine, measured at 254 nm (peak-area ratios) and 405 nm (peak-height ratios), represented in terms of x and y (see Fig. 3 legend) by the respective equations $y = 0.249\,x - 0.036$ ($r = 0.999$, $P<0.001$) and $y = 0.424\,x - 0.017$ ($r = 0.997$, $P<0.001$). The vertical bars denote ±1S.D.

References

1. Bodey, G.P. & Jacquillat, C., eds. (1983) *Amsacrine, Current Perspectives and Clinical Results with a new Anti-cancer Agent*, Communications Media for Education Inc., New Jersey, U.S.A., 107 pp.
2. Jurlina, J.L. & Paxton, J.W. (1983) *J. Chromatog. 276*, 367–374.
3. Malspeis, L., Khan, M.N. & Blat, H.B. (1979) *Proc. Am. Assoc. Cancer Res. 20*, 384 (Abs. C383).
4. Przybyski, M., Cysyk, R.L., Shoemaker, D. & Adamson, R.H. (1981) *Biomed. Mass Spectrom. 8*, 485–491.
5. Khan, M.N., Soloway, A.H., Cysyk, R.L. & Malspeis, L. (1980) *Proc. Am. Assoc. Cancer Res. 21*, 306 (Abs. 1227).
6. Denny, W.A., Cain, B.F., Atwell, G.J., Hansch, C., Panthanamickal, A. & Leo, A. (1982) *J. Med. Chem. 25*, 276–315.
7. Cysyk, R.L., Shoemaker, D. & Adamson, R.H. (1977) *Drug Metab. Disp. 5*, 579–590.
8. Gill, R., Alexander, S.P. & Moffat, A.C. (1982) *J. Chromatog. 247*, 39–45.
9. Wilson, W.R., Cain, B.F. & Baguley, B.G. (1977) *Chem. Biol. Interactions 18*, 163–176.

\#D-2

OVER-VIEW OF PROBLEMS IN DETERMINING TAMOXIFEN LEVELS IN BIOLOGICAL SAMPLES

H.K. Adam

Safety of Medicines Department
ICI Pharmaceuticals
Alderley Park, Macclesfield SK10 4TG, U.K.

As an introduction to the articles that follow (including a Note), early analytical work on tamoxifen and its metabolites is surveyed together with more recent work which has involved other laboratories. For exploratory studies TLC was advantageous. Sensitivity to single-figure ng/ml levels in blood is aimed at.

Any analytical chemist faced with a drug-assay need would start with a literature search. In this author's experience, the diffi-culty in setting up a published method rises in proportion to the square of the number of published methods! The later ones often use phrases such as "a more convenient", "a more robust", or "a more reliable". Typically, I suggest, these statements mean "We could not get the reported procedure to work, but here is one that does", or "Since we didn't have the relevant equipment we tried something else". [Cf. urinary metadrenalines! - #T-2 in Vol. 5.-*Ed.(E.R.).*]

This article does not seek to appraise the strengths and weak-nesses of the several methods reported for determining tamoxifen. It aims rather to set the scene with a short history of the drug in terms of analytical methodology, and to briefly discuss how some of the problems set by tamoxifen can be solved by the procedure devised in our laboratories. Tamoxifen citrate is the active drug ingredient.

tamoxifen
(trans isomer; 'Nolvadex'
- an ICI trademark - is
tamoxifen citrate)

$(CH_3)_2NCH_2CH_2O$

C_2H_5

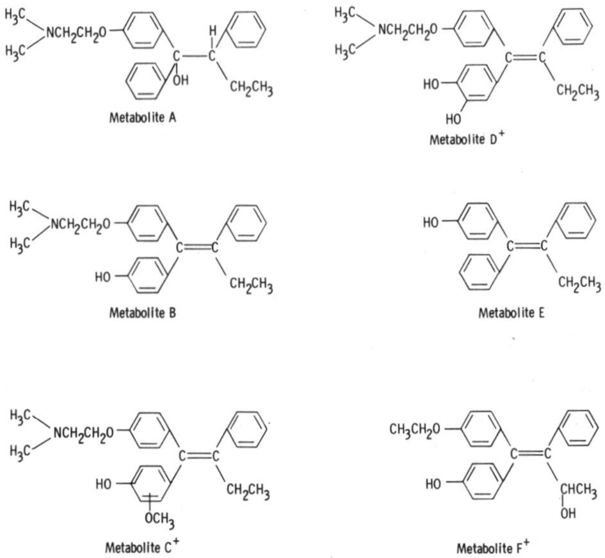

Fig. 1. Tamoxifen metabolites known [1] at the start of renewed analytical development work.

The drug is an anti-oestrogen of proved value in the treatment of hormone-dependent tumours, notably breast cancer. Because of its virtual lack of side-effects, compared to say cytotoxic agents, it is now in wide clinical use, as reviewed [2, 3]. Its pre-clinical development, through animal pharmacology and toxicological evaluation, was pre-1970. From the results of radiotracer work, it was clear that analytical methodology would have to be very sensitive to allow any significant pharmacokinetic input. Although the compound has a reasonable UV absorption, could be chromatographed by GC or TLC, and could be extracted without undue difficulty, the levels in biological samples were below the sensitivity of available detection techniques. Hence at that time sensitivity was the obstacle.

In the early 1970s the clinical usefulness was established but the question of pharmacokinetic evaluation remained unanswered. By the mid 1970s, with proven efficacy and safety in man, most subsequent work was aimed at analytical methodology for human samples. The renewed effort to devise analytical procedures hinged on work done in 1973 with ^{14}C-tamoxifen [1, 4] that clearly defined, for both animals and man, the targets of analytical method development in terms of specificity to begin with. Fromsom et al. [1] had isolated and identified 6 metabolites (Fig. 1). Evidently, initially at least, any method for tamoxifen had to distinguish parent drug from these metabolites.

Moreover, from the radioactivity profile shown in Fig. 2, evidently a single dose gives a peak level equivalent to ~100 ng/ml. This fell to 40 ng equivalents/ml by 24 h, and thereafter there was a slow decay such that radiotracer could still be detected in serum 14 days after drug administration. Because of the low levels of radio-activity, no indication of the nature of this long-lived material could be obtained, but at early times (~4 h) ~$\frac{1}{3}$ of the total ^{14}C content was unchanged drug and by day 8 less than 5% was tamoxifen. Moreover, by TLC metabolite B was identified in human serum.

In the late 1970s at least three groups were working independently to solve the problem of tamoxifen analysis in biological specimens. In summary, the requisite specifications are: (a) sensitivity to single-figure ng/ml levels for tamoxifen; (b) ability to distinguish tamoxifen from its known metabolites and (probably) a quantitative procedure for metabolite B since this had also been shown to be an anti-oestrogen [5]. The following articles outline how some of the groups working in the area have attacked the problem. I deal only with the analytical procedure devised at ICI.

OUR ANALYTICAL APPROACH

Given the structure of the compound, the most obvious choices for a sensitive assay to meet the specifications were ruled out, viz. GC-FID: insufficiently sensitive; GC-ECD: no easily derivatizable group; HPLC-UV: insufficiently sensitive; HPLC-fluorescence: no native fluorescence. However, we were interested in the possible conversion of

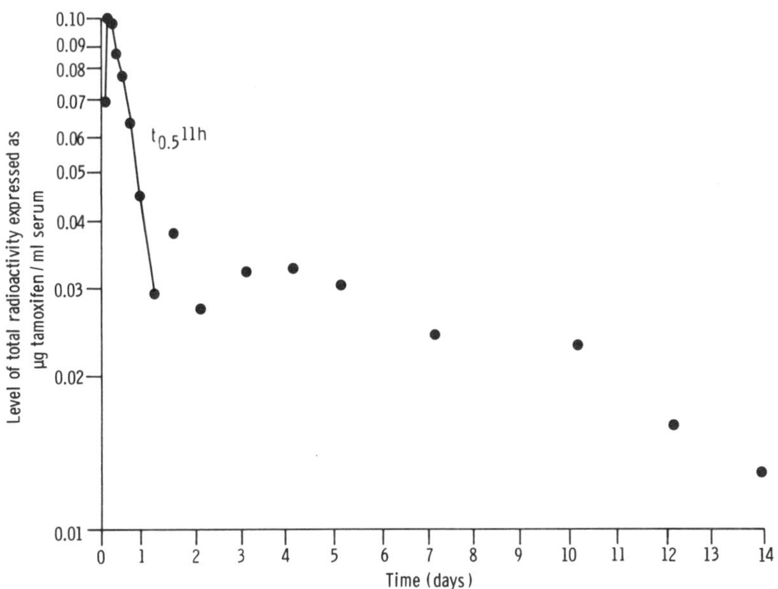

Fig. 2. Profile of total ^{14}C-concentrations after a single 20 mg dose of ^{14}C-tamoxifen in a female patient.

tamoxifen (ICI 46,474; trans isomer and anti-oestrogenic) to the oestrogenic cis isomer (ICI 47,699), and in UV-irradiation studies with TLC examination of the products it was discovered that a highly fluorescent product was formed. Subsequent work showed that this product was a phenanthrene, as exploited by two groups independently to devise a sensitive analytical procedure. Ours entailed TLC, and HPLC was used by Dr. L.A. Sternson, whose article in this volume (#D-4) shows the formula of the phenanthrene. We have published the TLC procedure and its subsequent use [6-8].

As one might expect for a relatively lipophilic base, the drug can be completely extracted into cyclohexane from aqueous buffers at all pH's above 3 (negligible at pH 0, 45% at pH 2, 98% at pH 3). However, in optimizing the extraction from serum we came across some unexpectedly subtle differences in the extraction. Fig. 3 shows not only that extraction may be slow [cf. J. Vessman, #F-4 in Vol. 10 – *Ed.*] but also that, despite the observed extraction efficiencies from aqueous buffer, extraction from alkalinized serum is inefficient with cyclohexane but better with hexane, and is better if the serum is at neutral rather than alkaline pH. For the final procedure the obvious choice was obviously neutral buffer and hexane.

We chose to convert tamoxifen to its fluorescent derivative by *in situ* irradiation of the TLC plate. We soon found that there was incomplete conversion of the total material in the spot. Fig. 4a demonstrates the effect of spotting at the origin, irradiation, chromatography and scanning by densitometer. Fig. 4b shows the results of re-irradiation and re-scanning the same post-chromato-

Fig. 3. Extraction of ^{14}C-tamoxifen (50 ng/ml) from human control serum (1 ml aliquots).
a) o: 1 ml 0.1 M NaOH added, and extracted with cyclohexane (10 ml) containing 1.5% v/v amyl alcohol.
b) ●: as for a) but using hexane instead of cyclohexane.
c) □: 1 ml 0.1 M phosphate buffer pH 7 added, and extracted with hexane containing 1.5% amyl alcohol.

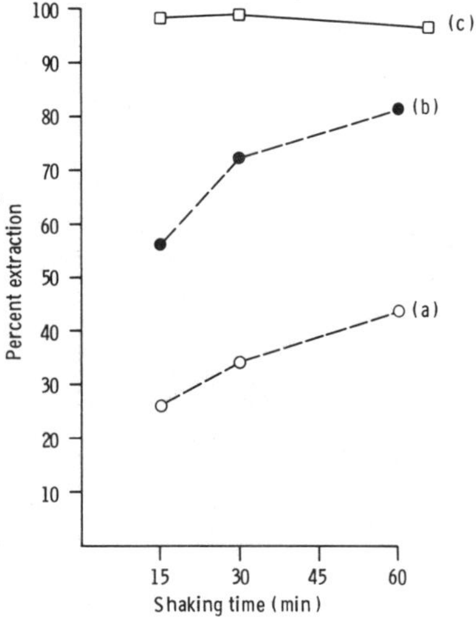

a

b

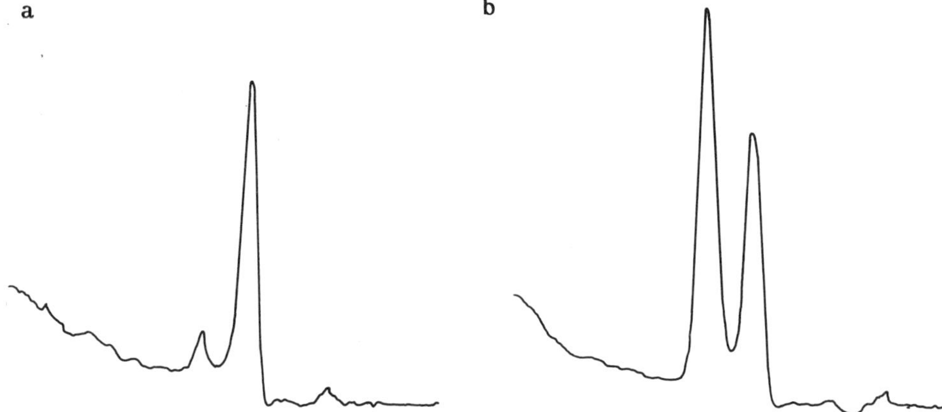

Fig. 4. Irradiation of tamoxifen on a TLC plate. a) Tamoxifen
(100 ng) spotted onto the plate, UV-irradiated, chromatographed
and scanned. b) Separated spot further irradiated and re-scanned.
Scanning was by fluorodensitometer.

Fig. 5. Comparison of results
by fluorodensitometry (●) and
^{14}C counting of TLC plate cut-
outs (▲), for ^{14}C-tamoxifen
analyzed by the final procedure:
calibration curves (100 nmol/l
corresponds to 37 ng/ml of
serum).

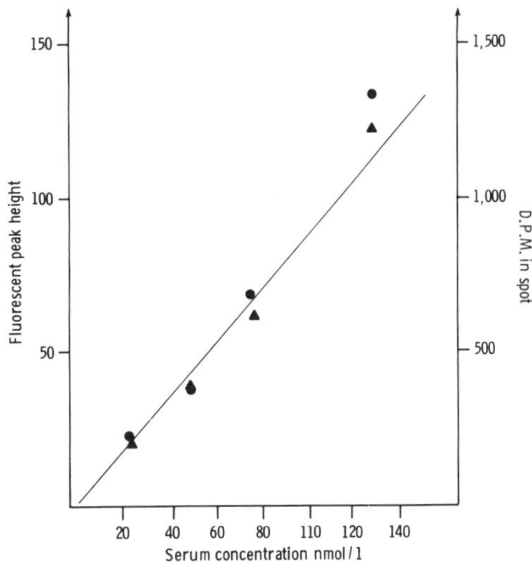

graphy spot. Evidently the phenanthrene derivative can be resolved
by TLC, but irradiation may not achieve complete conversion.   The
result was similar in a parallel study where the sample was chromatographed, irradiated, and then re-chromatographed in a second dimension at 90° to the first development. However, the reproducibility
of the TLC procedure was demonstrated by assaying samples containing
^{14}C-tamoxifen by the TLC procedure, with both densitometry and sec-
tioning and counting of the TLC plate (Fig. 5).

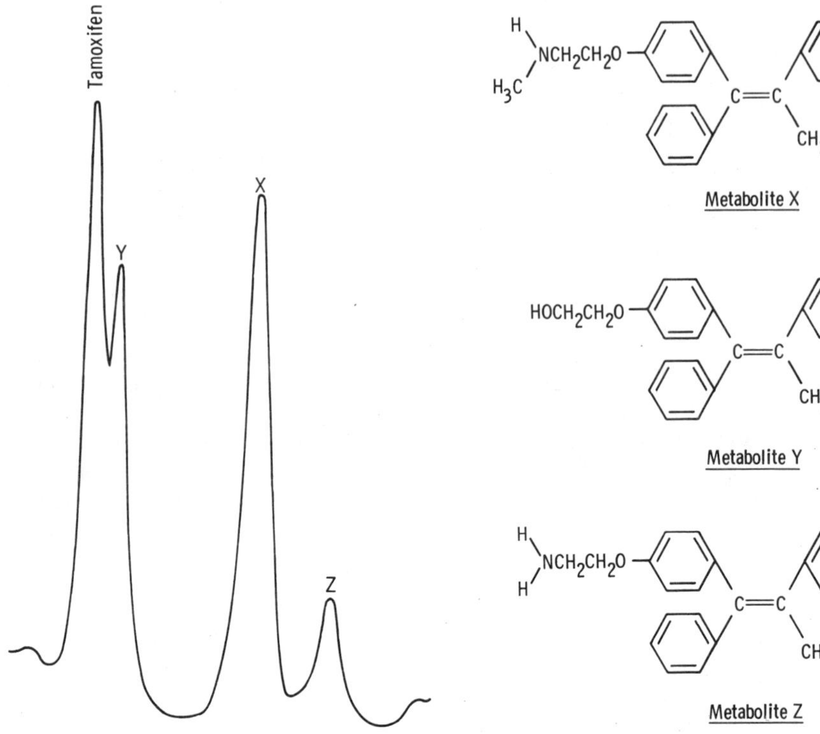

Fig. 6. HPLC fluorimetric trace
from a sample of human serum
after tamoxifen dosage, showing
three additional peaks due to
metabolites X, Y and Z.

Fig. 7. Structures of
metabolites X, Y and Z
(X being desmethyltamoxifen,
to which other articles in
this section refer).

All the known metabolites which retained the triphenylethylene
structure were resolvable from tamoxifen using a suitable TLC solvent and could be converted to fluorescent products. A sensitivity
of 2.5 ng/ml was attainable. Thus by this TLC procedure we had
apparently achieved our target. Two bonuses also emerged from this
TLC procedure. Firstly, since scanning of the plate *per se* led to
little conversion of the triphenylethylene structure to the fluorescent phenanthrene, there was a readily available check on any extraneous peaks in the trace, i.e. the appearance of an additional peak
only after irradiation was a strong indicator that such a peak was
tamoxifen-derived. Secondly we could, almost at will, alter solvent
systems to meet specific requirements. For example, by using a solvent specifically designed to resolve the cis and trans isomers we
were able to demonstrate that no conversion takes place when 'Nolvadex'
is dosed to humans.

The combination of the ability to alter the chromatography to meet specific needs and the convenient check that new peaks were indeed tamoxifen-derived were notably powerful tools when we started to use the assay in practice. When the procedure was applied to a sample from a patient on high-dose 'Nolvadex', three additional, previously unknown, metabolites were manifest besides tamoxifen (Fig. 6) with the structures shown in Fig. 7. This work has been reported in full elsewhere [9, 10].

References

1. Fromson, J.M., Pearson, S. & Bramah, S. (1973) *Xenobiotica* 7, 693-709.
2. Mourisden, H., Palshof, T., Patterson, J.S. & Battersby, L. (1978) *Cancer Treat. Rep. 5*, 131-141.
3. Patterson, J.S. (1981) in *Non-Steroidal Anti-Oestrogens* (Sutherland, R.L. & Jordan, C., eds.) Academic Press, Sydney, pp. 453-472.
4. Fromson, J.M., Pearson, S. & Bramah, S. (1973) *Xenobiotica* 7, 711-714.
5. Jordan, V.C., Collins, M.M., Rowsby, L. & Prestwich, W. (1977) *J. Endocrin. 75*, 305-311.
6. Adam, H.K., Gay, M.A. & Moore, R.H. (1980) *J. Endocrin. 84*, 35-42.
7. Patterson, J.S., Settatree, R.S., Adam, H.K. & Kemp, J.V. (1980) *Breast Cancer - Experimental and Clinical Aspects* (Mourisden, H. & Palsholf, T., eds.), Pergamon Press, Oxford, pp. 89-92.
8. Adam, H.K., Patterson, J.S. & Kemp, J.V. (1980) *Cancer Treat. Rep. 68*, 761-764.
9. Adam, H.K., Douglas, E.J. & Kemp, J.V. (1979) *Biochem. Pharmacol. 27*, 195-197.
10. Kemp, J.V., Adam, H.K., Wakeling, A.E. & Slater, S.R. (1983) *Biochem. Pharmacol. 32*, 2045-2052.

#D-3

ANALYSIS OF TAMOXIFEN AND ITS METABOLITES

V. Craig Jordan, Richard R. Bain, Stewart D. Lyman
and Raymond R. Brown

Department of Human Oncology
Wisconsin Clinical Cancer Center
University of Wisconsin
Madison, WI 53792, U.S.A.

Requirement   *Assay for serum tamoxifen, N-desmethyltamoxifen and metabolite Y, down to 5 ng/ml; tissue samples also.*

End-step   *Isocratic HPLC on silica, with different conditions for metabolite Y and a different internal standard (i.s.). Detection by post-column fluorescence activation.*

Sample preparation   *Serum (0.1 ml) extracted with 1 ml hexane containing 2% butanol (v/v) and evaporated to dryness; residue redissolved in HPLC solvent and injected.*

Comments   *Run times are <10 min. The methods can be used both for therapeutic monitoring of patients and for investigating drug interactions.*

Tamoxifen[*] is a non-steroidal anti-oestrogen used extensively in the treatment of breast cancer [1]. The usefulness of the drug and the interest in its mechanism of action have focused attention upon its pharmacokinetics and metabolism in animals and man (recently reviewed [2]). Several metabolites have been identified: metabolite B [3, 4], metabolite X [5] and metabolite Y [6, 7]. Convenient assay systems are required for clinical monitoring of patients and for pharmacological studies.

---

[*] *trans*-1-(*p*-β-dimethylaminoethoxyphenyl)-1,2-diphenylbut-1-ene.
Metabolites (X = *N*-desmethyltamoxifen) –
B: *trans*-1-(*p*-hydroxyphenyl)-1-(*p*-β-dimethylaminoethoxyphenyl)-2-phenylbut-1-ene;
X: *trans*-1-(*p*-β-methylaminoethoxyphenyl)-1,2-diphenylbut-1-ene;
Y: *trans*-1-(*p*-β-hydroxyethoxyphenyl)-1,2-diphenylbut-1-ene.

EXPLORATORY WORK

Serum samples (25–200 µl) from patients receiving tamoxifen (10 mg twice daily) was extracted with 2.0 ml hexane/n-butanol (96:4 v/v), dried under N_2 at 55°, dissolved in 0.1 ml mobile phase solvent, and analyzed immediately by HPLC using a 25 x 0.46 cm silica (10 µm) column. Detection was as fluorescent phenanthrenes after UV activation (cf. Sternson, #D-4) as shown in Fig. 1. Details are stated in [8]. Three isocratic solvent systems were used.-
- System A (methanol/diethylamine/acetic acid, 100:0.03:0.3 by vol.) separated tamoxifen, N-desmethyltamoxifen and the i.s., ICI 99311[*]: R_t's 3.7, 2.4 and 4.5 min respectively.
- System B (hexane/isopropanol, 96:4 by vol.) separated metabolites $E^†$ (3.1 min) and Y (6.2 min); metabolite E was used as an i.s. because it has not been detected in human serum.
- System C (hexane/ethanol/diethylamine, 70:30:0.004 by vol.) separated the i.s., enclomiphene[θ] (2.1 min), tamoxifen (3.6 min), metabolite B (monohydroxytamoxifen; 4.5 min) and N-desmethyltamoxifen (9.3 min).

Tamoxifen and four metabolites can be analyzed in patient sera by a TLC system with enclomiphene as i.s. (cf. Adam, #D-2, re TLC). Silica gel plates were developed with toluene/triethylamine/methanol (90:10:0.15 by vol.). After chromatography, plates were air-dried, coated with isooctane/paraffin oil (80:20), dried, and UV-irradiated for 45 sec. Plates were then analyzed with a scanning fluorescence densitometer. With 1% acetic acid in the coating solution the fluorescence intensity of metabolite E was increased 10-fold. R_f values were: metabolite B, 0.04; X (N-desmethyltamoxifen), 0.13; E, 0.18, Y, 0.25; tamoxifen, 0.41; and enclomiphene, 0.56.

The capacity of animal tissues to metabolize [^{3}H]tamoxifen, both *in vivo* and *in vitro*, was also studied. Two TLC systems were developed for analyzing tamoxifen metabolites on silica plates.-
- System A, toluene/triethylamine/methanol/piperidine/acetonitrile (90:10:1:1:0.5 by vol.), giving R_f values: tamoxifen-N-oxide, 0.00; bisphenol[φ], 0.05; metabolite B, 0.15; N-desmethyltamoxifen, 0.27; metabolite E, 0.35; metabolite Y, 0.41; and tamoxifen, 0.60.
- System B, chloroform/methanol/conc. ammonium hydroxide (95:5:0.25), giving R_f values: tamoxifen-N-oxide, 0.03; metabolite B, 0.18; desmethyltamoxifen, 0.24; bisphenol, 0.26; tamoxifen, 0.44; metabolite E, 0.52; metabolite Y, 0.57. Radioactivity was extracted from tissue homogenates or microsomal incubates using ethanol (recovery > 96%). The major metabolite formed *in vitro* by chicken liver microsomes was metabolite B.

---

[*] 1-(p-β-dimethylaminoethoxyphenyl)-1,2-diphenylacrylonitrile
[†] *trans*-1-(p-hydroxyphenyl)-1,2-diphenylbut-1-ene
[θ] E-2-[p-(2-chloro-1,2-diphenylvinyl)phenoxy]-triethylamine; E = *trans*
[φ] 1,2-(p-hydroxyphenyl)-2-phenylbut-1-ene

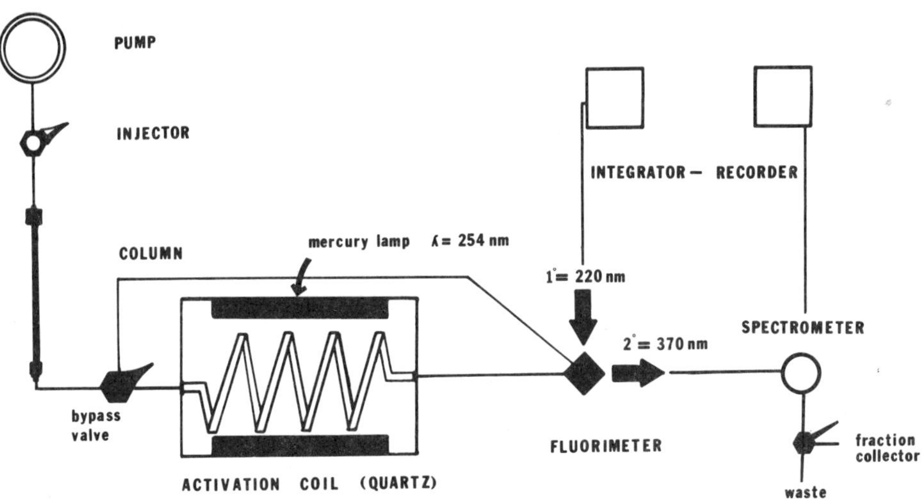

Fig. 1. Schematic representation of analytical and detection systems. Note bypass loop for detection of natively fluorescent compounds.

THE HPLC PROCEDURES FOR HUMAN SERUM AND ANIMAL TISSUES

Three isocratic solvent systems were developed to separate tamoxifen and metabolites on a silica column, as specified above. System **A** was used routinely to determine tamoxifen and desmethyl-tamoxifen in extracts of human serum. Sera (50 µl) from patients receiving tamoxifen therapy was spiked with 100 ng of the i.s. in 10 µl methanol, allowed to equilibrate for 10 min at room temperature, extracted as above with vortexing for 30 sec and centrifuged at 1000 g for 10 min at 4°. The supernatant was put in a conical vial, blown to dryness under N_2 at 55°, and redissolved in 0.1 ml mobile phase. Chromatography (Fig. 2, *upper left*) gave the R_t's stated above.

This system permitted extraction and analysis of 30 samples and 6 standards in a day. A weakness in this technique is that metabolite B co-elutes with tamoxifen. However, interference with tamoxifen measurement is minimized as levels of metabolite B are low (20 ng/ml) and the volume of serum analyzed is small (50 µl). Detection limits for metabolite X and tamoxifen were ~2 ng/ml.

System **B** (conditions and R_t's as stated above) was used for routine measurement of metabolite Y in human serum samples (100 µl) similarly extracted. The detection limit for Y is ~2 ng/ml. Analysis time is short and 20 samples may be analyzed conveniently in a day. The peak pattern is shown in Fig. 2 (*upper right*). Although metabolite E (i.s.) has never been reported in human serum, conceivably it could arise in certain individuals or under certain multi-drug regimens and thereby make this assay unreliable. A distinctive

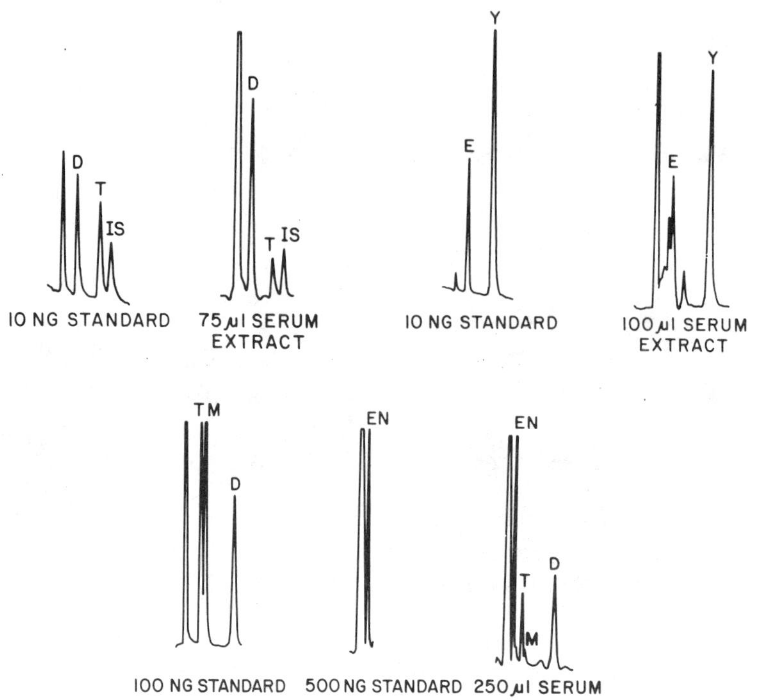

Fig.2. Normal-phase separation of tamoxifen (T) and metabolites
– E, Y, B (denoted M; monohydroxytamoxifen), and desmethyltamoxifen
(denoted D; metabolite X). IS = ICI 99311; EN = enclomiphene.
See text for conditions. The traces run from right to left. Area
measurements for standards led to linear regression curves.
Upper left: System **A**; flowrate 2.0 ml/min.   *Upper right:* System
B; 2.0 ml/min.   *Lower portion:* System **C**; 3.0 ml/min.

feature of this system is that all known metabolites of tamoxifen
that contain an aminoethoxy side-chain are retained indefinitely on
the column due to their higher polarities.

System **C** was used to determine tamoxifen, metabolite B and desmethyltamoxifen concentrations in human serum using enclomiphene as
an internal recovery marker. Due to the low levels of metabolite B,
larger volumes (250 µl) of serum are required. Serum was equilibrated with enclomiphene (500 ng in 50 µl ethanol), then made alkaline
by addition of 2 M Tris-acetate, pH 9.1 (50 µl), and extracted as
above. Extraction at pH 9.1 reduced the amounts of interfering substances eluting in the void volume. Chromatography (Fig. 2, *lower
portion*) was done under conditions stated above along with retention
times. Analysis time was somewhat longer than with Systems A and B,
but 15 serum extracts could be analyzed in a day.

The use of silica in this system confers greater selectivity than systems described by others. [Cf. use of RP–HPLC in accompanying articles:- Leith; with ion-pairing: Sternson (pre-column irradiation) and Stevenson.- *Ed*.]. In the C-18 RP method of Golander [9], large amounts of polar substances eluted in broad interfering peaks. The use of silica eliminates such interference and permits detection of small amounts of metabolite B, presumably by greater retention of polar compounds on the silica matrix. Detection limits for tamoxifen and desmethyltamoxifen are ~2 ng/ml, and the limit for metabolite B is ~4 ng/ml. All three systems give 88% average recovery from serum.

TLC PROCEDURES

TLC methods were used in parallel with HPLC in analysis of human serum, and disclosed an unidentified metabolite in serum of a patient receiving tamoxifen in high dosage (150 mg twice daily). The patient exhibited high serum levels of tamoxifen (2150 ng/ml) and desmethyltamoxifen (1270 ng/ml). The TLC method of Adam ([10], & #   ) was modified so that metabolites E and Y could be measured besides tamoxifen, desmethyltamoxifen and metabolite B. Sera from patients receiving tamoxifen were spiked with enclomiphene and extracted as above. The residue from drying down was dissolved in methanol (50 µl). Standards (10–500 ng) containing tamoxifen, desmethyltamoxifen, metabolites B, E & Y, and enclomiphene were spotted in a volume of 50 µl onto a silica plate with a mechanical spotter; test samples were spotted simultaneously. The plates were developed in the dark using toluene/triethylamine/methanol (90:10:0.15 by vol.), then air-dried, coated with isooctane/paraffin oil of b.p. 150° (80:20 by vol.), air-dried, UV-irradiated (at 254 nm) for 45 sec, and analyzed by scanning fluorescence densitometry [7]. The R_f values were: metabolite B, 0.04; desmethyltamoxifen, 0.13; metabolite E, 0.18; metabolite Y, 0.25; tamoxifen, 0.41; and enclomiphene, 0.56.

Metabolite E exhibited a low relative fluorescence intensity under these conditions. The above-mentioned 10-fold increase with acetic acid present was presumably due to neutralization of residual triethylamine in the silica matrix, thereby permitting protonation of the phenolic substituent of metabolite E. This proved to be a useful probe in the identification of the unknown metabolite. The unknown was not metabolite E because (1) its peak area did not rise when acetic acid was added to the coating solution, and (2) added metabolite E separated from the unknown peak (Fig. 3). The unknown was later identified by HPLC and GC–MS as metabolite Y [6].

METABOLISM AND CHROMATOGRAPHIC ANALYSIS OF [³H]TAMOXIFEN *IN VITRO*

Our method was based on the assay of Robertson [11]. The 250 µl reaction mixture contained [³H]tamoxifen (in 2.5 µl ethanol), hen liver microsomes (in 100 µl 0.25 M sucrose), buffer, and an NADPH gene-

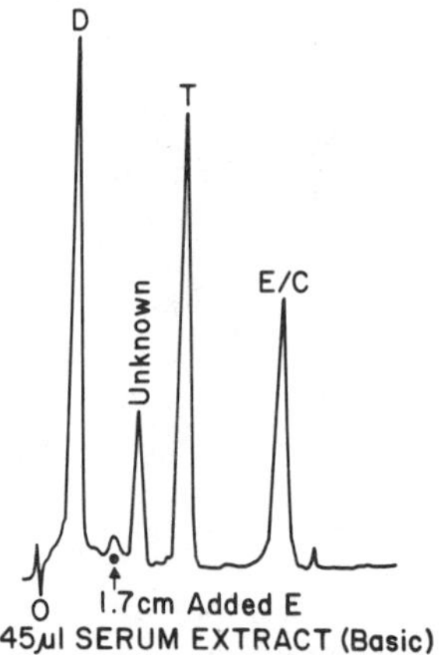

Fig. 3. TLC pattern for serum from a patient on high tamoxifen dosage, spiked with 50 ng of metabolite E. T, D as in Fig. 2; E/C denotes enclomiphene.

A basic coating solution was applied to the plate prior to analysis.

rating system; initial assay conditions were: 0.16 µM [³H]tamoxifen (1.8×10^6 dpm), 80 mM Na phosphate buffer (pH 6.9 at 37°), 10 mM $MgCl_2$, 1 mM nicotinamide, 10 mM glucose-6-phosphate, 0.5 units glucose-6-phosphate dehydrogenase, and 0.3 mM NADP. The reaction was started by adding the [³H]tamoxifen to the test tube, which was incubated in a water bath at 37° with shaking. The reaction was ended after 1 h by adding 1 ml ethanol. The contents of the tube were mixed, and the precipitate removed by centrifugation. At least 98% of the radioactivity was found in the supernatant, an aliquot of which was co-chromatographed with authentic standards on a TLC plate and analyzed as described below. The main metabolite formed (Fig. 4), metabolite B, has been shown to be the major metabolite formed *in vitro* by slices of male chicken liver [12]. This is of particular significance because metabolite B is a more potent anti-oestrogen than tamoxifen with a binding affinity for the oestrogen receptor equivalent to oestradiol [13].

In separating [³H]tamoxifen and its metabolites on Whatman LK5DF silica TLC plates, the following solvent systems were used:
- (A) toluene/triethylamine/methanol/piperidine/acetonitrile, 90:10:1:1:0.5 by vol.;
- (B) chloroform/methanol/conc. ammonium hydroxide, 95:25:0.25; modified from Ruenitz [14].
The R_f values in systems A and B respectively were: tamoxifen: 0.60, 0.44; monohydroxytamoxifen (metabolite B): 0.15, 0.18; *N*-desmethyltamoxifen (metabolite X): 0.27, 0.24; metabolite E: 0.35, 0.52; metabolite Y: 0.41, 0.57; bisphenol: 0.05, 0.26; tamoxifen-*N*-oxide: 0.00, 0.03.
Following chromatography, the standards were visualized by UV light, the silica gel was removed from the plate, and radioactivity was quantified by liquid scintillation spectrometry.

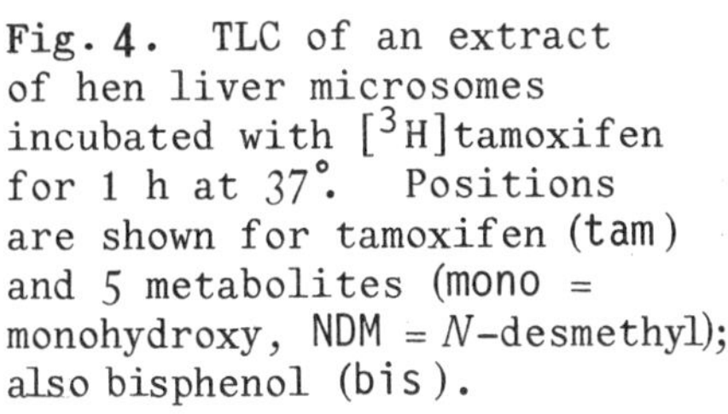 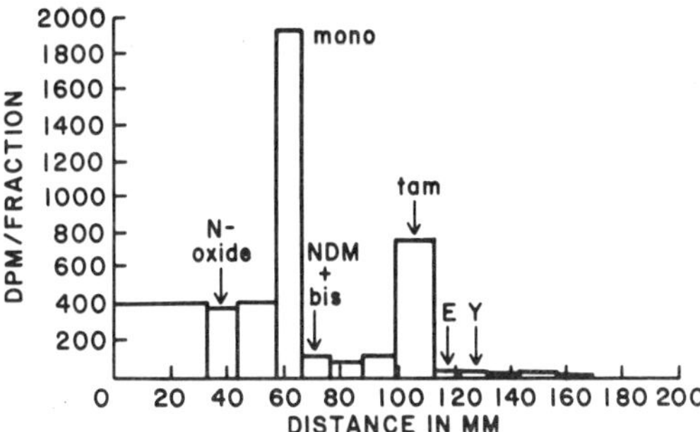

Fig. 4. TLC of an extract of hen liver microsomes incubated with [³H]tamoxifen for 1 h at 37°. Positions are shown for tamoxifen (tam) and 5 metabolites (mono = monohydroxy, NDM = *N*-desmethyl); also bisphenol (bis).

Acknowledgements

These studies were supported by grants P30-14520 and P01-20432 awarded to the Wisconsin Clinical Cancer Center by the National Institute of Health, U.S.A., and by grants from ICI plc (Pharmaceuticals Division), Macclesfield, U.K., and from Stuart Pharmaceuticals, Wilmington, DE. S.D.L. is a graduate student in the McArdle Laboratory for Cancer Research, University of Wisconsin.

References

1. Mouridsen, H., Palshof, T., Patterson, J. & Battersby, L. (1978) *Cancer Treat. Rev.* 5, 131–141.
2. Jordan, V.C. (1982) *Breast Cancer Res. Treat.* 2, 123–138.
3. Daniel, C.P., Gaskell, S.J., Bishop, H. & Nicholson, R.I. (1979) *J. Endocrinol.* 83, 401–408.
4. Fabian, C., Tilzer, L. & Sternson, L. (1981) *Biopharm. Drug Disp.* 2, 381–390.
5. Adam, H.K., Douglas, E.J. & Kemp, J.V. (1979) *Biochem. Pharmacol.* 27, 145–147.
6. Bain, R.R. & Jordan, V.C. (1983) *Biochem. Pharmacol.* 32, 373–375.
7. Jordan, V.C., Bain, R.R., Brown, R.R., Gosden, B. & Santos, M.S. (1983) *Cancer Res.* 43, 1446–1450.
8. Brown, R.R., Bain, R.R. & Jordan, V.C. (1983) *J. Chromatog.* 272, 351–358.
9. Golander, Y. & Sternson, L.A. (1980) *J. Chromatog.* 181, 41–49.
10. Adam, H.K., Gay, M.A. & Moore, R.H. (1980) *J. Endocrinol.* 84, 35–42.
11. Robertson, D.W., Katzenellenbogen, J.A., Long, D.J., Rorke, E.A. & Katzenellenbogen, B.S. (1982) *J. Steroid Biochem.* 16, 1–13.
12. Borgna, J.-L. & Rochefort, H. (1981) *J. Biol. Chem.* 256, 859–868.
13. Jordan, V.C., Collins, M.D., Rowsby, L. & Prestwich, G. (1977) *J. Endocrinol.* 75, 305–316.
14. Ruenitz, P.C. & Toledo, M.M. (1981) *Biochem. Pharmacol.* 30, 2203–2207.

#D-4

HPLC METHODOLOGY FOR THERAPEUTIC MONITORING OF TAMOXIFEN AND TWO MAJOR METABOLITES IN BLOOD AND TISSUES

Larry A. Sternson

Pharmaceutical Chemistry Department
University of Kansas, Lawrence, KA 66045, U.S.A.

Requirement	*Clinical assay of tamoxifen [I] and its N-desmethyl [II] and 4-hydroxy [III] metabolites in whole blood and various tissues.*
End-step	*Irradiate (UV) to form fluorescent phenanthrenes, then RP-HPLC (C-18) ion-pair separation, ending with fluorimetric detection.*
Sample preparation	*Tissues are first homogenized and digested with protease VIII. Homogenates and whole blood are extracted with diethyl ether, either directly or with trichloracetate ion-pair formation (only slight gain in analyte recovery). The residue from drying-down the extract is dissolved in the HPLC mobile phase.*
Comments	*The assay procedure allows quantitation down to analyte levels of 100 pg/ml of biological fluid or 100 pg/g tissue. Problems include a photocatalyzed degradation of the fluorescent product, minimized by optimized irradiation conditions so that the conversion to phenanthrene derivatives proceeds fast enough for convenient sample throughput.*

The non-steroidal anti-oestrogen, tamoxifen [I], is currently used extensively in treatment of metastatic breast cancer and other disseminated oestrogen-dependent tumours, as well as in the 'adjuvant' context. It is extensively metabolized in man to two major biologically active products: the *N*-desmethyl [II] and 4-hydroxy [III] derivatives [1-3] *(see overleaf)*. The former has oestrogenic activity while the latter is strongly anti-oestrogenic [4]. The relative binding affinities of I, II and III for oestrogen receptors [4] and relative blood levels in patients [2, 4, 5] who had received the drug

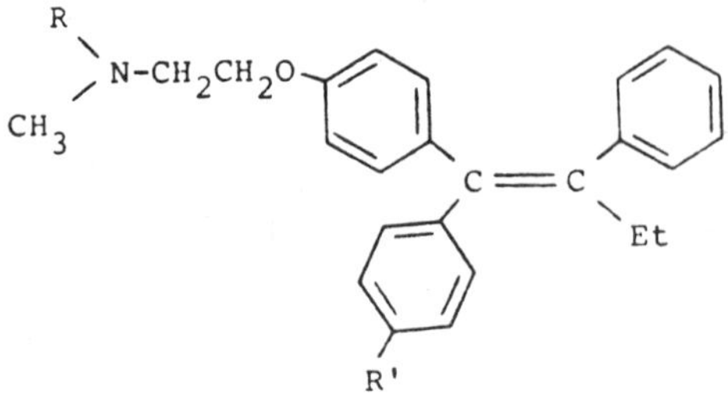

	R	R'
Parent drug, I	CH_3	H
Metabolite II	H	H
Metabolite III	CH_3	OH

over a period of time suggest that therapeutic monitoring should
include measurement of metabolites [2, 3] as well as parent drug. As
tamoxifen is now often given along with other drugs and their metabolism can be affected by tamoxifen and its metabolites [I-III], and
as tamoxifen is itself prone to metabolic alteration [6], dosage
guidance in combined therapy calls for specific monitoring of I-III
at the target tissues and/or in a compartment in equilibrium with
these sites, suitably blood. Early clinical studies [3] suggested
that sensitivity in the low ng/ml region had to be aimed at (cf. #D-2).

Besides high sensitivity, the development of a clinical assay
had to take account of tamoxifen's extreme hydrophobicity with the
risk of adsorptive loss from biological fluids onto storage-vessel
surfaces, of tamoxifen's chemical and photochemical instability that
may require special handling and storage of samples, and of extensive biotransformation calling for methodology with high selectivity.

RETRIEVAL OF TAMOXIFEN AND ITS METABOLITES FROM BLOOD

Tamoxifen and its metabolites can be efficiently extracted from
blood into diethyl ether, either directly or as their trichloroacet-
ate ion pairs [7, 8]. Recovery is ~90% by either approach. The
rationale for trying ion-pairing was to benefit extraction specificity, and also efficiency by minimizing adsorptive losses. Desorption of hydrophobic amines from glass surfaces was known [9] to be
favoured by lowering the pH and enabling ion-pairing to occur. However, with only ~2-3% gain in extraction efficiency, the extra time
needed to perform an ion-pairing step was not warranted.

ISOLATION OF TAMOXIFEN AND METABOLITES FROM TISSUE

The requisite efficient exposure of tissue material to the
extractant was generally achieved by homogenization of the sample in
pH 10.5 buffer [10]. With some tissues (e.g. adipose, uterus) this
procedure sufficed for good extraction of I-III (~90%; Table 1). For
most tissues, however, simple homogenization was inadequate, giving
~20% recoveries with ~ ±50% precision. Pre-incubation of tissue
homogenates with Protease VIII (Sigma) at 50° for 1 h gave, in the

Table 1. Recovery of tamoxifen and metabolites from blood and
tissue samples, spiked with each analyte (50 ng/ml of fluid or
homogenate) and analyzed as in ref. [10]. Generally the mean %
recoveries (±S.D.) are based on 6 spiked samples each analyzed
chromatographically in triplicate, from 200 g female rats except
that a human volunteer with diagnosed metastatic breast cancer
furnished the CSF and (sufficient for only single determinations)
the tumour. To assess recovery, analyte peak area was compared
with that for analyte spiked direct into the mobile phase.
Based, by permission of Marcel Dekker Inc., on Table 1 in [10].

Fluid or tissue	I (tamoxifen)	II (desmethyl)	III (4-hydroxy)
Whole blood	87.1 ±1.0	86.1 ±2.0	101.0 ±5.0
Plasma	90.1 ±3.0	85.3 ±1.7	96.1 ±2.3
CSF	96.3 ±1.2	88.5 ±3.1	97.7 ±1.9
Liver	88.7 ±5.1	87.7 ±4.2	94.8 ±1.0
Fat	85.8 ±3.2	90.7 ±6.0	87.5 ±7.4
Uterus	88.1 ±5.6	89.9 ±5.5	90.3 ±3.8
Tumour	85.8	84.4	90.0

subsequent extraction, high (~90%) and reproducible recovery of analytes (Table 1) [10].

STABILITY CONSIDERATIONS

Early studies in our laboratory suggested that tamoxifen undergoes *cis-trans* isomerization, especially at elevated temperatures.
Accordingly, samples were stored frozen; GC was ruled out as the
final step to get efficient separation. Photochemical instability
was observed too, especially for III. To filter out the light responsible for photodegradation, and also to minimize adsorptive loss,
samples were stored in plastic containers (polypropylene).

PHOTOCHEMICAL DERIVATIZATION OF THE ANALYTES

HPLC seemed the method of choice; but the physical properties
of the analytes precluded quantitation at therapeutic levels with an
ordinary detection mode. Although I is a polyaromatic conjugated
molecule, steric strain forces the phenyl rings out of plane, resulting in reduced resonance interaction [11] and a disappointing UV
absorbance [12] (λ_{max} = 238 nm, ε = 19,350). A derivatization reaction
was therefore sought that would convert the analytes to products
with improved detectability.

Stilbenes are known to cyclize under the influence of UV light
to yield phenanthrenes [13, 14] which often are highly fluorescent.

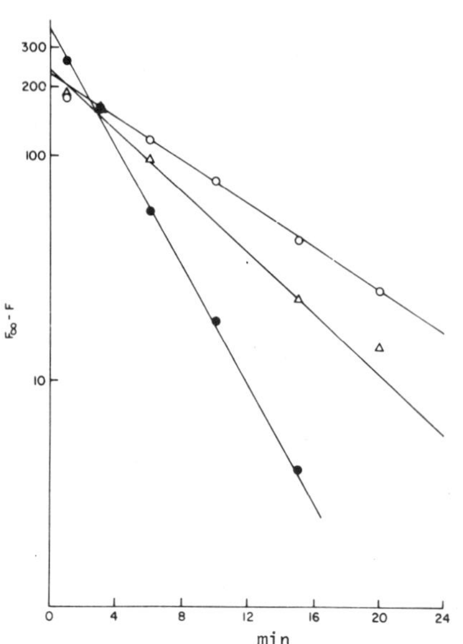

Fig. 1. Increase in fluorescence
when tamoxifen (I; Δ), desmethyltamoxifen (II; o) and 4-OH-tamoxi-
fen (III; ●) extracted from whole
blood are UV-irradiated (reaction
vessel 10 cm from light source,
viz. Hg vapour lamp, 15W, 253.7 nm
max.   *From [8], by permission.*
The plotted values (log scale;
$F_\infty - F$ in arbitrary units)
represent the lessening with
time of the difference in
fluorescence between the maximum
attainable and the observed
value (cf. Fig. 2).

Irradiation of acidic methanolic solutions of tamoxifen and metabolites II and III with UV light resulted in a rapid increase in
fluorescence (λ_{ex} = 255 nm, λ_{em} = 320 nm) [7, 8]. This photoactivated
increase in fluorescence could be plotted in a first-order manner
(Fig. 1).   Continued irradiation led to a gradual loss in fluorescence intensity and the appearance of other products.   However,
secondary irradiation products of I did not appear till nearly 10
½-lives of the first reaction had elapsed [7]. Thus it appeared that
a stable, highly fluorescent product was formed which could effectively be trapped by careful control of irradiation conditions.

Since irradiation intensity and, therefore, the rate of photolysis is inversely proportional to the square of the distance from
the light source, positioning of the reaction vessel is critical to
obtain a reaction rate slow enough to allow flexibility in irradiation time although not inconveniently prolonged.   Optimal kinetics
were achieved (Fig. 2) when the reactor was placed 10 cm from the
lamp [8].   Maximum fluorescence intensity was reached after irradiation for only 12.5 min, yet the fluorescence at 10 and 15 min was
near the peak value (98.6% and 98.2% respectively). First-order plots
of the appearance and disappearance of fluorescence of acidified
methanolic solutions of I irradiated under these conditions gave ½-
lives of 1.7 and 117 min respectively, confirming earlier speculation
that photolysis could be conveniently stopped at the fluorescence
maximum, before significant degradation occurred [7, 8].

The chemistry of tamoxifen photoactivation appears to parallel
that of diethylstilboestrol and other stilbenes [13, 14].   In the
presence of UV light, *cis-trans* isomerization is facilitated; ring

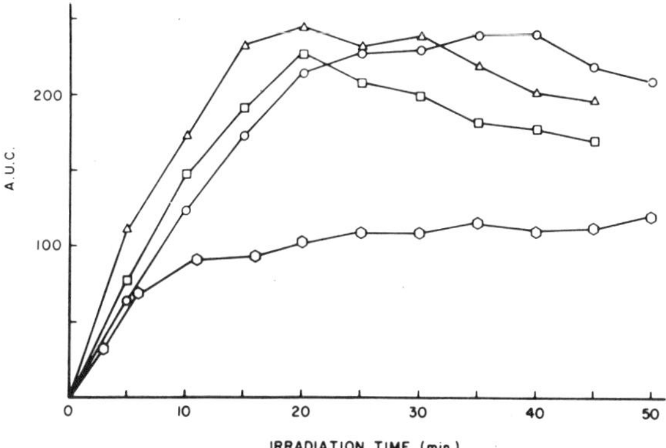

Fig. 2. Effect of irradiation time on the formation of the fluorescent derivative from tamoxifen. Distance between lamp and reaction
vessel: o, 20 cm; △, 15 cm; □,10 cm; ○, 25 cm. Reaction monitored by HPLC, and product yield determined as area under curve
(A.U.C.).

fusion of the isomers yields dihydrophenanthrenes - which are readily
air-oxidized to the corresponding phenylphenanthrenes (IV-VI or VII-IX).

Spectral data and elemental analysis confirm that fluorescent products are phenylphenanthrenes [12], apparently an unresolved mixture
of structurally isomeric products (viz. IV & VII; V & VIII; VI & IX).

The stability of the phenylphenanthrenes was greatly enhanced
by the presence of acid in the photolysis medium [8], allowing
greater latitude in the width of the time window chosen for irradia-

tion. This added flexibility is of particular importance when I, II
and III are being monitored simultaneously since the kinetics of
fluorophore formation differ significantly for the three (Fig. 1)
[8]. Degradation of fluorophore was found to proceed primarily with
formation of the hydroxyphenylphenanthrene X, apparently through
intramolecular aziridinium ion formation and concomitant expulsion
of the aziridine, XI [12].

The product X has spectral properties different from IV-VI, leading
to the observed reduction in fluorescence intensity. The stabili-
zation provided by the presence of acid in the solution apparently
results from protonation of the amine nitrogen (on IV-VI) reducing
its nucleophilicity and thereby its ability to attack the methylene
carbon to form X eventually [12].

The data presented here are consistent with a kinetic pathway
of the $A \to B \to C$ type, where A is the analyte (I-III); B is the
fluorophore (IV-IX), and C is the degradation product (X). The
linear, first-order plots for appearance of fluorescence suggest
$k_1 \gg k_2$. Solution of the appropriate differential equation for an
$A \to B \to C$ consecutive irreversible reaction gives an expression for B
in terms of k_1 and k_2. The fit of the theoretical curve generated by
solution of this equation:

$$\beta = \frac{A_o k_1}{k_2 - k_1} (e^{-k_1 t} - e^{-k_2 t}) \qquad \text{(Eq. 1)}$$

(using the experimentally determined values of k_1 and k_2) to the
experimental points was quite good, supporting the proposed kinetic
model (Fig. 3) [12].

CHROMATOGRAPHIC SEPARATION OF PRODUCTS

HPLC separation of the products of pre-column photochemical
activation, viz. IV-VI, could not be achieved by reverse- or normal-
phase modes. However, using ion-pair chromatography with a reversed-
phase (C-18) column, separation of tamoxifen species and corresponding phenanthrenes in biological samples was readily accomplished
[8, 10]. Using methanol-water mixtures and pentanesulphonate ion-

Fig. 3. Time course of
fluorescence change for
UV-irradiated tamoxifen
in acidic methanolic
solution. The points
are the experimental
values; the curve is
the theoretical line
drawn according to
Equation 1.

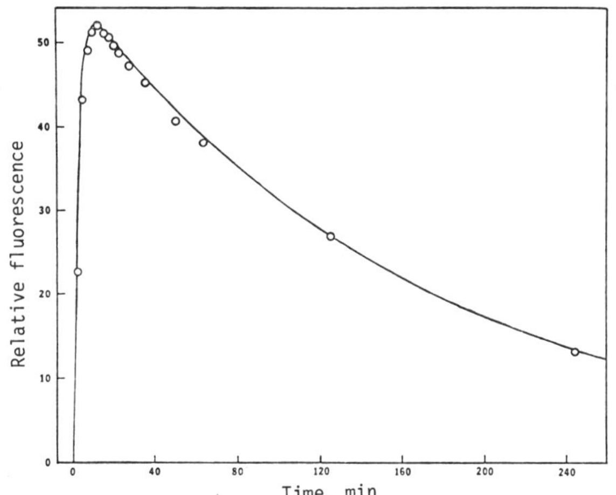

pairing agents, the capacity-factor (k') values were most sensitive
to change over the range 70-80% (v/v) methanol, varying from ~1.9 to
~5 for tamoxifen and over a similar range for V and VI.

The system was much less sensitive to the nature of the ion-
pairing agent. With methanol/water (80:20 by vol.) as mobile phase,
k' for tamoxifen was 1.9 with 1-pentanesulphonate (2.5 mM) and 2.3
with 1-decanesulphonate (2.5 mM). Optimal separation of analytes
from each other and from co-extracted contaminants as shown in Fig. 4
was obtained with the mobile phase specified in the legend. To aid
pre-column conversion to phenanthrenes, in later experiments irradiation was carried out directly in the HPLC mobile phase. Chromatography was significantly improved, although reactions were somewhat
slower: with the lamp and sample cuvette 10 cm apart, the respective
k values for the first-order reaction (and the $t_{\frac{1}{2}}$ values) were 0.15/
min (4.6 min) for I → IV, 0.29/min (2.4 min) for II → V, and 0.11/min
(6.5 min) for III → VI, the initial concentrations of I-III being 50 ng/ml.

Summarizing the outcome, the approach entailing irradiation,
HPLC and fluorescence detection afforded a sensitivity of 100 pg/ml
(at a signal:noise ratio = 3) in clinical samples containing tamoxifen and its desmethyl and 4-hydroxy metabolites. Thus methodology
is available to rapidly and easily monitor tamoxifen and its major
biologically active metabolites in blood and various tissues at
therapeutically relevant levels.

References

1. Adam, H.K., Douglas, E.J. & Kemp, J.V. (1979) *Biochem.*
    Pharmacol. 28, 145-147. [Cf. #D-2, this vol.-*Ed.*]
2. Fabian, C., Sternson, L.A. & Barnett, M. (1980) *Cancer Treat.*
    Rep. 64, 765-773.

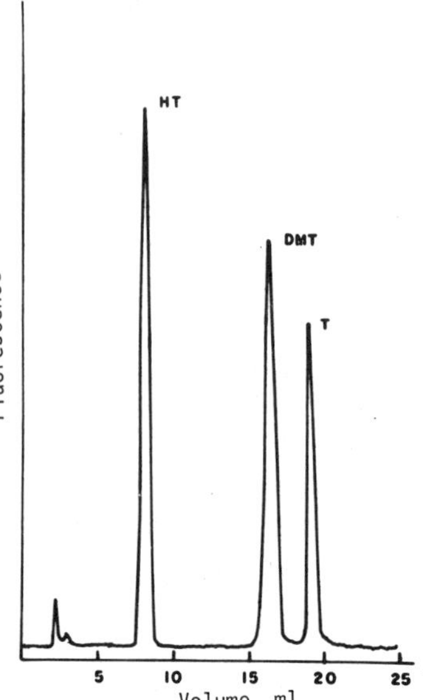

Fig. 4. HPLC of I, II and III extracted from blood (each 1 ng/ml) and UV-irradiated to form phenanthrene derivatives, run with ion-pairing.
Column: 10 µm C-18, 30 cm 4.6 mm i.d., ambient temp.
Eluent: methanol/water (73:27 by vol.) containing 0.5% (by vol.) acetic acid and 2.5 mM sodium pentanesulphonate.
Flow-rate: 2 ml/min.
Detection: fluorescence; 256 nm excitation; 340 nm cut-off filter for the emission.
Injection: 20 µl, from 250 µl of solution of residue.

3. Fromson, J.M., Pearson, S. & Braman, S. (1973) *Xenobiotica* *3*, 711-714.

4. Fabian, C., Tilzer, L. & Sternson, L.A. (1981) *Biopharm. Drug Dis. 2*, 381-390.

5. Fabian, C. & Sternson, L.A. (1981) *Reviews on Endocrine Related Cancer, Suppl. 9*, ICI Pharm. Div., pp. 158-168.

6. Meltzer, N., Stang, P. & Sternson, L.A. (1984) *Biochem. Pharmacol. 33*, 115-123.

7. Mendenhall, D.W., Kobayashi, H., Shih, F.M.L. & Sternson, L.A. (1978) *Clin. Chem. 24*, 1518-1524.

8. Golander, Y. & Sternson, L.A. (1980) *J. Chromatog. 181*, 41-49.

9. Thaaker, K., Higuchi, T. & Sternson, L.A. (1979) *J. Pharm. Sci. 63*, 93-95.

10. Sternson, L.A., Meltzer, N. & Shih, F.M.L (1981) *Anal. Lett. 14*, 583-600.

11. Bedford, G.R., Walpole, A.L. & Wright, B. (1974) *J. Med. Chem. 17*, 147-149.

12. Mendenhall, D.W. (1978) Ph.D. Thesis, Dept. of Pharmaceutical Chemistry, University of Kansas, Lawrence, KA.

13. Stermite, F.R. (1967) in *Organic Photochemistry*, Vol. 1 (Chapman, O.L., ed.), Dekker, New York, pp. 247-282.

14. Doyle, T.D., Filipescu, N., Benson, W.R. & Banes, D. (1970) *J. Am. Chem. Soc. 92*, 6371-6372.

#D-5

ANALYSES FOR TAMOXIFEN, ITS METABOLITES AND
ENDOGENOUS OESTROGENS IN BLOOD PLASMA AND TUMOUR TISSUE

Heather M. Leith, C. Paul Daniel*,
Robert I. Nicholson and Simon J. Gaskell

Tenovus Institute for Cancer Research
Welsh National School of Medicine
Heath, Cardiff CF4 4XX, U.K.

Requirement	*(i) 'Reference' analyses, to the log-ng level, for tamoxifen, N-desmethyltamoxifen and 4-hydroxytamoxifen in blood plasma and tumour tissue from treated patients. (ii) 'Routine' techniques for the same analytes in blood plasma only.*
End-step	*(i) GC-high resolution MS/selected ion monitoring (SIM) of molecular ions of tamoxifen, N-heptafluorobutyryl-N-desmethyltamoxifen and 4-hydroxytamoxifen TMS ether. (ii) HPLC of underivatized analytes using a reverse-phase (RP) system and detection by UV absorption (254 nm).*
Sample preparation	*Initial internal-standard (i.s.) addition: pentene analogues of the 3 analytes. Solvent extraction of plasma (diethyl ether) and homogenates (acetone /ethanol) is followed by cation-exchange chromatography on derivatives of Sephadex LH-20 (Scheme 1).*
Comments	*GC-MS data are useful for reference purposes and initial determination of concentration ranges. Sample work-up is similar for both procedures, but the use of HPLC obviates derivatization and is less dependent on complex instrumentation and expertise.*

Tamoxifen [trans-1-(4-β-dimethylaminoethoxyphenyl)-1,2-diphenylbut-1-ene] is now widely used in the treatment of advanced breast cancer. The development in this laboratory of methodology for the analysis of tamoxifen and related compounds has had two principal objectives: (a) an assessment of correlations between dose, concen-

---

*now at Dept. of Endocrinology, Royal Postgrad. Med. Sch., London W12 0HS

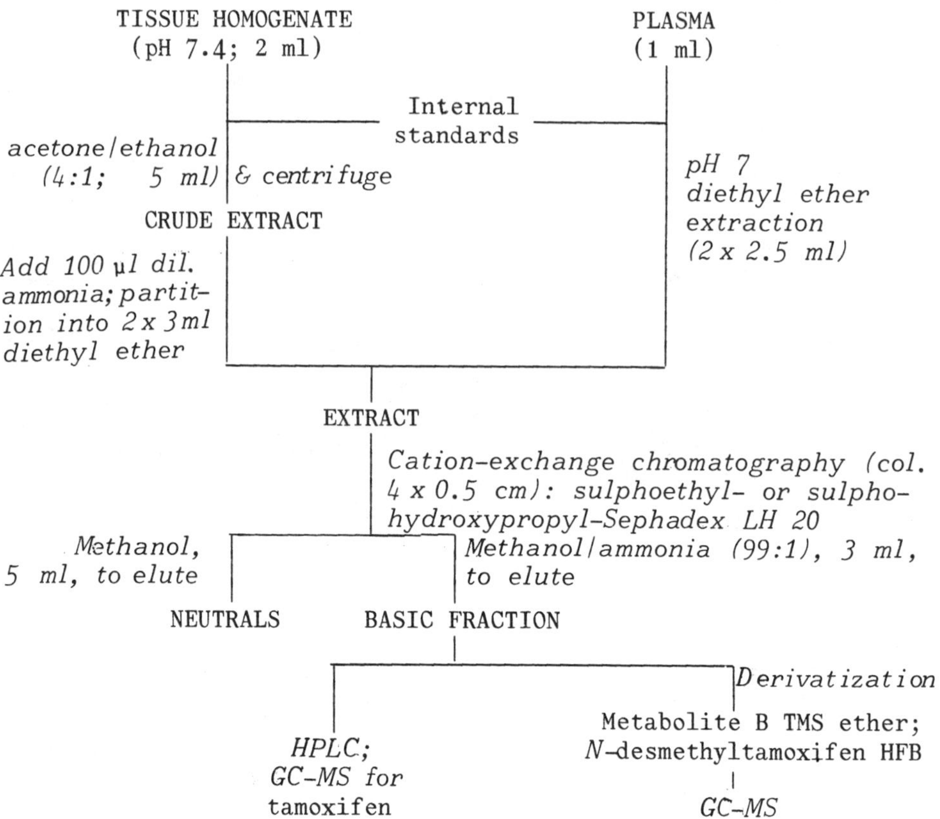

Scheme 1. Flow diagram for the analysis of tamoxifen and metabolites in plasma and tissue.' TMS = trimethylsilyl; HFB = heptafluorobutyryl. Ammonia strength: s.g. 0.88; 'dil.' = 10-fold dilution.

trations of the drug and its metabolites in blood plasma and target
tissue, and response to treatment; (b) elucidation of the mechanism
of action. The early assays for tamoxifen and its metabolites in
plasma, based on highly specific techniques using GC-high resolution
MS, have been followed by development of a more routine plasma assay
using HPLC and extension of MS studies to tissue extracts.

ANALYSES FOR TAMOXIFEN AND METABOLITES IN PLASMA AND TUMOUR TISSUE

Analyses of human blood plasma and tumour tissue from treated
patients, using GC-MS with SIM, have been described in detail elsewhere [1, 2]. The incorporation of internal standards, added to the
plasma or the tissue homogenate, improves the accuracy and precision
of quantitative determination. The compounds used were the pentene
analogues of tamoxifen and metabolite B, viz. 4-hydroxytamoxifen,

Table 1. Representative data for concentrations of tamoxifen (tam), N-desmethyltamoxifen (des), metabolite B and oestradiol-17β (E_2-17β), in plasma and tumour tissue from treated patients [2].

Patient (& days of treatment)	Plasma, ng/ml				Tumour homogenate, ng/mg prot.			
	tam	des	B	E_2-17β[*]	tam	des	B	E_2-17β[*]
QB (21)	160	356	4.5	0.037	18.7	24	0.37	$<1.3 \times 10^{-4}$
ESt (26)	410	534		0.028	42.5	21.1	0.29	$<2.6 \times 10^{-4}$
MS (30)	306	391	8.2	0.059	14.3	14.3		
FS (34)	146	299	6.0	0.037	13.8	25		
FS (60)	248	396	5.4	0.037	18.4	35.6		
JC (75)	149	416	5.8	0.048	5.4	92.3	1.13	4.7×10^{-4}

[*] measured by RIA

both gifted by ICI Pharmaceuticals, and of N-desmethyltamoxifen, prepared by desmethylating tamoxifen's pentene analogue. Cation-exchange derivatives of Sephadex LH-20, first developed for steroid separations [3, 4], provide a rapid and selective means of isolating tamoxifen and related compounds.

During GC-MS, molecular ions were monitored in each case, giving detection limits of ~1 ng analyte in plasma and tissue extracts. Other performance characteristics have been reported elsewhere [1,2]. Intra- and inter-assay C.V.s were generally 5-10%. Typical analytical data for plasma and tissue analyses are given in Table 1, together with estimates of oestradiol-17β concentrations obtained by radio-immunoassay [2]. The latter analyses have yet to be validated by comparison with GC-MS values; nevertheless the data suggest that a greater than 1000-fold excess of tamoxifen and desmethyltamoxifen over oestradiol-17β is achieved in the plasma and tumour tissue of treated patients, consistent with effective competition with endogenous oestrogens for binding to the oestrogen receptor. Indeed, the tissue concentrations far exceed those of the latter (typically <2 pmol/mg protein). Fractionation of tumour samples to permit the analysis of subcellular compartments indicated concentration of tamoxifen and desmethyltamoxifen in a crude nuclear fraction [2].

Amounts of 4-hydroxytamoxifen (metabolite B) were much lower than for tamoxifen or desmethyltamoxifen (Table 1). A contributory role to the anti-oestrogenic effect of tamoxifen treatment is nevertheless indicated in view of the greater affinity of the 4-hydroxy metabolite than the parent drug for the oestrogen receptor [5].

Analyses incorporating the highly specific technique of GC-MS with SIM at high MS resolution have provided reliable data on the plasma and tissue concentrations of tamoxifen and its metabolites in

Fig. 1. HPLC analysis of a
plasma extract for tamoxifen
(peak **b**) and *N*-desmethyltam-
oxifen (peak **a**) (peak **c** =
tamoxifen i.s.).

Column: ODS-silica, 12.5 cm.
Solvent: methanol with n–butyl-
amine (0.8%)/water (82:18).
Detection at 254 nm.

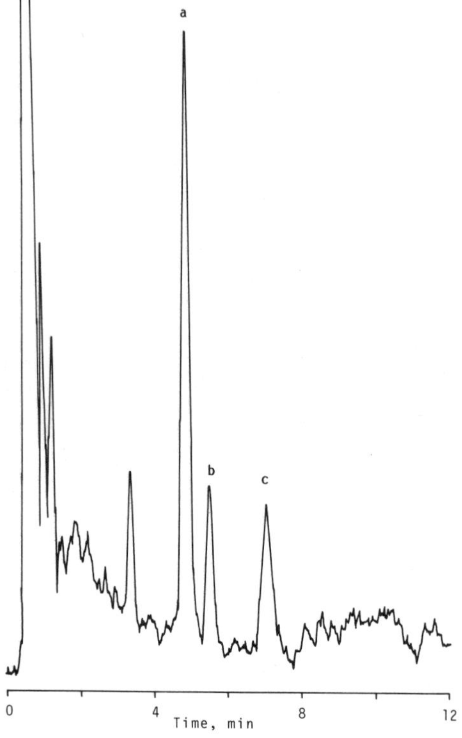

treated patients. The substantial expense and technical complexity
of analyses using GC–MS, however, have encouraged the development of
a plasma assay using HPLC. The sample work-up procedure is similar to
that employed for GC–MS analyses (Scheme 1). The high concentrations of tamoxifen and desmethyltamoxifen in the plasma of patients
(Table 1) permit the use of fixed-wavelength (254 nm) UV absorbance
detection.

Fig. 1 shows a typical analysis of a patient plasma. UV detection has been preferred, largely on the grounds of instrument availability, to the more sensitive fluorescence detection described in
accompanying contributions from the laboratories of L.A. Sternson
(#D-4) and V.C. Jordan (#D-3). The UV method does, however, have
the advantage of needing no pre- or post-column modification of the
analyte. Another laboratory has also developed a HPLC–UV procedure
[D. Stevenson, #NC(D)-1].

Analytical data obtained so far have not revealed any correlation
between tamoxifen and metabolite levels in plasma or tissue and response to treatment. In rats, however, the reduction in tumour (DMBA)
volumes in treated animals showed a correlation with the tamoxifen
concentrations in cytosol fractions [6].

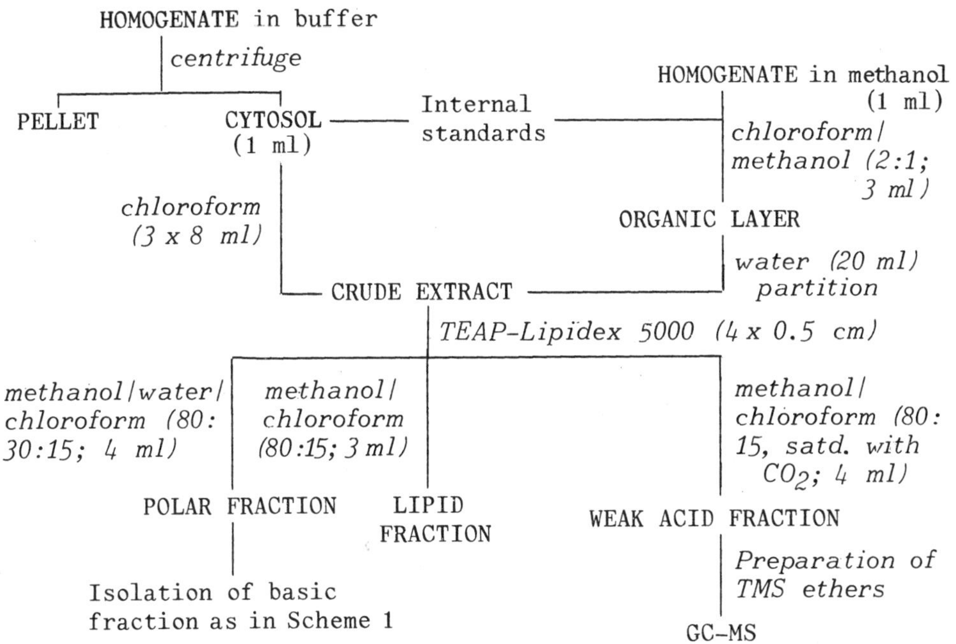

Scheme 2. Flow diagram for the analysis of endogenous oestrogens in tumour tissue. *TEAP*, triethylammoniohydroxypropyl.

ANALYSES FOR ENDOGENOUS OESTROGENS IN TUMOUR TISSUE

Reliable data concerning relative concentrations of tamoxifen and its metabolites and of endogenous oestrogens in tumour tissue are of particular value in elucidating the mechanism of action of the drug. The levels of oestradiol-17β and other oestrogens are extremely low, placing severe demands on the analytical methodology. Techniques based on GC-MS have been developed to provide the required sensitivity and selectivity for the analysis of oestradiol-17β, oestrone and oestriol.

Scheme 2 summarizes the procedures for the analysis of total tissue homogenates or subcellular fractions. [^{13}C]- and [^{2}H]labelled analogues are used as internal standards. The weakly acidic nature of the phenolic oestrogens permits their isolation by anion-exchange chromatography, using *(see legend)* TEAP-Lipidex 5000 [3]. The polar fraction obtained in this chromatographic step may be further fractionated to permit analysis of tamoxifen and its metabolites. Fig. 2 shows a GC-MS/SIM analysis for oestradiol in a cytosol extract, demonstrating the notably high sensitivity and selectivity achieved. Detection limits for oestradiol, oestrone and oestriol are 1-5 pg/g of tissue.

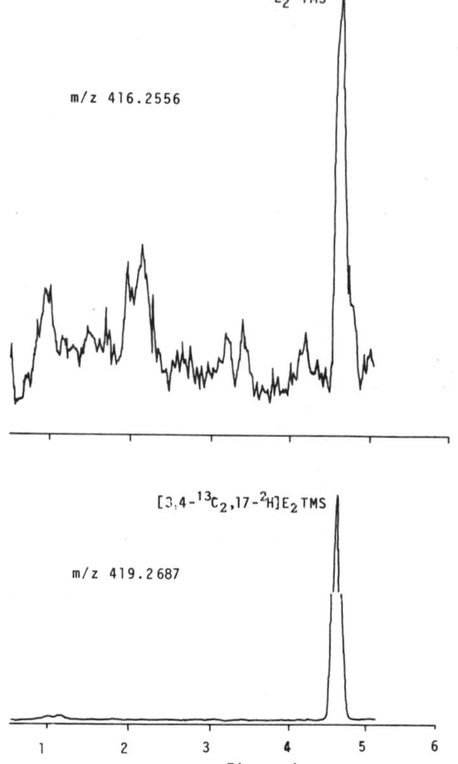

Fig. 2. GC-MS/SIM(EI) assay of a tumour cytosol extract for oestradiol-17β (E₂) as its TMS derivative (m/z 416.2556). For the ¹³C- and ²H-enriched i.s. *(lower portion)*, m/z is 419,2687. MS instrument: VG 7070 HS; MS resolution 2000 (10% valley). GC column: 12 m DB-1 (bonded phase) fused silica open-tubular, 235°. Splitless injection; 'falling needle' mode.

Acknowledgements

The continuing support of the Tenovus Organization is gratefully acknowledged. Additional financial assistance was received from the Medical Research Council and the Jane Hodge Foundation. We thank C.J. Collins for technical assistance.

References

1. Daniel, C.P., Gaskell, S.J., Bishop, H. & Nicholson, R.I. (1979) *J. Endocrin. 83*, 401-408.
2. Daniel, C.P., Gaskell, S.J., Bishop, H., Campbell, C. & Nicholson, R.I. (1981) *Eur. J. Cancer Clin. Oncol. 17*, 1183-1189.
3. Sjövall, J. & Axelson, M. (1979) *J. Steroid Biochem. 11*, 129-134.
4. Axelson, M. & Sjövall, J. (1979) *J. Chromatog. 186*, 725-732.
5. Nicholson, R.I., Syne, J.S., Daniel, C.P. & Griffiths, K. (1979) *Eur. J. Cancer 15*, 317-329.
6. Daniel, C.P., Gaskell, S.J. & Nicholson, R.I. (1984) *Eur. J. Clin. Oncol. 20*, 137-143.

#NC(D)

NOTES and COMMENTS relating to

Tamoxifen and other anti-cancer drugs

Comments related to particular contributions:

#D-1 to #D-5, p. 255; also #D-5, p. 256, & #D-4, p. 257

#NC(D)-1

A Note on

DETERMINATION OF TAMOXIFEN AND *N*-DESMETHYLTAMOXIFEN
IN PLASMA BY HPLC

D. Stevenson

Robens Institute of Health and Safety
University of Surrey, Guildford GU2 5XH, U.K.

Requirement | *Assay for tamoxifen and its desmethyl metabolite in plasma, sensitive down to 25 ng/ml.*

End-step | *Ion-pair HPLC on C-8 column with UV detection.*

Sample preparation | *The plasma aliquot (1.0 ml), made alkaline with ammonia (1 M, 1.0 ml), is extracted into 10 ml of diethyl ether. The residue from drying down the extract is redissolved in the mobile phase (0.1 ml) and a 20 µl aliquot is injected onto the HPLC column.*

Comments | *1. The HPLC run times are short (9 min), and sample preparation is simple. 2. HPLC separation does not depend on the conventional ion-pairing mechanism. 3. The method allows determination of plasma levels in patients undergoing tamoxifen therapy.*

Tamoxifen is an anti-oestrogenic agent widely used in the treatment of breast cancer. Dose levels are typically 20 mg/day, sometimes after an initial loading dose of 40 mg. Two main metabolites are observed, *N*-desmethyltamoxifen and 4-hydroxytamoxifen. (For formulae, see #D-2.) The stated dosage gives tamoxifen and *N*-desmethyltamoxifen levels of 150–250 ng/ml plasma, and 4-hydroxytamoxifen levels of <8 ng/ml [1]. This Note describes an HPLC method for the determination of tamoxifen and desmethyltamoxifen down to 25 ng/ml plasma. 4-Hydroxytamoxifen would need a more sensitive method (cf. #D-4).

Several columns were evaluated, with the following results.-
ODS: too lipophilic; peaks well retained even with conditions chosen to promote ionization.
Zorbax CN: ion suppression gave good separation of tamoxifen and its metabolite, but peaks very broad.

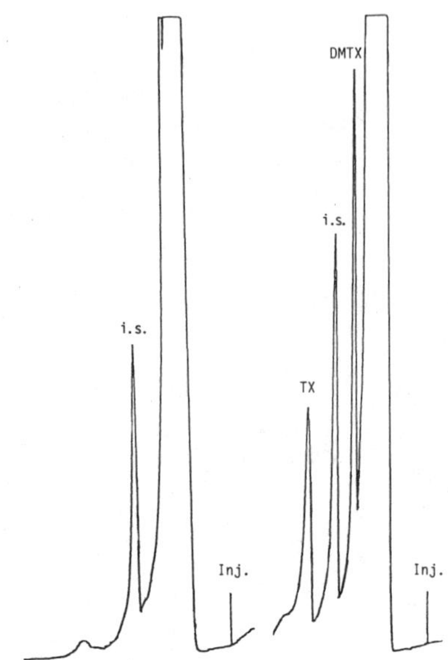

Fig. 1. HPLC analysis of extract from plasma spiked with tamoxifen (TX) and des-methyltamoxifen (DMTX), each 250 ng/ml plasma. The internal standard (i.s.) was 'ICI 129817', a structural analogue of TX. For load sample, see text.
Column: 5 μm C-8 silica; length 20 cm, i.d. 4.6 mm.
Eluent: 50 mM heptane sulphonate in methanol containing 20% v/v 0.1 M phosphate buffer, pH 3; flow-rate 1.5 ml/min.
UV detector setting: 240 nm.
Typical retention times (min): DMTX, 5.4; i.s., 6.4; TX, 7.9.

C-5 silica: ion pairing gave good peak shape and near-baseline resolution of TX and DMTX.
C-8 silica: ion pairing gave good peak shape and baseline resolution of TX, DMTX and 4OH-TX; this packing was adopted as the best choice.

For sample preparation a single ether extraction [1] proved to be suitable. The legend to Fig. 1, which shows the peak pattern, specifies the ion-pairing system that proved effective, the conditions for HPLC operation, and the chosen i.s. - a compound similar in t_R to 4-hydroxytamoxifen; but this is undetectable and hence assay precision is unaffected. The method has been used to monitor levels of tamoxifen and desmethyltamoxifen in patients undergoing tamoxifen therapy.

The mechanism of the HPLC separation is unconventional. The addition of a pairing-ion (e.g. heptane sulphonate) to an aqueous mobile phase usually increases t_R for basic solutes, whereas under the above conditions t_R decreased. Although not easily explained, this phenomenon proves useful when chromatographing compounds such as tamoxifen, since besides shortened t_R's peak shape is improved. A similar effect has been observed when chromatographing chlorpromazine and metabolites on a silica column [2] (mentioned in Vol. 12, #E - *Ed.*).

References

1. Golander, Y. & Sternson, L.A. (1980) *J. Chromatog. 181,*
2. Stevenson, D & Reid, E. (1981) *Anal. Lett. 14B,*

#NC(D)-2

A Note on

IDENTIFICATION OF ADRIAMYCIN AND ITS METABOLITES
IN HUMAN AND ANIMAL TISSUE AND BLOOD

J. Cummings, J.F.B. Stuart, *C.S. McArdle and K.C. Calman

Department of Clinical Oncology   *Department of Surgery
University of Glasgow   Phase 1 Glasgow Royal
1 Horselethill Road   Infirmary
Glasgow G12 9LY, U.K.   Glasgow G4 0SF, U.K.

Adriamycin, an anthracycline antibiotic with a broad spectrum of anti-tumour activity, is metabolized to at least 8 different products in man, all of which have been extracted from patient urine and identified using TLC [1]. These are adriamycinol (AOL), the major metabolite; 5 non-polar aglycones of adriamycin (ADR) and AOL, and two conjugated aglycones. The kinetics of several of the aglycones has been followed in 14 patients using TLC [2].

Recently, simplified HPLC methods coupled with sensitive spectrofluorimetric detection have been reported for the separation and quantitation of standard mixtures of ADR and up to 8 metabolites [3-6]. When applied to the analysis of serum or plasma from patients receiving ADR, various extraction techniques have been employed: liquid-solid extraction with C-18 material in cartridges [7], acid/alcohol extraction [6] and ammonium sulphate precipitation with chloroform/propan-2-ol extraction [5]. With all these extraction methods only one metabolite, AOL, was seen in measurable amount.

A new simple extraction procedure for plasma or serum which can also be applied to tissues, and a new rapid HPLC method, as described below, can detect aglycone metabolites as well as AOL.

HPLC: performed isocratically using a C-18 μ-Bondapak column (250 x i.d. 4.6 mm) at 2 ml/min flow-rate with either phosphoric acid (5 mM)/propan-2-ol, 80:20 by vol., pH 3.2, or phosphoric acid (5 mM) /acetonitrile/methanol/propan-2-ol, 62.5:15:15:7.5 by vol., pH 3.2. Fluorescence was measured with 480 nm excitation and 560 nm emission. The method separates ADR and 5 metabolite standards: AOL and 4 aglycones.

Rapid extraction from plasma or serum: achieved by direct mixing with 5 vol. chloroform/propan-2-ol (2:1), only briefly. The lower organic phase obtained by centrifugation (1000 g, 15 min) is evaporated to dryness, ready for HPLC. Extraction efficiency for all 6 analytes is 85 ±5%. Daunorubicin serves as i.s. ADR, AOL, ADR- & AOL-7-deoxyaglycone have thereby all been identified in patient sera. Most tissues (including human tumour biopsy) are amenable to the rapid extraction (60-90% efficiency expected) provided that the tissue is well homogenized. However, we reckon that only free and reversibly bound drug and metabolites are extracted, not covalently bound.

Extraction of bound ADR and metabolites from tissue: based on the method of Schwartz [8] who showed that $AgNO_3$ released ADR bound to DNA and RNA. Homogenized tissue is treated with 33% $AgNO_3$ (0.2 ml per ml homogenate; 10 min, 4°) with very rapid mixing. Immediately, 5 vol. of the above solvent mixture is added and the rapid extraction procedure applied after a further 30 min of vigorous shaking. Extraction efficiency was 60-90% from the Ridgway osteogenic sarcoma (AKR mice, s.c. growth; 5 mg/kg ADR given i.v.). We found ADR, AOL, AOL- & ADR-7-deoxyaglycone, and an unidentified metabolite whose peak was close to ADR and thus was thought not to be an aglycone. Recoveries compared with the rapid method alone were 5-fold greater for ADR and 15-fold for AOL, but no different for the aglycones which, it is inferred, do not bind to DNA or RNA. Human tissue homogenization: by Potter-Elvehjem, 5 strokes: for liver biopsy (and animal tissues), and tumour biopsy but with preceding Ultra Turrax treatment (30 sec).

Human biopsies (gastric and breast carcinomas) were collected 30 min after trace-dosage with ADR i.v. at operation, and analyzed using the silver and rapid methods. Only ADR was present, and ~80% appeared to be DNA/RNA-bound as judged by recovery comparisons.

References

1. Takanshi, S. & Bachur, N.R. (1976) *Drug Metab. Disp.* 4, 79-87.
2. Benjamin, R.S., Riggs, C.E. & Bacchur, N.R. (1977) *Cancer Res.* 37, 1416-1420.
3. Israel, M., Pegg, W.J., Wilkinson, P.M. & Garnick, M.B. (1978) *J. Liq. Chromatog.* 1, 795-809.
4. Pierce, R.N. & Jatlow, P.I. (1979) *J. Chromatog.* 164, 471-478.
5. Andrews, P.A., Brenner, D.E., Chou, F.T.E., Kubo, H. & Bachur, N.R. (1980) *Drug Met. Disp.* 8, 152-156.
6. Bolanowska, W., Gessner, T. & Preisler, H. (1983) *Cancer Chemother. Pharmacol.* 10, 187-191.
7. Robert, J., Iliadis, A., Hoerni, B., Cano, J.P., Durand, M. & Lagarde, C. (1982) *Eur. J. Cancer Clin. Oncol.* 18, 739-745.
8. Schwartz, H.S. (1973) *Biochem. Med.* 7, 396-404.

Nomenclature note: adriamycin has now been re-named doxorubicin – *Ed.*

#NC(D)-3

A Note on

ANTIBODIES TO DEOXYCYTIDINE TRIPHOSPHATE (dCTP)
SUITABLE FOR ITS RADIOIMMUNOASSAY

E. Piall, G.W. Aherne and V. Marks

Department of Biochemistry
University of Surrey, Guildford GU2 5XH, U.K.

Deoxyribonucleoside triphosphate (dNTP) concentrations in cancer cells are of particular interest because of their regulation of DNA synthesis and repair. Alterations in concentration following anti-metabolite treatment are also of interest in connection with understanding the drug's mode of action and mechanisms of resistance.

Well-established methods of measuring dNTPs include microbiological assay, isotope dilution in intact cells, HPLC and enzymatic assay [1]. These methods all suffer from some disadvantage such as lack of sensitivity, extreme tediousness or difficulty of performance. The development of a specific and sensitive RIA for dCTP was undertaken as a simple alternative to these methods, and one which could be applied to other dNTPs.

dCTP is thought to be present in elevated concentrations in leukaemic cells of some patients resistant to cytosine arabinoside (Ara.c.), an important drug in the treatment of acute myelogenous leukaemia. Ara.c., an analogue of 2'-deoxycytidine, is cytotoxically active when phosphorylated to cytosine arabinoside triphosphate. Resistance to ara.c. is a frequent occurrence, and altered enzyme levels [2, 3] and raised dCTP concentrations [4, 5] are possible causes shown in cell lines. dCTP exerts inhibitory feedback on pyrimidine phosphokinases as well as stimulating dCTP deaminase, thereby blocking phosphorylation of ara.c. and inactivating - by deamination - any ara.c. monophosphate already formed to uracil arabinoside monophosphate [6]. To date, dCTP concentrations have not been measured in leukaemic cells from patients sensitive and resistant to ara.c.

Development of RIA. - The radiolabel used was [5-^{3}H]dCTP (New England Nuclear Corpn.). Standard compounds were obtained from Sigma. Antibody was raised in a New Zealand White rabbit to a conjugate of

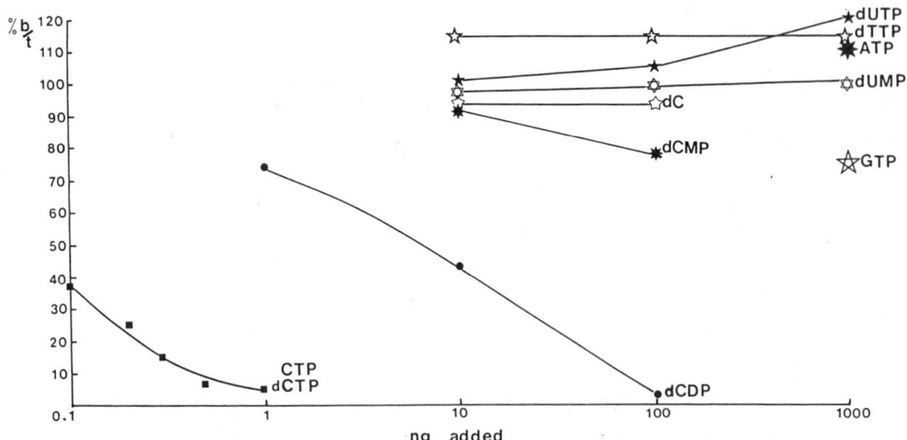

Fig. 1. Cross-reactivity of antibody with dCTP and structurally similar endogenous compounds. b = bound, t = total radioactivity.

dCTP and ovalbumin, made using a carbodiimide as coupling reagent [7]. The assay diluent was ammonium tetraborate, 0.1 M, pH 6.2. The assay was done with all reagents kept in ice. A 2 h incubation at 4° was found optimal. After phase separation by dextran-coated charcoal liquid scintillation counting was done on aliquots of the supernatant.

Using these conditions, the affinity constant (K) of the antibody for dCTP was found by a Scatchard plot to be 7.083×10^{10} 1/mol. Cross-reactivity (Fig. 1) was 100% with dCTP and CTP, 4% with dCDP and CDP, and <1% with the other endogenous compounds shown. Sensitivity of the assay was 4.2 pM, and the C.V.s between and within batches (reflecting reproducibility) were below 12% over the range 30-300 pM (10-100 pg added). Checking of overall recovery will include extraction of nucleotides from blast cells and a simple chromatographic separation of CTP from dCTP.

References

1. Hunting, D. & Henderson, J.F., (1982) in *Methods in Cancer Research*, Vol. 20 (Busch, H. & Yeoman, L.C., eds.), Academic Press, New York, pp. 245-289.
2. Chu, M.Y. & Fischer, G.A. (1964) *Biochem. Pharmacol. 14*, 333-341.
3. Coleman, C.N., Stoller, R.G., Drake, J.C. & Chabner, B.A. (1975) *Blood 46*, 791-803.
4. Tattersall, M.H.N., Ganeshagaru, K. & Hoffbrand, A.V. (1974) *Br. J. Haematol. 27*, 39-46.
5. de Saint Vincent, B.R., Dechamps, M. & Buttin, G. (1980) *J. Biol. Chem. 255*, 162-167.
6. Chabner, B., Hande, K.R. & Drake, J.C. (1979) *Bull. du Cancer, Paris 66*, 373-378.
7. Halloran, M.J. & Parker, C.W. (1966) *J. Immunol. 96*, 373-378.

#NC(D)-4

A Note on

COMBINED RIA/HPLC OF METHOTREXATE AND ITS METABOLITES

G.W. Aherne, D.J. Lawson, M. Quinton and V. Marks

Department of Biochemistry
Univesity of Surrey
Guildford GU2 5XH, U.K.

At the present time methotrexate (MTX) is probably the only cytotoxic drug for which therapeutic drug monitoring is important. Radioimmunoassay (RIA), HPLC and enzyme inhibition assays have all been used for this purpose [cf. arts in this vol. by S.H. Curry, #E-3, and J.W. Paxton, #NC(D)-5]. Comparative studies have shown that immunoassay techniques tend to slightly overestimate plasma MTX concentrations especially at late time points [1, 2], although in another study [3] it was concluded that for monitoring purposes any of the techniques was suitable.

In this laboratory, the concentration of MTX in the plasma of patients receiving MTX in high dosage (1-4 g/m^2) has been measured by RIA [4] and HPLC [5]. Whilst RIA gave consistently higher values overall, a close correlation between the methods was found (n = 52, r = 0.99; y = 0.77x + 0.99 where y values = HPLC and x values = RIA). The values obtained by RIA could be affected by the presence of cross-reacting metabolites. 7-Hydroxy-MTX, a major metabolite of MTX in humans, cross-reacts by <1% in the RIA and has been measured in plasma samples using HPLC. Fig. 1 shows the concentration range of both MTX and 7-OH-MTX in plasma samples from 4 patients receiving 3-4 g/m^2 MTX (4.8-6.4 g over 6 h). By 6-24 h after the end of drug infusion the concentration of metabolite exceed that of MTX itself.

Chromatographically pure 2,4-diamino- N^{10}-methylpteroic acid (DAMPA) cross-reacts in the RIA by 83%. DAMPA has been shown to be a metabolite of MTX in some animal species and has been detected in human plasma following MTX in high dosage [1, 2]. Combined RIA/HPLC was therefore used in an attempt to detect and quantitate DAMPA in plasma following MTX administration in man. HPLC was performed on 7 plasma samples collected from patients 10-40 h following high-dose therapy. Fractions (2 ml) of HPLC eluate were collected and each fraction assayed by RIA. Immunoreactivity was detected in those

Fig. 1. Ranges of concentration
of MTX and 7-OH-MTX in plasma,
following MTX in high dosage,
measured by HPLC with assay of
fractions by RIA.

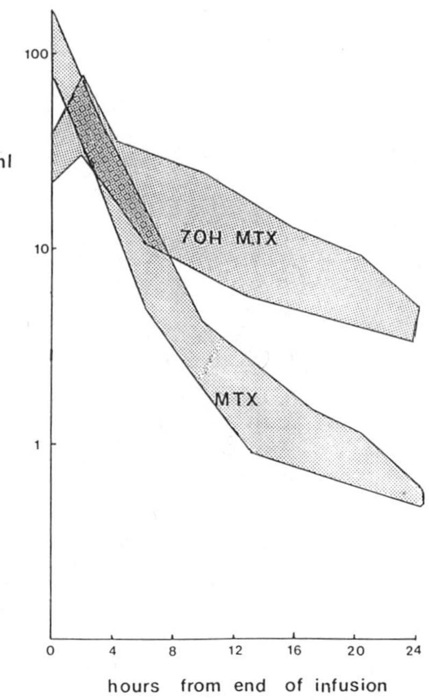

fractions corresponding in retention to DAMPA standard chromatographed under identical conditions.

In these samples at least, the amount of DAMPA present (if any)
would have had little effect on the RIA results. The relatively
high concentrations of 7-OH-MTX at late time points may be a source
of error even though the percentage cross-reaction is low.

Acknowledgement

G.W.A. and M.Q. are grateful to the Leukaemia Research Fund for
financial support.

References

1.   Donehower, R.C., Hande, K.R., Drake, J.C. & Chabner, B.A. (1979)
     Clin. Pharmacol. Ther. 26, 63–72.
2.   Howell, S.K. Wang, Y., Hosoya, R. & Sutow, W.W. (1980) *Clin.
     Chem. 26*,   734–737.
3.   Buice, R.G., Evans, W.E., Karas, J., Nicholas, C.A., Sidhu, P.,
     Straughn, A.B., Meyer, M.C. & Crom, W.R. (1980) *Clin. Chem. 26*,
     1902–1904.
4.   Aherne, G.W., Piall, E.M. & Marks, V. (1977) *Br. J. Cancer, 36*,
     608–617.
5.   Lawson, G.T., Dixon, P.F. & Aherne, G.W. (1981) *J. Chromatog.
     223*, 225–231.

\#NC(D)-5

A Note on

RADIOIMMUNOASSAY OF METHOTREXATE

J.W. Paxton

Department of Pharmacology and Clinical Pharmacology
University of Auckland School of Medicine
Auckland, New Zealand

The development of a radioimmunoassay (RIA) for methotrexate (MTX) proved to be straightforward. No major problems were encountered in the production of a suitable antiserum or radiolabelled drug.

PRODUCTION OF SUITABLE ANTISERUM

The MTX molecule does not require any modification before covalent linkage to a macromolecule to form an immunogen, since it contains two carboxyl groups. This allows direct formation of a peptide bond between either of these groups and an amino group on the carrier protein. The reagent used was the water-soluble carbodiimide 1-ethyl-3-(3-dimethylaminopropyl)carbodiimide, with direct conjugation of MTX to bovine serum albumin (BSA). The resulting immunogen was calculated to contain 34 MTX residues per BSA molecule [1]. This immunogen (0.25 mg in 2 ml emulsion containing saline, Tween 20 and Freund's adjuvant) was injected at various i.p., i.m. and s.c. sites in each of 5 rabbits. Booster injections were given at 4-weekly intervals, with harvesting of blood 10-14 days after each booster injection. One rabbit gave a consistently high titre allowing a final dilution of 1/4000 for 50% binding of 1.69 pmol ^{3}H-MTX [1].

PRODUCTION OF RADIOLABELLED DRUG

Three isotopically labelled drug derivatives, ^{3}H-MTX, ^{125}I-MTX and ^{75}Se-MTX, are available for this assay. The commercially available tritiated form, ^{3}H-MTX, is perhaps best for the initial testing for binding and the assessment of antisera. However, assays using β-emitting isotopes suffer from several disadvantages in that the overall cost (including the label, scintillation fluid and counting vials) is relatively high and the procedures of decanting, adding scintillation fluid and capping vials are time-consuming. Quenching and chemiluminescence may also cause problems during counting.

In contrast, the γ-emitting isotopes of MTX offer many advantages. In particular, ^{125}I-MTX may be synthesized very simply and at relatively low cost by radio-iodination of the *p*-hydroxyphenyl ring of the tyrosine methyl ester derivative of MTX, using Na[125]I and the chloramine-T reaction [2]. The MTX-tyrosine methyl ester was prepared using the same carbodiimide reaction as used for conjugation of MTX to BSA [3]. After purification, a high specific activity ^{125}I-MTX derivative was obtained which allowed counting times of <60 sec in the assay. One disadvantage of the ^{125}I label is its short half-life of 60 days, which means that fresh label must be synthesized at least every 2 months. A doubled shelf-life is obtainable by use of ^{75}Se which has a half-life of 121 days. This radio-labelled MTX derivative is prepared by conjugation of [^{75}Se]methyl-selenocysteine with 4-amino-4-deoxy-N[10]-methylpteroic acid using isobutyl chloroformate [3].

Apart from the rapidity of the counting of the γ-labelled derivatives, these labels are better suited to automation, as the supernatant can be decanted to waste, and the precipitated antibody-bound label counted in the incubation tubes directly.

Similar standard curves over the range 22-2200 fmol per tube were obtained for all three labels [3]. When using γ-labels, a 20-fold increase in the BSA concentration in the assay buffer was necessary to achieve acceptably low non-specific binding (viz. <5%).

SEPARATION OF ANTIBODY-BOUND AND FREE DRUG

A major disadvantage of RIA compared to enzyme-immunoassay or fluorescence polarization immunoassay systems is the necessity for a procedure to separate bound and free antigen. For the MTX assay several methods were examined, including absorption of free fraction by coated charcoal, and precipitation of bound fraction by polyethylene glycol or ammonium sulphate, or by a second antibody raised against rabbit IgG. The latter method, employing Wellcome's donkey anti-rabbit precipitating sera, was chosen after consideration of completeness of separation, the practicality, and whether affected by serum or plasma.

ASSAY PROTOCOL

Reagent **A** consists of ~1 nM labelled MTX plus anti-rabbit precipitating serum, final dilution 1/100. Reagent B contains anti-MTX serum plus normal rabbit serum, final dilutions 1/4000 and 1/1000 respectively. The steps are as follows:
- 200 μl each of reagents A and B are added to 100 μl of appropriately diluted serum sample or standard. Sorensen's phosphate buffer, 0.05 M, pH 7.4, containing 0.1% BSA (or 2% BSA for γ-emitting labels) and 0.1% sodium azide is the diluent for all assay reagents.
- After vortexing and incubating for 1 h at 4°, assay tubes are cent-

rifuged at 4° for 30 min at 2500 g.
- With tritium-labelled MTX, the supernatant is decanted into counting
vials, scintillation fluid added, and the vials counted for 4-10 min
in a β-counter. Otherwise the supernatant is aspirated to waste and
the incubation tubes containing the precipitated bound γ-labelled
drug are counted for <60 sec in a γ-counter.

CHARACTERISTICS OF THE ASSAY

A unique characteristic of most RIA methods is their very high
sensitivity. In our system, 50 fmol of MTX was sufficient to produce
significant displacement, which allowed concentrations as low as
1.6×10^{-10} M to be assayed in 250 µl of biological fluid. This sensitivity allowed MTX pharmacokinetic studies to be undertaken in small
sample volumes (<50 µl) in laboratory animals such as the rat [4].

The main disadvantage of RIA in pharmacokinetic study of MTX is
the possible interference of 4-amino-4-deoxy-N^{10}-methylpteroic acid
(APA) which shows 15% cross-reactivity with this antiserum [5]. This
compound has been identified in the urine and faeces of mice and rats
and also in some patients [6, 7]; however, significant amounts are
not observed in plasma even after high-dose therapy [Aherne et al.,
#NC(D)-4,   this vol.]. Other naturally occurring folates, and also
7-hydroxy-MTX, showed lesser degrees of cross-reactivity (Fig. 1).
It has been suggested that the latter metabolite may contribute to

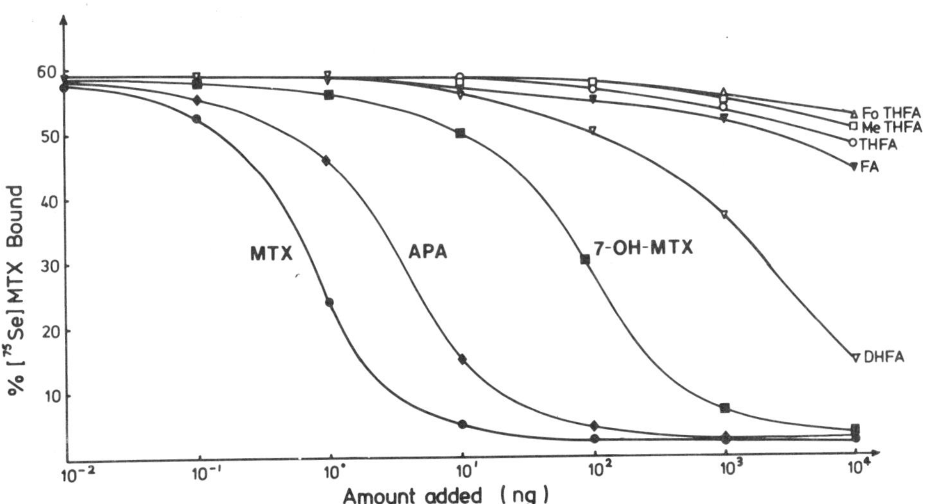

Fig. 1. Inhibition of antiserum binding of ^{75}Se-MTX by increasing
amounts of MTX, APA, 7-hydroxy-MTX, dihydrofolic acid (DHFA), folic
acid (FA), tetrahydrofolic acid (THFA), N^5-methyltetrahydrofolic
acid (MeTHFA) and 5-formyltetrahydrofolic acid (FoTHFA).

the nephrotoxicity of the parent drug [8], and monitoring of both
parent drug and 7-hydroxy- MTX may be clinically useful in patients
receiving high-dose therapy. In such situations, an HPLC method
which could separately quantitate both compounds would have considerable merit. However, the clinical significance of MTX metabolites
has not yet been adequately established.

CONCLUSIONS

The high sensitivity and throughput of RIA as applied to MTX
are offset by disadvantages including radioisotope health risk, the
need to separate bound and free drug, and equipment cost. These
disadvantages could be eliminated by a method based on an enzyme- or
fluorescent-labelled antigen. Where it is desirable to measure metabolites such as APA and 7-hydroxy-MTX, HPLC would appear to be the
method of choice.

References

1. Paxton, J.W. & Rowel, F.J. (1977) *Clin. Chim. Acta 80*, 563–
    572.
2. Paxton, J.W., Rowel, F.J. & Ratcliffe, J.G. (1977) *Clin. Chim.
    Acta 79*, 81–92.
3. Paxton, J.W., Rowel, F.J. & Cree, G.M. (1978) *Clin. Chem. 24*,
    1514–1538.
4. Paxton, J.W. (1979) *J. Pharm. Meth. 2*, 203–212.
5. Paxton, J.W. (1979) *Clin. Chem. 25*, 491–492.
6. Valerino, D.M. (1972) *Res. Comm. Chem. Path. Pharmacol. 4*,
    529–542.
7. Howell, S.A., Wang, Y-M., Hosoya, R. & Sutow, W.W. (1980) *Clin.
    Chem. 26*, 734–737.
8. Jacobs, S.A., Stoller, R.G., Chabner, B.A. & Johns, D.G. (1976)
    J. Clin. Invest. 57, 534–538.

Comments on material in #D

Comments on #D-1, J.W. Paxton – AMSACRINE ASSAY

L.A. Sternson *commented:* AMSA may have fluorescent properties which could represent an alternative means of HPLC detection. *Reply to a question:* the assay has not so far been applied to metabolism studies (J.W. Paxton).

Comments on #D-2 to -5, H.K. Adam; V.C. Jordan et al.;
L.A. Sternson; H.M. Leith et al.- TAMOXIFEN

H.K. Adam, *answering* G.E. von Unruh: In our studies with radio-labelled tamoxifen (#D-2), the ^{14}C was in the phenyl group attached to the same carbon as that carrying the basic side-chain. *Reply to* A.P. Beresford: each metabolite has some bioactivity, but 4-hydroxy-tamoxifen is the most active (H.K. Adam). Metabolite X (present in serum at twice the concentration of the parent drug) may be important for anti-cancer activity since in high concentration it has a very specific action, surpassing tamoxifen, on breast cancer cells.- V.C. Jordan (cf. #D-3), *who also replied to* G.P. Mould: we are trying to obtain an animal model to investigate the possibility that drugs used in treating the patients may affect the metabolic pattern of tamoxifen. The patients, whose selection takes into account their medical status, tend to be pre-menopausal (*reply to* B.S. Thomas). H.J. Eggar (*remark to* V.C. Jordan): taking account of work by Henschler (Würzburg) on the toxicity of stilbenes, it would be of interest to know whether tamoxifen gives rise to reactive intermediates such as epoxides. In fact (V.C. Jordan) tamoxifen is remarkably non-toxic; we have not observed any retinopathy with normal-dose treatment, but any teratogenicity is difficult to demonstrate because the drug inhibits implantation in laboratory animals.

Questions by W.J. Ellis *and* D. Newell.- Is there any particular reason for assaying whole blood rather than plasma? or any difference in correlation with patient response? H.K. Adam, *in reply* (L.A. Sternson did not have relevant data).- It seems not to matter which is assayed, although erythrocytes bind the drug strongly. *In reply to* Z.H. Siddik.- Urinary excretion of tamoxifen and metabolites is very small. Plasma methodology generally applies to bile, which is the drug's main excretion route, largely as conjugated metabolites. *Replies to* I.S. Krull (L.A. Sternson, H.M. Leith).- Spiking with labelled drug and metabolites has confirmed that there is good recovery from tissues or from patients' plasma (>75%).

EDITOR'S INTERPOLATION
*Trial of a radioreceptor approach for assay of tamoxifen in blood
(pers. comm. to Senior Editor: B. Scales and colleagues)*

One approach tried in the early 1970's at ICI Pharmaceuticals
for assaying tamoxifen and its metabolites hinged on competition
with oestrogens for a specific oestrogen receptor present in a
protein fraction from rabbit uterus. The method was a modification
of a published assay [1] for plasma oestradiol, and entailed incubation of the drug and [³H]oestradiol with the fraction; unbound
radioligand was removed by use of dextran-charcoal, and the bound
label in the supernatant assayed by liquid scintillation counting.

Comparable calibration curves, for concentrations up to 10 ng/
ml, with receptor preparations differing in activity (e.g. through
a fall-off with storage), were obtained by use of a normalization
procedure: all measurements were related to a 100% value, viz. the
amount of [³H]oestradiol bound to uterine protein in the absence of
any competitor. The metabolite 4-hydroxytamoxifen was much more
active than the parent drug, especially if isolated (as glucuronide)
from dog bile rather than chemically synthesized (which gives a 1:1
cis-trans mixture). The receptor-based approach to tamoxifen assay
was not pursued in view of encouraging progress with chromatographic
approaches (cf. #D-2, by H.K. Adam, this vol.). In pilot studies with
tamoxifen-spiked serum, 10 ng/ml was barely detectable; there is a
significant blank, although naturally occurring oestrogens are eliminated by the extraction procedure. The latter entailed a cyclohexane
extraction at alkaline pH, back-extraction into 0.1 M HCl, extraction
into benzene, and drying down; its efficacy for metabolites was not
checked.

An unexplained phenomenon was the higher extraction efficiency
(attained after 1 h) with alkaline plasma or serum compared with
water (cf. #D-2). It was mainly in the early work that adsorptive
losses were encountered [cf. a query by R. Schmid at the Forum concerning such losses when extracts were blown dry - *Ed.*].

1. Korenman, S.G., Perrin, L.E. & McCallum, T.P. (1969) *J. Clin.
   Endocrin. 29*, 879-893.

In recovery experiments *(discussion continued from previous p.)*
no differences were seen between tissue samples giving a positive
result in the oestrogen receptor assay and those giving a negative
result (H.M. Leith, *answering* D. Newell; cf. #D-5). *In reply to*
H. de Bree: there is indeed a structural similarity between tamoxifen and diethylstilboestrol (DES; a cancer promoter), such that bonding occurs to the oestradiol receptor; but the extra ring prevents
further unwanted actions in the target cell. *Added by* V.C. Jordan.-
The mechanism of action of tamoxifen and other anti-oestrogens is

dependent upon the aminoethoxy chain extending away from the stilbene-like portion of the molecule. It is this side-chain that causes the anti-oestrogenic actions of tamoxifen when bound to the oestrogen receptor. Absence of the side-chain leads to oestrogenic effects.

Questions and replies on photochemical 'derivatization' (L.A. Sternson, #D-4).- A non-chemist (Z.H. Siddik) wondered why, taking account of the phenanthrene formation resulting from UV irradiation of tamoxifen, a similar bridge cannot be formed between the two phenyl rings of the same C atom in tamoxifen. *Reply.*- The close proximity of the phenyls differing in C-attachment aids bridge formation that leads to phenanthrene formation, whereas the distance between the two rings on the same C-atom is greater, ~1.4 Å, and this in fact hinders bridge formation. *Query by* A.P. Beresford: In view of the thermal and photolytic instability, exploited for the assay, is it risky to subject the compounds to pre-treatments such as incubation at 37° or a 2-h equilibration with i.s.? *Reply.*- Tamoxifen and its *N*-desmethyl metabolite are reasonably stable, but not the hydroxy metabolite; as a precaution we store samples in the refrigerator in wrapped vials. *Comment by* H. J. Eggar.- Convincing information on the validity of the assay would be welcomed, e.g. concerning the blank, sensitivity, accuracy and precision.   H. M. Leith, *replying to* W.J. Ellis: the GC-MS assay can detect 1 ng/ml.

Supplementary citations (by Editor) on anti-cancer drugs

Tamoxifen.- In a useful survey [1] that overlaps with #D-5, points mentioned by S.J. Gaskell et al. include assay of patient plasmas for diethylstilboestrol (also testosterone) by RIA, tallying well with GC-MS(SIM) assay which involved measuring molecular ions from derivatized analyte; derivatization was performed for *N*-des-methyltamoxifen and metabolite B but not for tamoxifen itself. HPLC-UV sufficed for routine assay of plasma containing tamoxifen and metabolites, but GC with thermionic detection (N- & P-specific) was also feasible. The citation of [1] below includes a Corrigendum. Related to ref. [10] in the foregoing article by H.K. Adam (#D-2), there is a paper in a later issue of the journal (pp. 2045-2052; same authorship etc.) titled "Identification and biological activity of tamoxifen metabolites in human serum".

Daunorubicin and doxorubicin (daunomycin and adriamycin).- Normal and neoplastic tissues have been assayed, with ~95% recovery, by a simple extraction procedure (non-degradative) followed by HPLC with fluorescence detection [2]. *Daunorubicin   and   metabolites* (the aglycone - not seen in a patient's plasma - and daunorubicinol) have been assayed in plasma and other biological fluids with no ext-raction other than use of a loop column linked to the HPLC column; oxidative EC detection surpassed fluorescence in sensitivity (10 ng/ ml) [3]. *(Note altered nomenclature; cf. #NC(D)-2 - Ed.)* The anti-

cancer drug bisantrene (ADCA) is likewise amenable to HPLC–EC (C. Riley
et al.), also to HPLC–UV [Y-M. Deng et al. (1981) *Life Sci.* *29*, 361-369].

   A tricyclic nucleoside 5'-phosphate.- The anti–cancer drug TCNP
(structural features of which include a primary amino group), and
TCN as formed in plasma by dephosphorylation, can be assayed in
blood, urine and bile from rabbits,   by RP–HPLC with UV detection
using a gradient; alternatively, for TCNP in urine, anion–exchange
HPLC can be performed [4].   Sample pre-treatment in the case of
blood consisted of deproteinization by perchloric acid, extraction
onto ODS-silica, and elution by methanol. *Celiptium*, an elliptici-
nium derivative, gives rise to urinary conjugates (GSH, *0*-glucuro-
nide, cysteine and *N*-acetylcysteine) which have been investigated by
RP-HPLC [5]; high water-solubility precluded solvent extraction.

1. Gaskell, S.J., Daniel, C.P. & Nicholson, R.I. (1983) *Anal. Proc. 20*,
   34-35 *and Corrigendum* 137.
2. Straus, J.F., Kitchens, R.L., Patrizi, V.W. & Frenkel, E.P.
   (1980) *J. Chromatog. 221*, 139-144.
3. Akpofure, C., Riley, C.A., Sinkule, J.A. & Evans, W.E. (1982)
   J. Chromatog. 232, 377-383.
4. Basseches, P.J., Durski, A. & Powis, G. (1982) *J. Chromatog.*
   233, 227-234.
5. Monsarrat, B., Maftouh, M., Meunier, G., Dugué, B., Bernadou, J.,
   Armand, J-P., Picard-Fraire, C., Meunier, B. & Paoletti, C.
   (1983) *Biochem. Pharmacol. 32*, 3887-3890.

   Methotrexate metabolites (cf. NC(D)-4 & -5; also NC(B)-2, Vol. 12*).* -
The conversion of MTX to 7-OH-MTX (inactive) and to poly-γ-glutamyl
derivatives (active) has been studied, with confirmatory assays on
plasma, in incubated cells (human acute lymphoblastic leukaemia cell
line) and the incubation medium [6]. Polyglutamate formation from
MTX was decreased by 7-OH-MTX, possibly acting on the membrane transport carrier.   The RP-HPLC assays, with TBAN as ion-pairing agent,
were performed after extracting the analytes (from centrifugal supernatants in the case of lysed cells) onto C-18 silica, eluting with
acetonitrile, and drying down.

6. Fabre, G., Cano, J.P., Catalin,J. & Just, S. (1983) *Int. J. Clin.*
   Pharm. Res. 3, 475-484.

   Further anti-cancer drug refs.- 6-Mercaptopurine, HPLC, even 5 ng
/ml: acetonitrile protein-free supernatant washed with dichloromethane and dried down [7]. Chloroethylnitrosoureas (lipophilic): ether-
extract, and decompose (base-catalyzed) to *0*-methylcarbamates for GC-
AFID assay [8].

7. Narang, P.K., Yeager, R.L. & Chatterji, D.C. (1982) *J. Chromatog.*
   230, 373-380.
8. Weinkmam, R.J. & Liu, T-Y. (1982) *J. Pharm. Sci. 71*, 153-157.

Section #E

LIGAND METHODS FOR DRUGS IN FORENSIC AND OTHER CONTEXTS

#E-1

RADIOIMMUNOASSAY IN FORENSIC SCIENCE

R.N. Smith

Metropolitan Police Forensic Science Laboratory
109 Lambeth Road, London SE1 7LP, U.K.

The use of radioimmunoassay (RIA) for drug analysis in the Metropolitan Police Forensic Science Laboratory (MPFSL) is here described. The particular advantages of RIA in forensic toxicology are outlined, and the assays in routine use are described with brief details including sources of antisera and radiolabelled drugs. The advantages and disadvantages of ^3H and ^{125}I radiolabels are discussed, and the use of disposable silica cartridges for purifying radiolabelled drugs is described. Finally, the likely advantages of monoclonal antibodies in forensic RIA are considered.

The only routine application of RIA in British forensic science laboratories at present is the analysis of drugs in body fluids. The most frequently analyzed samples are blood and urine; but bile, stomach contents, ocular fluid, saliva, tissue extracts or syringe washings may also be encountered [cf. R.L. Williams, #F-4 – *Ed.*]

We first used RIA in the MPFSL in 1976 when a mere 43 samples were analyzed, mainly for opiates. The number of assays that we have available has since increased and so have our sample numbers, to the point where virtually every body fluid submitted for drug analysis is examined by RIA as well as by other techniques. Our present case-load is 800-1000 samples per year, about one-quarter of which are taken under the Road Transport Act and subsequently found to contain less than the maximum legal alcohol content. The other samples are from cases of murder, suicide, poisoning, rape, etc., in which drug use may be a contributing factor.

On average, each sample is analyzed by three different RIAs and one or more positive results are obtained in about two-thirds of our cases. About one-third of the cases examined for cannabinoids and more than half of those examined for benzodiazepine tranquillizers prove to be positive.

ADVANTAGES OF RIA

Our case-load is small compared with that of a clinical chemistry laboratory, but our use of RIA on a limited scale is justified by its particular advantages for our work. The size and quality of the samples that we receive are dictated by circumstances outside our control; hence any routine method of analysis must be able to cope with the worst of our samples, exemplified by 2-3 ml of haemolysed, decomposing, semi-solid blood from an unidentified body for which no obvious cause of death has been established. Decomposition products, co-extractives and plasticizers in such a sample have relatively little effect on a typical RIA, whereas they are likely to cause gross interference in a routine chromatographic drug-screen.

A second advantage of RIA in analyzing this type of sample is sensitivity. Only 50-75 μl of sample are required for a complete assay, and so several assays can be done on a small sample and yet leave enough for positive results to be confirmed by alternative means. Much smaller sample volumes could be used in most assays if the detection of therapeutic or toxic levels of drugs were the only consideration; but in forensic toxicology the absence of a drug may be as significant as its presence, and so the assay conditions are a compromise between sensitivity and sample availability.

Another important feature of RIA in forensic toxicology is group specificity: an antiserum may cross-react with a number of drugs having structural features in common. This is in contrast to the clinical application of RIA where specificity for a single compound is desirable [cf. S.H. Curry, #E-3 -*Ed*.]. Advantageously, a negative result excludes a range of drugs from further investigation while a positive result indicates the class of compound present and gives an approximate estimate of the amount, so simplifying the choice of method for confirming the RIA result. More specific assays are also useful in forensic work; yet some degree of cross-reaction with major metabolites is advantageous since it increases the assay sensitivity.

Additional advantages of RIA in drug-screening are speed, reliability and the ease with which samples may be analyzed in batches. Economy is also important. The running cost (excluding labour) of an 'in-house' RIA is only a few pence per tube, whereas the cost of an equivalent commercial RIA may be £1 or more per tube. Were we to use commercially available RIA kits wherever possible in accordance with the manufacturers' instructions, our expenditure would increase by ~£15,000 p.a.

PARTICULAR ASSAYS

The following assays are in current use in the MPFSL.-
Routine: amphetamine, barbiturates, benzodiazepines (2 assays), benzoylecgonine, cannabinoids (2 assays), LSD, methadone (2 assays), opiates, tricyclic antidepressants.

Non-routine: digoxin, fentanyl, haloperidol, insulin, methaqualone, paraquat, phenyclidine, phenytoin, ricin.

The assays in routine use are kept in stock; some of the radiolabelled drugs are provided by courtesy of the Central Research Establishment (CRE) of the Home Office Forensic Service while others are prepared in our laboratory or bought from Amersham International plc. Antisera are bought from commercial sources or obtained as gifts. The non-routine assays, which are needed infrequently, are purchased as complete kits or otherwise acquired when necessary (see later).

The *amphetamine* RIA is the only routinely-used assay that we buy as a complete kit (Amphetamine Abuscreen; Roche Products Ltd., Welwyn Garden City, Herts.). Abuscreen kits are currently marketed with a second antibody rather than an ammonium sulphate precipitation of the bound fraction; so we no longer modify the protocol as previously reported [1]. Instead we use 25 µl of standard or 1:2 dilution of sample in phosphate buffer (0.067 M, pH 7.4, containing 0.2% bovine γ-globulin and 0.1% sodium azide) with one-quarter of the recommended reagent volumes; 400 assay tubes may thus be readily obtained from a 100-tube kit. The dose-response curve is sensitive enough for the assay to be applied to blood as well as urine (50% displacement of label by ~40 ng/ml unlabelled amphetamine), and the kit may be used for a period of weeks following the stated expiry date.

It is difficult to define a 'cut-off' value for the assay since β-phenethylamine, a matrix decomposition product, cross-reacts with the antiserum: the cross-reactivity is small and less than that of alternative amphetamine assays that we have tried, but high concentrations (mg/ml) of β-phenethylamine in putrid samples result in significant false positives. Fresh or preserved blood samples containing no amphetamine give an apparent response of ~0-10 ng/ml for the drug, and so it is prudent to consider a result >10 ng/ml as a possible positive and to examine the sample by alternative means. Positive results are thus of limited value, but time and effort are saved by eliminating negative samples from further investigation.

The *barbiturate* assay was developed by Mason et al. [2]. The label is 3-[^{125}I]iodo-4-hydroxyphenobarbitone. It is used with a 1:30 dilution of the antiserum from an EMIT-dau barbiturate enzyme immunoassay kit (Syva UK, Maidenhead). Commonly prescribed barbiturates and their metabolites cross-react well in the assay. Blood samples must be diluted 10-fold to reduce interference by some samples that results in erroneously low values.

We screen for *benzodiazepine tranquillizers* with two assays [3, 4]. [^{3}H]Diazepam is used as the label for one assay in conjunction with an anti-diazepam serum (Guildhay Antisera, Biochemistry Dept., Univ. of Surrey, Guildford). Only diazepam and temazepam (3-

hydroxydiazepam), but not their metabolites, cross-react well, and so the assay can be used to measure either compound. The other assay uses [^{3}H]flunitrazepam as the label with the antiserum from an EMIT-tox serum benzodiazepine kit (Syva) at a dilution of ~1:350. Therapeutic or sub-therapeutic levels of numerous benzodiazepines can be detected by this assay. Our most recent development is the production of an anti-benzodiazepine serum in rabbits using the drug-enzyme conjugate from a number of EMIT-tox benzodiazepine kits as the immunogen. This convenient stratagem, first adopted by Kamel et al. [5] for preparing anti-phenytoin serum, is a simple and economical way of producing anti-drug sera.

Cocaine is monitored by assaying benzoylecgonine, the principal breakdown product of cocaine. The label is radioiodinated *p*-hydroxy-benzoylecgonine, and the antiserum from an EMIT-dau cocaine metabolite kit (Syva) is used at ~1:600 dilution [6].

Cannabis metabolites are analyzed by two assays. Samples are analyzed by one assay and positive results are checked by the other. One assay uses [^{3}H]tetrahydrocannabinol as the label with an anti-tetrahydrocannabinol serum (Guildhay Antisera) and is a modified version of the assay originally developed by Teale et al. [7]. The assay mixture contains 25% (v/v) methanol to solubilize the canna-binoids [8], and polyethylene glycol of mol. wt. 6000 (PEG-6000) is used to separate the free and bound fractions. The other assay [9] uses 2-[^{125}I]iodohistamine-labelled Δ^8-tetrahydrocannabinol-11-oic acid with a 1:100 dilution of the antiserum from an EMIT-dau cannabinoid kit (Syva).

The assay for *lysergic acid diethylamide, LSD* [10], has [^{3}H]-LSD as the label and two alternative antisera (provided by courtesy of the CRE). The antisera were prepared using two bovine serum albumin conjugates, one with the LSD linked to the protein via the indolic N atom and the other with the linkage via the amide side-chain. The use of two antisera directed against different portions of the LSD molecule reinforces confidence in the results. The assay was originally used on urine or serum, but we have found that whole blood may be assayed after deproteinization with 3 vol. of methanol. The methanolic extract is evaporated to dryness, and the residue taken up in the assay buffer. The recovery from spiked samples is ~80%.

For *methadone* we have developed two assays [11], one with a [^{3}H]methadone label and one with an ^{125}I-conjugate of 4-dimethyl-amino-2,2-diphenylpentanoic acid and tyrosine methyl ester. The antiserum from an EMIT-dau methadone kit (Syva) is used in both, at dilutions of ~1:30 and ~1:50 respectively. Pre-extraction of blood samples is required with the ^{3}H assay but not with the ^{125}I assay. Urine can be analyzed directly by either assay.

Fig. 1. Opiate cross-reactions.
B = bound activity.
B_0 = activity bound in absence
       of unlabelled drug.

1, levallorphan;
2, hydrocodone;
3, morphine-3-glucuronide;
4, dihydromorphine;
5, dihydrocodeine and
   morpholinylethylmorphine;
6, ethylmorphine;
7, codeine;
8, morphine;
9, codeine-6-glucuronide.

No cross-reaction with:
dextromethomorphan,
dextrorphan, etorphine,
morphine-*N*-oxide, myrophine,
nalorphine, oxycodone.

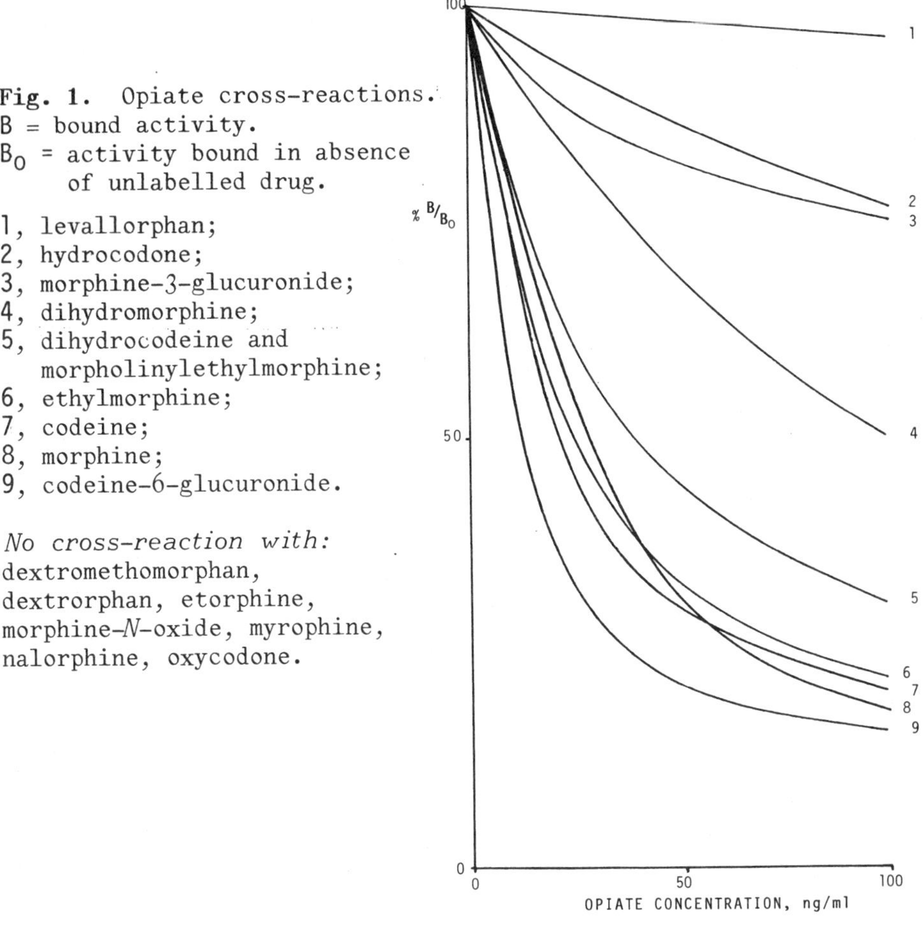

For *opiates* we initially used the morphine Abuscreen (Roche) in
a modified form [12].   More   recently,   we   devised a   simple   assay
using 2-[^{125}I]iodomorphine [13] and an anti-morphine serum (Guildhay
G/G/1).   Samples and reagents are all diluted with 0.067 M phosphate
buffer, pH 7.4, containing 0.2% bovine γ-globulin and 0.1% sodium
azide.   The samples (25 µl; diluted 1:2) or standards (25 µl; range
0–100 ng/ml morphine) are mixed with 50 µl each of label (10^4 cpm/
50 µl) and antiserum (diluted 1:750), and incubated for 1 h at room
temp.   PEG-6000 (23%, 400 µl) is then added, the tubes are centrifuged, the supernatants are aspirated, and the precipitates γ-counted.

Fig. 1 shows the dose–response curves for a number of opiates.
Morphine is used as the standard unless another opiate is suspected.
The 'cut-off' is 20 ng/ml for both blood and urine (mean response of
blank samples *plus* 3 standard deviations).

Tricyclic antidepressants are assayed with [³H]imipramine as
the label and an anti-nortriptyline serum (Guildhay) at 1:600 dilution. We have published brief details of the procedure [1].

Non-routine assays entail purchase of the following kits:-
digoxin & phenytoin, Amersham International; fentanyl & haloperidol,
IRE UK Ltd. (19 Castle St., High Wycombe); insulin, Pharmacia; methaqualone & phenycyclidine Abuscreens, Roche Products. Reagents for
the paraquat [14] and ricin [15] assays were kindly donated by Dr.
B. Widdop (Poisons Unit, New Cross Hospital, London) and Dr. A. Godal
(Norsk Hydro's Inst. for Cancer Res., Oslo) respectively. Most of
the non-routine assays are readily applicable to forensic samples,
but care must be taken in the interpretation of digoxin and insulin
results [1, 16, 17].

GENERAL COMMENTS

Our overall aim is to provide a comprehensive and reliable
assay service at minimum cost. The simplest and cheapest way to
devise an assay is to buy both an antiserum and a ³H-labelled drug.
The development cost is minimal. However, the running cost exceeds
that of an ¹²⁵I-labelled assay because liquid scintillation counting
is necessary. For example, the running costs, excluding labour, of
our methadone assays are 6.5 p and 3p per tube respectively. The
protocols of the two assays are compared in the following outline,
where the primary description is for the ³H-assay and procedural
differences in respect of the ¹²⁵I-assay are *italicized*.

SAMPLE PREPARATION.- **Blood** buffered to pH 9 is extracted twice with
diethyl ether, and residue obtained after evaporation is dissolved
in buffer; **urine** is assayed directly. *OR blood and urine are
each diluted 1:2 with buffer and assayed directly.*

PROTOCOL.- Sample as above, or standard (50 µl *or 25 µl*) + label
(50 µl) + antiserum (50 µl) are vortex-mixed.
After incubation at room temp. for 1 h *or 30 min*, 475 *or 400* µl of
23% (w/v) PEG-6000 are added; vortex-mix, and centrifuge for
2 min at 12,000 g.
Supernatant (400 µl) is transferred to a 5 ml liquid scintillation
tube; add 4 ml scintillant, shake 10 min, let phases separate, and
count. *OR Supernatant is aspirated, and precipitate counted.*
For TOTAL LABEL.- Label (50 µl) + buffer 100 µl are treated as
above and counted. *OR label (50 µl) is counted.*

An initial extraction of blood samples is necessary in a ³H-
labelled assay to prevent colour quenching by haemolysis products
and to eliminate non-specific binding of the label by blood proteins.
We also incorporate a second extraction in most of our ³H-labelled
assays to avoid the use of expensive 1,4-dioxane-based scintillants

that would otherwise be required to count the PEG-containing supernatants in a single-phase system. We use instead a low-cost scintillant consisting of 4 g/l 2,5-diphenyloxazole (PPO) and 0.2 g/l
1,4-bis[2-(5-phenyloxazolyl)]-benzene (POPOP) in reagent-grade
toluene. The ^{3}H-labelled drug in the supernatants is extracted into
the scintillant by shaking, and is counted after ~15 min when the
two phases have separated. The disadvantages of extra steps in the
protocol are obvious, although they are minimal with our relatively
small number of samples; but a ^{3}H-labelled assay can be a valuable
interim measure until an alternative has been developed. A ^{3}H-
labelled assay, however, has the advantage that the label may be
stored for long periods (^{3}H half-life = 12.3 years; degradation by
radiolysis of ^{3}H-labelled drugs is generally <10% p.a.), a useful
consideration in the case of infrequently used assays.

^{125}I-labelled assays undoubtedly have methodological advantages
for routine use, but they too have their disadvantages, notably the
difficulty of synthesizing a derivative of the drug that can be
radioiodinated while retaining the capacity to bind to the antibody.
A similar problem is encountered in the preparation of labelled
drugs for non-isotopic immunoassays in which the labels may be enzymes or fluorescent compounds. The relatively short half-life of
^{125}I (60 days) compared with ^{3}H is a minor disadvantage since, in
general, radioiodination of a suitable drug derivative is easily
carried out [18] and many radioiodinated drugs are usable for at
least two half-lives of the isotope.

A crucial step in the preparation of radioiodinated derivatives
is the separation of the derivative from unused radioiodide and other
reaction products. Chromatographic, ion-exchange or solvent extraction methods are generally employed for this purpose. Recently we
have found that silica Sep-Pak cartridges (Waters Associates) are
useful for isolating radioiodinated p-hydroxybenzoylecgonine and 2-
[^{125}I]iodomorphine. In each case, the radioiodination reaction mixture is applied to the dry cartridge and unreacted radioiodide is
eluted with acetone. Radioiodinated p-hydroxybenzoylecgonine is
eluted with methanol and 2-[^{125}I]iodomorphine is eluted with 25% dimethylformamide in methanol. The advantages of this procedure are
speed, simplicity, minimal exposure of the operator to radiation,
and the production of only a small volume of radioactive waste. Possible solvent combinations for a particular application can be tested
by TLC.

Non-isotopic immunoassays are viable alternatives to RIA in
clinical chemistry (cf. #E-3 -*Ed.*). In forensic science, however,
RIA remains the best option, because no alternative immunoassay so
far developed can match its sensitivity and versatility in analyzing
the samples encountered in forensic toxicology.

The antiserum is just as important as the label in an immunoassay. Alternatives to conventional polyclonal antisera such as tissue
receptors have been used, but it was not till the advent of monoclonal
antibody (MAB) technology that any significant advance in antibody
production occurred following the early pioneering work. To date,
however, MABs have been little used in small-molecule immunoassays.
A polyclonal antiserum contains families of antibodies directed
against the various antigenic determinants of the target molecule,
whereas the antibodies produced by a single clone are all identical
and directed against a single determinant. Thus MABs to a drug molecule are unlikely to be more specific than a polyclonal antiserum.
However, a lack of specificity can be advantageous in forensic RIA,
and MABs directed against a determinant common to all drugs of a
particular group could be invaluable in developing an assay with
broad specificity.

References

1. Smith, R.N. (1980) in *Forensic Toxicology* (Oliver, J.S., ed.),
    Croom Helm, London, pp. 34-47.
2. Mason, P.A., Law, B., Pocock, K. & Moffat, A.C. (1982) *Analyst*
    107, 629-633.
3. Rutterford, M.G. & Smith, R.N. (1980) *J. Pharm. Pharmacol. 32*,
    449-452.
4. Robinson, K., Rutterford, M.G. & Smith, R.N. (1980) *J. Pharm.
    Pharmacol. 32*, 773-777.
5. Kamel, R.S., Landon, J. & Smith, D.S. (1978) *Clin. Chim. Acta 84*,
    403-405.
6. Robinson, K. & Smith, R.N. (1984) *J. Pharm. Pharmacol. 36*,
    156-162.
7. Teale, J.D., Forman, E.J., King, L.J., Piall, E.M. & Marks, V.
    (1975) *J. Pharm. Pharmacol. 27*, 465-472.
8. Williams, P.L., Moffat, A.C. & King, L.J. (1978) *J. Chromatog.
    155*, 273-283.
9. Law, B., Mason, P.A., King, L.J. & Moffat, A.C. (1983) *J. Anal.
    Toxicol. 7*, in press.
10. Ratcliffe, W.A., Fletcher, S.M., Moffat, A.C., Ratcliffe, J.G.,
    Harland, W.A. & Levitt, T.E. (1977) *Clin. Chem. 23*, 169-174.
11. Robinson, K. & Smith, R.N. (1983) *J. Pharm. Pharmacol. 35*,
    566-570.
12. Smith, R.N. (1981) *J. Immunoassay 2*, 75-84.
13. Mason, P.A., Law, B. & Ardrey, R.E. (1981) *J. Labelled Comp.
    Radiopharm. 18*, 1497-1505.
14. Levitt, T. (1979) *Proc. Anal. Div. Chem. Soc. 16*, 72-76.
15. Godal, A., Olsnes, S. & Pihl, A. (1981) *J. Toxicol. Environ.
    Health 8*, 409-417.
16. Fletcher, S.M. (1983) *J. Forensic Sci. Soc. 23*, 5-17.
17. Smith, R.N. (1983) *Anal. Proc. 20*, 417-419.
18. Seevers, R.H. & Counsell, R.E. (1982) *Chem. Rev. 82*, 575-590.

#E-2

SPIN IMMUNOASSAY FOR DRUG DETERMINATION

Mark R. Montgomery

Research Service, Veterans Administration Hospital
and Department of Pharmacology and Therapeutics
University of South Florida Medical Center
Tampa, FL 33612, U.S.A.

Immunoassay for various drugs and foreign chemicals in biological fluids has become a routine, sensitive and reliable technique. Its principal drawback in routine clinical and forensic laboratories is the requirement for multiple sample manipulation and ultimate separation of the labelled from the unlabelled chemical. The use of electron spin resonance (ESR) as the detection system allows a greatly simplified procedure.-
1. An antibody-labelled antigen complex is prepared (the 'label' is an unpaired electron).
2. This complex is mixed with the sample (the only mixing step).
3. The ESR absorption of the displaced label is determined.
The main advantage is achieved in step (3): no separation of antibody-labelled antigen complex from displaced labelled antigen is required (nor radioactivity measurement); only the unbound form resonates and is easily quantitated from a standard curve. In another application, binding to endogenous protein in blood or a tissue homogenate may be quantitated directly, by first measuring the resonance absorption of the free, labelled chemical and then the decrease on adding the test material.

The technique has been evaluated for morphine, phenytoin, amphetamine, methadone, secobarbital and benzoyl ecgonine, with checking for cross-reactivity. With appropriate automation and computerization, reliable, precise and accurate analysis, in a few sec, of numerous chemicals in the nanomolar range should be achievable.

With increasingly sophisticated analytical tools, the need for sensitive assays and information on spatial relationships has led to the examination of chemical groups and, in turn, of individual molecular species, atoms within the molecule and, finally, subatomic particles. It is to this most recent development, the quantitation of molecules by probing protons and electrons, that the physical science of magnetic resonance is applied.

The principles of physical resonance are well known and described in numerous review sources. For chemical magnetic resonance, the principle is straightforward: when an applied magnetic field is in frequency resonance with a magnetic moment placed within that field, energy is absorbed. Close examination of the absorption characteristics is informative concerning the environment of the magnetic moment under investigation. For example, protons may be examined (nuclear magnetic resonance, NMR) with magnetic fields in the radio-frequency range ($\simeq 10^8$ Hz) utilizing magnetic field strengths of 10^4-10^5 gauss. Electrons may be examined (electron spin resonance, ESR) with magnetic fields in the microwave frequency ($\simeq 10^{10}$ Hz) utilizing magnetic field strengths of 10^3-10^4 gauss.

Combining the physical principle of ESR with the biological principle of immunoassay has permitted the development of a simplified, rapid and sensitive assay for chemicals which is called 'spin immunoassay'.   It relies upon the inclusion of a magnetic moment (an unpaired electron) into the structure of the target chemical. Most frequently, drugs are 'labelled' with a nitroxide free radical which contains a stable, delocalized electron that resonates within the –N–O bond.   Fig. 1 exemplifies such labelling for phenytoin.

ASSAY PROCEDURES

The spin immunoassay procedure (details in [1-4]) is exemplified by phenytoin assay.-
a. Labelled drug (phenytoin) is prepared as outlined in Fig. 1.
b. Phenytoin conjugate with bovine serum albumin is prepared with 3-*N*-(methoxycarbimidomethyl)-phenytoin. The hapten number is determined by reference to a standard UV calibration curve.
c. The specific drug-sensitive γ-globulin is obtained from sheep given multiple injections of phenytoin conjugate in complete Freund's adjuvant.   The binding site concentration and binding constant are obtained by titration against phenytoin spin label.

Sample preparation for the ESR spectrometer is then simple.-
1. The drug-specific antibody-labelled drug complex (from steps b & c above) is mixed with the unknown sample.
2. The unlabelled drug ('unknown') competitively displaces the labelled drug from the antibody-antigen complex.
3. The energy absorption of the displaced, nitroxide-labelled compoint is determined *without* separation of the free and residually bound label (obviously an advantage).
3. The observed absorption is quantitated by comparison with a standard curve.

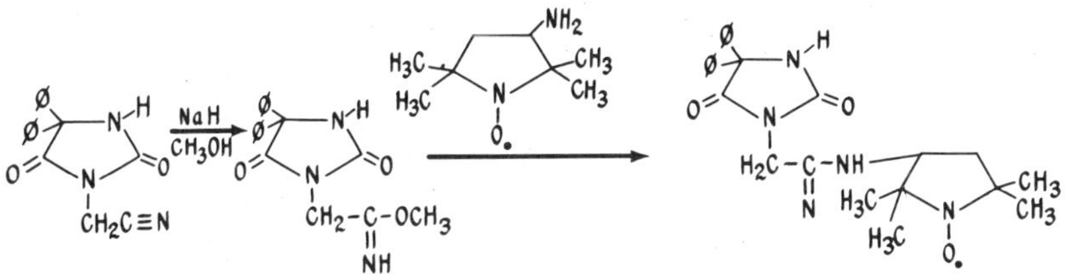

Fig. 1. Preparation of phenytoin-nitroxide spin label. For details of reagents and conditions, see [2].

One important characteristic of ESR is that only the free, unbound electron will tumble randomly in three dimensions and, therefore, will demonstrate absorption. This contrasts with the use of radioactive labels, which are counted equally whether antibody-bound or displaced and thus require tedious separation procedures.

For quantitation, samples are mixed in a 50 µl quartz capillary tube which is placed directly into the magnetic cavity of the ESR spectrometer (Varian E-4 was used in these studies). Typical ESR settings for determining phenytoin in blood are: scan rate 20 G/min, time constant 1 sec, modulation amplitude 4 G, modulation frequency 100 KHz, microwave power 12 mW, microwave frequency 9,538 GHz. A typical spectrum for both bound and unbound nitroxide-labelled phenytoin is reproduced in Fig. 2. The low-field peak (H = 3377) is quantitated routinely by the peak-to-peak difference between the free and bound signals (signal-to-noise). This peak-to-peak difference is then referred to a standard curve prepared in the same biological fluid.

APPRAISAL OF THE APPROACH

The accuracy, precision, sensitivity and reproducibility of the technique have been closely evaluated for both phenytoin and morphine [1, 2]. In one study, serum samples were obtained from 27 patients receiving phenytoin for seizure control. The phenytoin concentration was detrermined in each sample by spin immunoassay and by standard GC technique. The linear regression of the correlation had a slope = 0.94 ±0.04 ($\bar{x}$ ± 1 S.D.) and a correlation coefficient r = 0.97. Selected samples were assayed in triplicate by spin immunoassay and typically had <10% difference between highest and lowest values.

In separate experiment, both morphine and ^{14}C-labelled phenytoin were injected i.v. into female rabbits. Disappearance from the serum was then followed by spin immunoassay, GC and, for phenytoin, ^{14}C-counting (Fig. 3). Evidently spin immunoassay provides an accurate assessment of serum drug concentrations in an *in vivo* model.

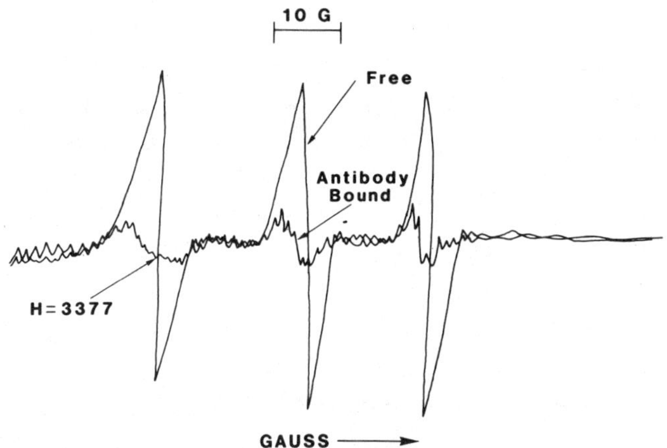

Fig. 2. Typical ESR absorption spectra for a spin-labelled drug, phenytoin, in saline without specific antibody ('free' state) and with antibody present ('antibody bound'). ESR settings as in text.

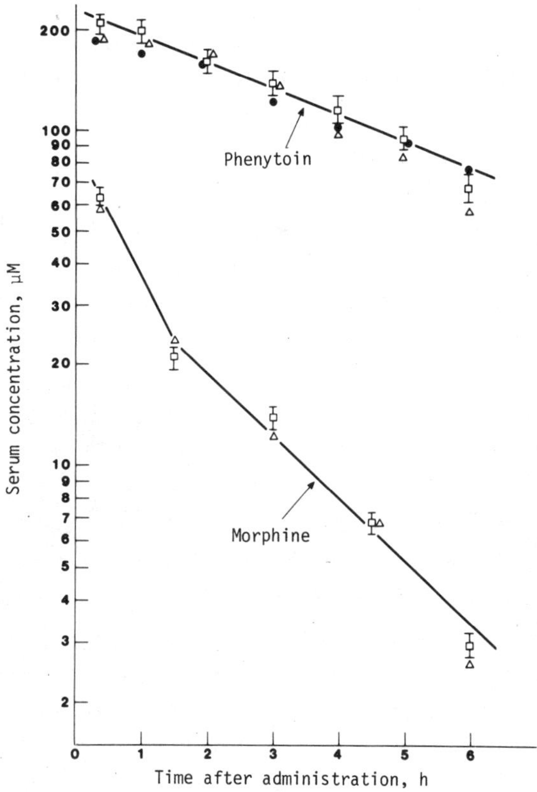

Fig. 3. Serum decay curves for phenytoin and *(lower curve)* morphine after i.v. administration to a rabbit.

□, ESR absorption assay.
△, GC assay.
● (for phenytoin), ^{14}C assay.

Note use of log scale for concentrations.

In forensic and other contexts, any cross-reactivity is important. The potential for false results must be considered not only
for structural analogues but also for compounds which are commonly
administered in combination with the drug in question. As with all
immunoassays, the spin immunoassay is subject to this limitation
which depends on the specificity of the antibody and not the detection system. [For metabolites *vs.* parent drug, see p. 259 in Vol.12,
this series.-*Ed.*] With the particular antisera used for the 6 drugs
investigated, a range of compounds relevant to each drug showed <1%
cross-reactivity. Only α-acetylmethadol cross-reacted (100%) to
methadone, and ecognine to benzoyl ecognine (8%). Elevated responses
were found for codeine *vs.* morphine, and mephobarbital *vs.* secobarbital (500% and 250% respectively). Examples of >10% cross-reactivity included heroin *vs.* morphine, pyrimidone *vs.* phenytoin, metham-
phetamine or phenteramine *vs.* amphetamine, and several barbiturates
(amongst a number) *vs.* secobarbital.

It was concluded that several structural analogues, or a metabolic product such as *p*-hydroxyphenytoin, may cross-react sufficiently to yield a false positive result if the assay were interpreted
on a 'presence' or 'absence' basis. Furthermore, significant
interference may be present with selected combinations, yielding
artificially elevated concentrations. This potential complication
must be recognized and evaluated for all immunoassays used in
forensic toxicology, particularly when the results have medico-legal
implications.

Finally, the question of assay sensitivity and limits of detection requires discussion. Utilizing a flat cell of 0.30 ml volume
at present allows assay down to 20-30 nM. Manipulations available
to increase sensitivity have been thoroughly discussed elsewhere [4].
The most direct and easily accessible improvement would be computerization of the ESR spectrometer to scan only two frequencies: that
corresponding to the peak and that corresponding to the trough (see
Fig. 1). By rapidly determining multiple peak-to-peak heights and
comparing with similar measurements on the blank sample, the signal-
to-noise ratio could be enhanced significantly. Such improvements
should allow concentrations below 10 nM to be assayed, thus making
the system available for routine monitoring of such drugs as cardiac
glycosides and hormones.

References

1. Montgomery, M.R. & Holtzman, J.L. (1974) *Drug Metab. Dispos.*
   2, 391-395.
2. Montgomery, M.R., Holtzman, J.L., Leute, R.K., Dewees, J.S. &
   Bolz, G. (1975) *Clin. Chem. 21*, 221-226.
3. Montgomery, M.R., Holtzman, J.L. & Leute, R.K. (1975) *Clin.
   Chem. 21*, 1323-1328.

4. Montgomery, M.R., Holtzman, J.L. & Leute, R.K. (1977) in *Drug Fate and Metabolism*, Vol. 1 (Garrett, E.R. & Hirtz, J.L., eds.), Dekker, New York, pp. 243-268.

#E-3

ADVENT AND PRESENT-DAY STATUS OF LIGAND METHODS

Stephen H. Curry

Division of Clinical Pharmacokinetics
J. Hillis Miller Health Center (Box J-4)
University of Florida
Gainesville, FL 32610, U.S.A.

Biological methods have been crucial in pharmacological studies since the earliest days of quantitation. Disciplined use of bioassay allowed the discovery and quantitation of such endogenous compounds as acetylcholine and noradrenaline long before chemical assay became possible. In contrast, little was done in attempts to measure exogenous compounds introduced into the body until key developments in chemical analysis took place, notably with the Bratton-Marshall reaction, absorption and fluorescence spectroscopy and, more recently, GC and HPLC. Specificity in chemical drug assays became a rigorous aim, often hinging on both work-up procedures and detection modes.

As the demand grew for extensively determining drugs in large sets of clinical samples, calling for rapid methods, many investigators reverted to biological assays in the form of radio-immunoassays (RIAs) and, more recently, various kit techniques such as the EMIT and fluorescence types. No-one can deny the contribution of these assay methods to knowledge, but they need careful assessment especially for specificity. Drugs produce metabolites similar to themselves. Specific and separate assay of drug and metabolites is essential. Biological methods that lack the requisite specificity are of no value in pharmacokinetics and find application only in very limited areas of research.

As Edward T. Maggio has said, "Novus Ordo Seclorum" [1].* This remark was made in 1979 in the light of the fact that, with RIA only 20 years old, and the Nobel Prize recognizing its importance having been awarded only two years previously, a "new order" was already emerging. This new order is epitomized by the, now very important, and newer, enzyme immunoassay systems.

---

* May be rendered as "A new order is upon us" (from Preface of [1]).

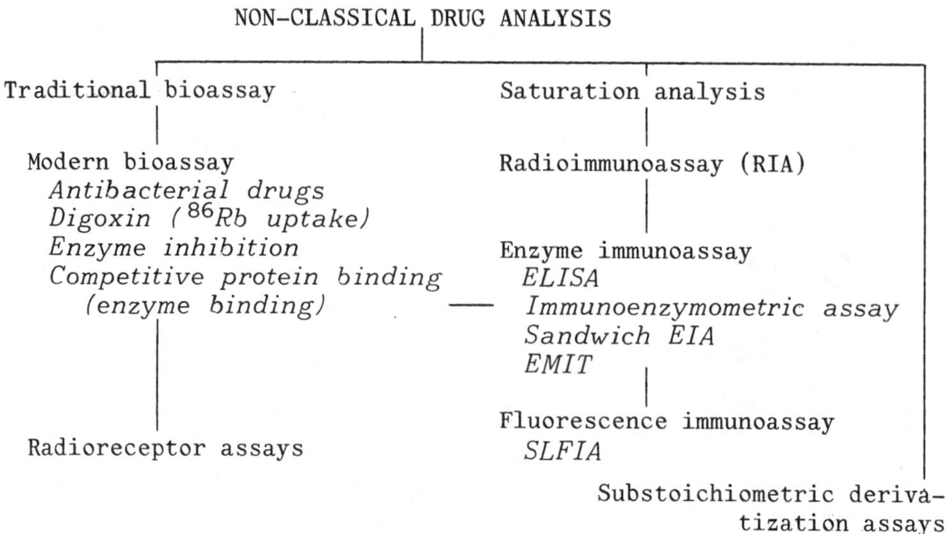

Fig. 1.   Some ligand binding and other related assay aystems.

In this survey I plan to relate current developments in biological assay methods for drugs to the historically important concept of biological assay in pharmacology. This is not an exhaustive review. Rather it is an attempt to focus attention on the development, over the last 50 years, of certain concepts in drug analysis, and to stimulate discussion and evaluation of biological methods in relation to present-day needs. As is indicated in Fig. 1, which attempts to relate the different methods in a sequence of thinking, I shall consider bioassay – classical and later forms (e.g. microbiological, radioreceptor) – and saturation analysis (especially ELISA and EMIT).

CLASSICAL BIOLOGICAL ASSAY

Classical bioassay was the foundation of modern pharmacology, although at first sight it seems to be of little relevance to modern drug analysis. The classical pharmacologists needed to quantify responses to potential drugs, which included compounds extracted from mammalian tissues, and they needed to use the resulting standardized systems to determine drug levels in other solutions. Diverse drugs were thus assayed, e.g. adrenaline, acetylcholine and histamine as examples of endogenous substances quantified by inducing responses, but also including antagonists of these materials, e.g. atropine, antihistamines and, more recently, blockers of catecholamine actions. Similar principles were also applied to truly exogenous therapeutic agents, e.g. dioxin [2]. The range of strategies employed included threshold doses for effect measured on individual animals, the practice of repeatedly measuring responses and equalizing responses to

Fig. 2. The classical 2 x 2 bioassay
in which two doses of an unknown (U)
are compared with two doses of a
standard (S) in order to generate a
potency ratio (M).

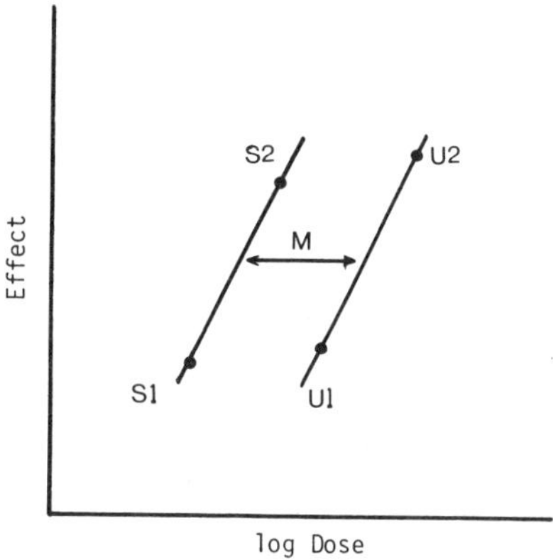

solutions of different strengths by an iterative process, the well-
known 2 x 2 procedure (principle shown in Fig. 2), and the measurement of positive incidents of criterion responses (quantal assay).

The principles of classical biological assay were a model for
methods used in later years. Like was compared with like, in that
solutions to be assayed were tested against known concentrations of
the same compound. Specificity was assessed. Sampling and replica-
tion were carefully controlled to ensure a result which could be
assessed by a statistician.

While classical bioassay is now relegated to a secondary role
more concerned with characterization of drug responses, chemical
methods having taken over the measurement of known substances, I
develop below the view that one group of biological assays (radioreceptor assays) is bringing us back 'full circle' to a re-enactment
of some of the elements in the classical drama.

BIOASSAYS SUBSEQUENTLY DEVELOPED

Certain later bioassay types combine the principles of chemical
assay with the use of biological systems. An obvious example is the
biological assay (with cultures) of antimicrobial drugs, still widely
practised as the method of choice for measuring not only the type of
activity in a plasma sample but also its concentration [3] (cf. #T-10
in Vol. 5, this series – *Ed.*). Again, the emphasis is on quantitation of a specific compound, in that potential interfering compounds
and metabolites of the drug in question are assumed or proved to have
no activity in the test system. Quantitation requires the replica-
tion and statistical review characteristic of classical bioassay.

In theory, a biological response can act as the detection function for any assay. This is exemplified in the rubidium uptake assay
which was once·popular for digitoxin [4]. Observed uptake of radio-

active rubidium into red cells had nothing to do with digoxin pharmacology, nor was it a chemical assay technique. The assay entails the following steps.-
1. Extract digoxin from plasma into dichloromethane; separate and evaporate extract to dryness.
2. Re-dissolve residue in Krebs-Ringer phosphate buffer or saline-glucose solution.
3. Add 0.5 ml packed washed red blood cells, and incubate at 37° for 2 h; then cool.
4. Add 3 µCi ^{86}Rb, incubate at 37° for 2 h, then cool.
5. Centrifuge, remove supernatant, and wash red cells.
6. Measure radioactivity (γ) of the red cells.
7. Plot % inhibition of radioactivity uptake from the ratio [(uptake in control tube – uptake in experimental tube)/uptake in control tube] x 100 *vs.* digoxin concentration.

The movement of rubidium serves the same function as that of a colorimeter in a classical chemical drug assay involving extraction of the drug into a solvent and assessment of the extract by optical density measurements at a chosen wavelength.

SATURATION ANALYSIS AND RADIOIMMUNOASSAY (RIA)

Quite obviously, the advent of RIA provided a new lease of life for biological assay, as much for drug analysis as for assay of endogenous materials [5]. It should be noted that RIA is really a specialized form of saturation analysis – the term originally proposed to describe a general analytical method relying on progressive saturation of a specific reagent by a test compound. The general principle is illustrated in Fig. 3, for Compound X (non-radioactive) to be measured in a biological sample. To X is added a known quantity of radioactive X (X^*), increasing the total quantity of X present by a known amount. Material Y as subsequently added (Fig. 3) reacts chemically or physically with the mixture ($X + X^*$); the utilization of Y must be such that <100% of ($X + X^*$) is used. The reaction product ($X + X^*$)Y must differ in properties from X, such that the extent of reaction can be assessed from the proportional change in radioactivity. Then comparison with X in standards gives the unknown value.

In RIA – where most drugs must be converted to haptens to help prepare the requisite antiserum – the calibration graph is by definition non-linear. There are a number of mathematical methods for generating linear calibration graphs from curves, but it must be realized that they all involve weighting. Thus the variance may not be similar at all points on the calibration graph, and the graph may be more accurate in the middle region or at one end depending on the transformation used. This need not be a problem with classical chemical methods.

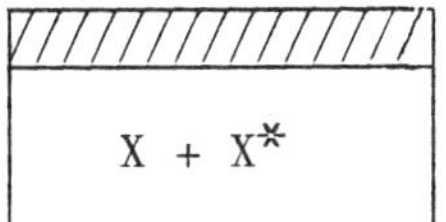

Biological sample containing X
(non-radioactive)

Add a known (trace) quantity
of radioactive X *(*X*)*

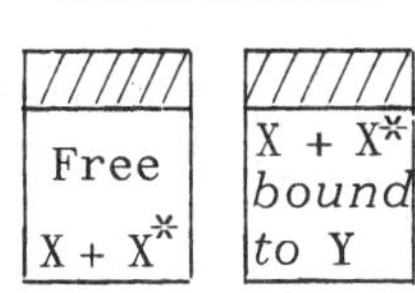

Mix to ensure homogeneity
with respect to X *and* X*

Purify if necessary, e.g. by extraction;
some loss of both X *and* X* *can be*
expected, and is noted from the radio-
label recovery (total X*; cf. the final*
distribution, = X* *diluted by* X*)*

A material Y *is now added to the*
system; it reacts chemically or physically
*with (*X + X*), <100%, resulting in two*
fractions one of which contains the
*reaction product (*X + X*)* Y : *the propor-*
tional change in radioactivity gives the
extent of reaction (vs. X *in standards)*

Fig. 3. Principles of saturation analysis. Y is typically an
antibody or other protein, or charcoal; it alters the properties of X.

THE USE OF PROTEIN- AND ENZYME-BINDING REACTIONS

The discovery of RIA was more than the discovery of a new method
of analysis. It was the demonstration of the potential of immunological methods, and so was the catalyst for the development of a
considerable number of alternatives [6, 7]. The detractors of RIA
pointed out that RIAs required the use of isotopes, with implications
for handler safety, involved unstable chemicals, and also required
separations which could be difficult to achieve. Enzyme immunoassay
(EIA) has the potential to eliminate the first of these two problems.
The step from RIA to EIA was not difficult conceptually, but it was
actually the discovery of homogeneous EIA, overcoming the third problem - the need for a separation - that set the ball rolling, leading
to the development of the EMIT system.

Before considering EIA, the use of protein-binding and enzyme-
binding assays warrants brief mention. They are illustrated by
methotrexate assay methods, which have involved chromatographic as
well as fluorescence, bacteriological, radioimmunological, direct
ligand binding, enzyme inhibition and competitive protein binding

strategies. The enzyme-inhibition strategy relies on the fact that methotrexate inhibits dihydrofolate reductase, which is obtained from *Lactobacillus casei* [8]. The enzyme reaction entails NADPH utilization, the fall in which (monitored by spectrophotometry) shows the inhibition, related to methotrexate concentration.

The competitive protein-binding assay for methotrexate relies on the binding of the drug to dihydrofolate reductase [9]. The patient sample is mixed with a trace of [^{3}H]methotrexate, and incubated with the enzyme. Free and bound methotrexate are separated by adsorption of free drug onto dextran-albumin coated charcoal. There is a single homogeneous class of binding sites with a Ka (association constant) of 2.1×10^8 M^{-1}. A logit plot of the ratio: (Drug bound in experimental determination)/(Drug bound in a control determination) is used, the control tube being from a drug-free human sample. [For RIA assay of methotrexate, see #NC(D)-4.- *Ed.*]

ENZYME IMMUNOASSAYS (EIA)

Several different types of EIA exist, including competitive EIA systems for antigens, immunoenzymometric assays for antigens, sandwich EIA systems for antigens, antibody EIA systems, and homogeneous EIA for haptens (EMIT). The ELISA system is the competitive EIA. It was originally devised for measuring antibodies. It is a direct analogue of RIA. 'Labelled' and 'unlabelled' antigens compete for attachment to a limited quantity of solid-phase antibody. The labelling is achieved by complexing the hapten and an enzyme. The drug to be measured competes with the drug-enzyme complex for attachment to the antibody. A separation is involved, and the enzyme activity remaining available after the separation relates to the amount of drug. 'ELISA' signifies enzyme-linked immunosorbent assay.

In the immunoenzymometric procedure, an unknown quantity of antigen is reacted with an excess of labelled antibody. Solid-phase antibody is then added. Centrifugation removes unreacted enzyme-linked antibody molecules. Enzyme activity associated with the solid phase is measured. This activity is then an expression of the antigen concentration in the unknown sample.

EMIT

The discovery of the EMIT system (Enzyme-Multiplied Immunoassay Technique) may well come to be recognized in the future as an event at least as important as the discovery of RIA. EMIT can be traced to a modest paper published in 1972 by K.E. Rubinstein et al. [10] describing *'Homogeneous' Enzyme Immunoassay. A New Immunochemical technique.* One of the earliest applications was to morphine. The addition of morphine antibodies to a conjugate of morphine and lysozyme was found to inhibit the activity of lysozyme [10]. The addition of free morphine to a mixture of the conjugate

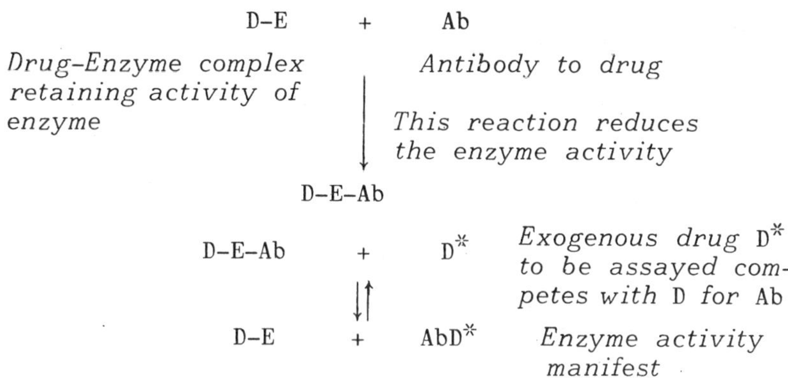

EMIT system: E is glucose-6-phosphate
dehydrogenase (G6PDH), and reduction of
NAD^+ or $NADP^+$ is followed spectrophotometrically

Fig. 4.   Principles of an EMIT assay. See text for another
Stage 3 example, where lysozyme displaced from a complex with
morphine causes cell-wall lysis in suspended bacteria, as is
monitored by the increase in UV absorption.

and morphine antibodies reduced the inhibition of enzyme activity in
proportion to the quantity of free morphine added.   The method was
sensitive to 10^{-9} M morphine. Fig. 4 summarizes the assay principles,
with use of glucose-6-phosphate dehydrogenase (G6PDH) rather than
lysozyme, and of a final coenzyme reduction step.   The latter may be
$NAD^+ \rightarrow NADH$, or $NADP^+ \rightarrow NADPH$.

The EMIT system has the fundamental advantages of needing no
radioisotopes, nor separation of bound and free drug.   The enzyme
G6PDH serves for a wide range of assays, with specificity introduced
at the stage of competition between free and conjugated drug for the
antibody.   In the comercial EMIT system, plasma (containing D^*, Fig. 4)
is mixed with a reagent which contains antibodies to the drug to be
assayed, together with G6PDH substrates. The drug binds to its antibodies (giving AbD^*).   A quantity of the same drug labelled with
G6PDH (D-E) is then added.   It combines with any unfilled antibody
binding sites, and enzyme activity is proportionately reduced.   The
residual activity is directly related to the concentration of the
drug in the plasma.   Besides G6PDH (e.g. for digoxin) and lysozyme
(e.g. for barbiturates and cocaine metabolites), enzymes used in
currently available EIAs include malate dehydrogenase (e.g. for morphine and codeine), alkaline phosphatase and acetylcholinesterase.

Besides advantages indicated above, others that are   claimed in
propaganda style involve intangible factors such as cheap equipment,
long reagent shelf-life, simplicity and potential for automation. What
really matters is specificity.

Enzymic reaction:

$$\text{F-D} \xrightarrow{\text{enzyme}} \textbf{fluorescent product}$$

Antibody-binding reaction:

$$\text{F-D} + \text{Ab} \rightleftharpoons \text{F-D-Ab} \left[\xrightarrow{\text{enzyme}} \textit{no product}\right]$$

Competitive binding reaction:

$$\left.\begin{array}{c} \text{F-D} + \text{Ab} + \text{D} \\ \Updownarrow \\ \text{F-D-Ab} + \text{D-Ab} \end{array}\right\} \xrightarrow{\text{enzyme}} \begin{array}{l} \textbf{fluorescent product} \\ \text{(reaction rate related to D)} \end{array}$$

Fig. 5. Principles of reactant-labelled fluorescence immunoassay.
F-D is drug/dye conjugate; D is the drug to be assayed; Ab is
antibody to the drug. This technique can be termed substrate-
labelled fluorescence immunoassay (SLFIA).

FLUORESCENCE IMMUNOASSAY

The principles of fluorescence immunoassay are exemplified in a
description, published in 1977 [11], of a system for gentamicin. A
conjugate of drug and a fluorescent dye serves as substrate for an
enzyme, whereby the non-fluorescent conjugate yields a fluorescent
product. When the conjugate interacts with an antibody to the drug,
a complex is formed which prevents conjugate-enzyme interaction [11].
In competitive binding systems, the drug in a test sample competes
with the conjugate for antibody binding sites [12]. The amount of
conjugate available for reaction with the enzyme is therefore controlled by the amount of drug in the sample. Thus the rate of
increase in fluorescence intensity is also related to this amount.
Because the conjugate bound to antibody is inactive as a substrate,
no separation steps are needed (Fig. 5). In the gentamicin system,
just 1 µl of sample is added to a reagent containing antiserum and
bacterial β-D-galactoside galactohydrolase, after which the drug-dye
conjugate is added and the rate of fluorescence production is followed.

The drug-dye conjugate is a complex molecule, β-galactosyl-umbelliferone-sisomycin (Fig. 6). Gentamicin is a mixture of 3 components. [See Vol. 5, this series: art. #T-12 on radioenzymatic assays. –
Ed.] The conjugate is formed from sisomicin, which differs from
gentamicin C_{1a} by the presence of one double bond. Sisomicin is
immunochemically similar to the gentamicins. [Cf. NC(E)-2.–*Ed.*]

ISOTOPE DILUTION WITH SUBSTOICHIOMETRIC DERIVATIZATION

As far back as 1970, this author proposed a general method of
analysis which was akin to saturation analysis (Fig. 1) but involved
substoichiometric derivatization of the drug with a reagent rather
than binding to a solid phase [13]. In the method, fX mol of the drug

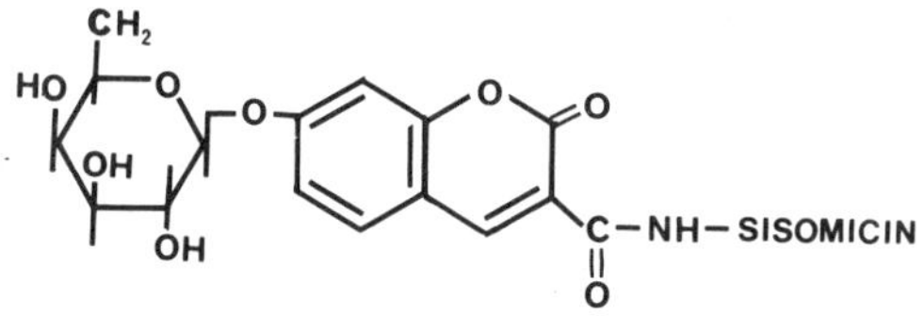

Fig. 6. β-Galactosyl-umbelliferone-sisomycin.

would first be mixed with 0.01X mol of radioactively labelled drug.
The mixture would then be derivatized with 0.01X mol of a reagent
which changed the physicochemical properties of the drug, such that
the derivative could be separated from the underivatized drug by
solvent extraction, TLC or any available method. Since f is always
>0.01, the extent of reaction of the labelled drug would depend on f.
The proportional conversion of the labelled drug to its derivative
could serve to generate a signal related systematically to f.

This method is limited only by the specific activity of the
labelled drug and by the ingenuity of those seeking to apply it. It
is, however, dependent on the specific reaction of the analyte with
another compound, and this problem has prevented its use.

RADIORECEPTOR ASSAYS

Radioreceptor assays for drug molecules came to prominence in
1977 with the publication by Creese & Snyder [14] of a method for
neuroleptic drugs. These drugs are thought to exert their therapeutic effects by blocking brain dopamine receptors. It was believed
that the radiolabelled butyrophenone neuroleptics bind selectively
and with high affinity to dopamine receptor sites in mammalian brain.
This specific binding was believed to be reduced by other neuroleptics in close parallel with the clinical potencies of these other
drugs. The inhibition of haloperidol binding occurs at concentrations lower than those found in blood.

In the assay, serum is used without treatment. The serum is
mixed with a sample of a homogenate of membranes of the corpus stri-
atum to which labelled haloperidol or spiroperidol is bound. The
extent of displacement of the labelled drug indicates the amount
of neuroleptic present. This method obviously has the advantage of
phenomenal simplicity, and it can be likened to classical biological
assay, in that it is believed that something representing the required pharmacological response is used for assay of the drug. This
relates to my earlier comment about 'turning full circle'.

The problem with radioreceptor assays is that they are inherently non-specific. They detect both parent drug and metabolites. Furthermore, they detect mixtures of drugs. This is claimed to be advantageous, in that for patient monitoring 'total neuroleptic activity' in blood is a valuable measure. This may be so, but such a measure is of no value in pharmacokinetics, or for any of the purposes for which those of us not involved in monitoring analyze drugs. Apart from this, the basis of the belief in the 'close parallel' of clinical potency and ability to displace labelled butyrophenone is tenuous. The correlation coefficients involved are too low to say the least, and they are based on very poor estimates of clinical doses, which in any case omit the influence of first-pass effects and other pharmacokinetic variables on our ideas of the potency of the drugs. [Vol. 13 features RRA principles, correlations, etc.- *Ed*.]

Furthermore, it is assumed that the metabolites of the drugs have 'clinical potencies' related to their ability to displace the butyrophenone. In fact, it has been shown that their ability to affect dopamine receptors closely parallels their lipid solubilities. The clinical potencies of the neuroleptic drug metabolites have never been measured, so the assumption is clearly highly suspect.

EVALUATION

The apprehensiveness of this author about ligand binding methods concerns specificity primarily, and the mathematical derivations necessary in producing calibration graphs secondarily. I accept that many available immunoassay methods have been proved to be highly specific for the drug in question in comparison with the known metabolites of the drug, other drugs likely to be given at the same time, and endogenous constituents of the biological fluid. I believe that there is no answer to the risk that a particularly interesting result may arise when a new metabolite or interacting factor comes into play, and the investigator is left trying to decide whether he is reading unusual concentrations of the drug he expected to assay, or a combination of the drug and a second, hitherto unknown, material of interest. I believe that radioreceptor assays can never meet this criticism.

As to quantitation, our usual requirement of unweighted linear calibration is not available with ligand-binding assays, and this means that great care must be exercised in handling numbers at the extremes of the range of enquiry, which is just where the interesting data are. Altogether, the trend towards use of ligand-binding assays rather than classical chemical assays warrants close scrutiny.

How do RIA and the EMIT systems compare? Both are dependent on antibodies to the drug in question, but only RIA requires radioactivity. Thus, the equipment needed, and procedures involved, are a major factor. Special conditions are required for radioactivity handling; none for EMIT. EMIT systems are more amenable to automation, in that EMIT spectrophotometers can retain calibration curves and work in 'black-box' style. The quality of the methods still depends on specificity, sensitivity, precision and accuracy, which are functions of the individual systems for particular drugs, not of the basic principles.

References

1. Maggio, E.T., ed. (1980) *Enzyme Immunoassay*, CRC Press, Boca Raton, Florida, 295 pp.
2. Burgen, A.S.V. & Mitchell, J.F. (1978) *Gaddum's Pharmacology*, 8th edn., Oxford Univ. Press, Oxford. - See pp. 229-247.
3. Alcid, D.V. & Seligman, S.J. (1973) *Antimicrobial Agents & Chemotherapy 3*, 559-561.
4. Berther, A. & Redfors, A. (1970) *Clin. Pharmacol. Ther. 11*, 665-673.
5. Berson, S.A. & Yalow, R.S. (1968) *Clin. Chim. Acta 22*, 51-69.
6. Schall, R.F. & Teroso, H.J. (1981) *Clin. Chem. 27*, 1157-1164.
7. Wisdom, G.B. (1976) *Clin. Chem. 22*, 1243-1255.
8. Falk, L.C., Clard, D.R., Kalman, S.M. & Long, T.F. (1976) *Clin. Chem. 22*, 785-788.
9. Myers, C.E., Lippman, M.E., Eliot, H.M. & Chabner, B.A. (1975) *Proc. Nat. Acad. Sci. 72*, 3683-3686.
10. Rubenstein, K.E., Schneider, R.S. & Ullman, E.F. (1972) *Biochem. Biophys. Res. Comm. 47*, 846-851.
11. Burd, J.F., Wong, R.C., Feenery, J.E., Carrico, R.J. & Bogulaski, R.C. (1977) *Clin. Chem. 23*, 1402-1408.
12. Visor, G. & Schulman, S.G. (1981) *J. Pharm. Sci. 70*, 469-475.
13. Curry, S.H. (1970) *Br. J. Pharmacol. 39*, 250P.
14. Creese, I. & Snyder, S.H. (1977) *Nature 270*, 180-182.

#E-4

RADIORECEPTOR ASSAY OF ANTICHOLINERGIC DRUGS
IN BIOLOGICAL FLUIDS

K. Ensing and R.A. de Zeeuw

Department of Toxicology, State University
A. Deusinglaan 2, 9713 AW Groningen, The Netherlands

Requirement   *An assay for quaternary anticholinergic drugs sensitive enough for pharmacokinetic studies after oral and i.m. administration as well as inhalation. The assay should discriminate between biologically active and inactive compounds or enantiomers.*

End-step   *Extract containing the analyte is incubated with 2 nM [^{3}H]dexetimide and muscarinic receptors (0.5 nM). After 10 min at 37°, ice-cold buffer is added. Bound radiolabelled ligand is collected on glass-fibre filters and, after dispersion in Plasmasol, determined by liquid scintillation counting.*

Sample preparation   *Ion-pair extraction from 1–4 ml plasma with 0.1–1.0 mM Na picrate into 5 ml dichloromethane, 4.5 ml of which is re-extracted with 2 ml of 1 M tetrapentylammonium iodide, displacing the drug from the ion-pair into the aqueous phase. Aliquots (3 x 0.5 ml) of this phase are freeze-dried to remove traces of the organic solvent, and assayed after adding 500 μl of water.*

Comments   *Plasma detection limits are 100 pg racemic oxyphenonium, 50 pg D-oxyphenonium and 200 pg ipratropium respectively. The choice of isolation procedure depends on its influence on the binding properties of the receptor.*

Anticholinergic drugs are used in the treatment of chronic aspecific respiratory affections in order to obtain bronchodilation. These drugs antagonize the bronchus-obstructing effect of acetylcholine, mediated by an interaction with muscarinic receptors, and may also be useful as a protective against certain irritants. For a good understanding of the pharmacological action of these drugs and to obtain optimum therapeutic efficacy, it is necessary to determine pharmacokinetic profiles after i.m. and oral administration as well as after inhalation, and to relate the findings to the observed lung function data.

Because most of the antimuscarinic drugs used clinically are highly potent, therapeutic doses result in plasma levels in the low ng/ml range. The GC method of Greving [1] allows plasma levels down to 1 ng/ml to be assayed, but is time-consuming and prone to interferences. Because of the low levels expected after inhalation and the limited sensitivity of chemical methods, we investigated the potential of a radioreceptor assay (RRA). Such assays, offering stereoselectivity, apply wherever there is a drug-receptor interaction. Basic factors in RRA are briefly considered [see also accompanying article #E-5, also #E-3. -*Ed.*], and illustrated with assay-development data for anticholinergic drugs.

PRINCIPLES OF RADIORECEPTOR ASSAYS, AND ASSAY DEVELOPMENT

An RRA hinges on competitive displacement of a certain amount of labelled ligand from the receptor by the drug, depending on drug concentration and equilibrium dissociation constant, Kd^*. The radioactivity of the bound fraction as collected by rapid filtration will be inversely related to the amount of competitive drug present (Fig. 1).

RRA development calls for several precautions. When only the parent drug molecule is active *in vivo* the composition of the incubation medium and the incubation time can be chosen to increase the sensitivity of the assay. When more than one active form is present in the assay material, the composition of the incubation medium has to be comparable to the physiological medium; otherwise no meaningful quantification of the total biological activity can be obtained. The influence of any small departure from the near-physiological conditions has to be established to determine the stability of the binding of the radiolabelled drug and the competitive drug to the receptor [2, 3]. [Notably relevant to RRA: #A-3 & A-4 in Vol. 13.-*Ed.*]

Receptor preparation

The receptor 'solution' used in the RRA has to be characterized before use. The source of tissue used for the preparation of a receptor suspension has to be chosen carefully:
- the receptor (sub)type must have identical properties to the receptors in human tissue;
- a high receptor density per mg protein, giving low non-specific binding, is preferable;
- the tissue must be easily obtainable at low cost;
- homogenization and conservation of the tissue should be simple.

---

* altered from authors' K_D; see Preface to Vol. 13, this series.-*Ed.*

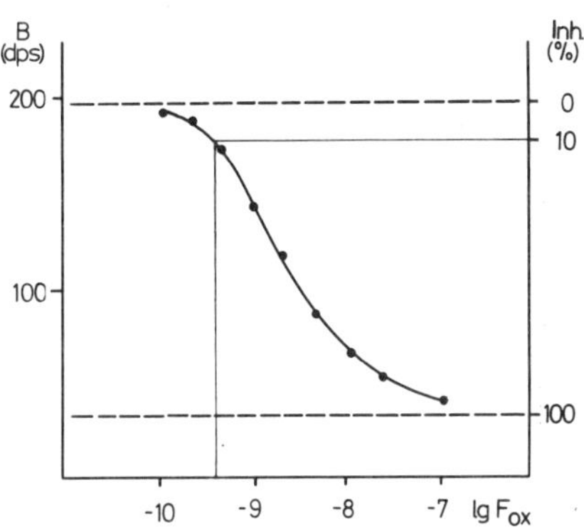

Fig. 1. Inhibition of [3H]dexetimide (DEX) binding by oxyphenonium (OX): F_{OX} is the concentration of free OX; B is bound radiolabelled ligand; Inh. is inhibition of receptor-bound radiolabelled ligand. 100% inhibition is amount of aspecific bound radiolabelled ligand; 10% is reckoned to be the determination limit of the RRA (0.4 nM OX).

In our experiments we used striata from calf brain. Fractionation of the receptor suspension may increase the receptor concentration per mg protein and the homogeneity of the material, which may improve the sensitivity and selectivity of the assay. However, fractionation should not cause a change in the binding properties of the receptors [4]. (*Note:* receptor M values as cited derive from radioligand binding.)

Radiolabelled ligand

This ligand must have the following properties:
- chemically stable, no hydrolysis or oxidation;
- radiochemically stable, minimal radiolysis (which depends on the choice of isotope and its position in the molecule);
- high purity;
- sufficiently high specific activity so that counting can be done within 10 min with a variation coefficient of 0.5% (40,000 counts);
- fast binding to one (sub)type with high affinity and, consequently, a low dissociation rate;
- reversibility of the binding, and low non-specific binding to membrane fragments, proteins, glassware and utensils.

Carbon-14 is preferable to tritium, whose short half-life (12.8 years), whilst environmentally advantageous in respect of disposal, causes ~0.5% loss of specific activity per month. There is a relatively large difference in isotope mass which can influence the rate and type of chemical reactions. Further, liquid scintillation counting efficiency for tritium is <50%. Yet tritium is used in most cases because of low cost, low toxicity and relatively simple synthetic incorporation to give desired labelled compounds.

In binding experiments on the muscarinic type of acetylcholine receptor, commonly there have been two different labelled ligands. Single tritium-labelled dexetimide (^{3}H-DEX), the active dextro-isomer of the anticholinergic drug benzetimide, and double tritium-labelled quinuclidinyl benzilate (^{3}H-QNB).   Both compounds have high and selective affinities for the muscarinic receptor, but ^{3}H-DEX is more suitable for a quantitative assay than ^{3}H-QNB as the latter shows some irreversible binding and a much longer association time.

The binding of a ligand can be pH-dependent, as shown in Fig. 2 for dexetimide and pirenzepine. The affinity of dexetimide is optimal and constant around the physiological pH, but the affinity of pirenzepine increases at lower pH. This is due to protonation of the drug (pKa 8.20); the protonated form has a higher affinity for the muscarinic receptor than the unprotonated form. Thus, pirenzepine is unsuitable as a radiolabelled ligand, not because of its lower affinity but because small changes in the pH of the incubation medium cause changes in the position and magnitude of the inhibition curve. However, if pirenzepine is the drug that has to be assayed, a lower detection limit can be obtained when the pH of the incubation medium is decreased.

The influence of ionic strength on the affinity of a ligand to the receptor is reported for most agonists to be more important than for antagonists [5].

Incubation time, temperature, and separation of free and bound ligand

Binding is usually performed at 0°, 25° or 37° and the incubation time is chosen so as to allow equilibrium of bound and free fractions of radiolabelled and competitive ligand. Depending on the separation technique, the separation is carried out either under equilibrium conditions (e.g. dialysis or centrifugation of the membranes) or under non-equilibrium conditions (e.g. filtration, or centrifugation after addition of charcoal, which binds all free radiolabelled ligand). The difference between total added radioactivity and the bound fraction represents the free fraction.

The bound ligand held on glass fibre filters along with receptor (in membrane fragments) is washed free of unbound label with the incubation-medium buffer. Because there is no equilibrium during the filtration, the receptor-ligand complex can dissociate, depending on the temperature and time of the washings. Therefore the dissociation rates of the receptor-ligand complex have to be determined at the incubation and filtration temperature. The dissociation of dexetimide from the muscarinic receptor at 0° and 37° depends on the pH (6.2-8.2 in our experiments). The half-life of the ligand-receptor complex increases with increasing pH. It can be calculated that the loss of specific binding at 0° during the filtration procedure, which takes 15 sec, is <0.1%.

In the receptor assay we are interested only in the amount of radiolabelled ligand bound to the receptor, which is inversely related to the amount of competitive ligand present.   Therefore we prefer

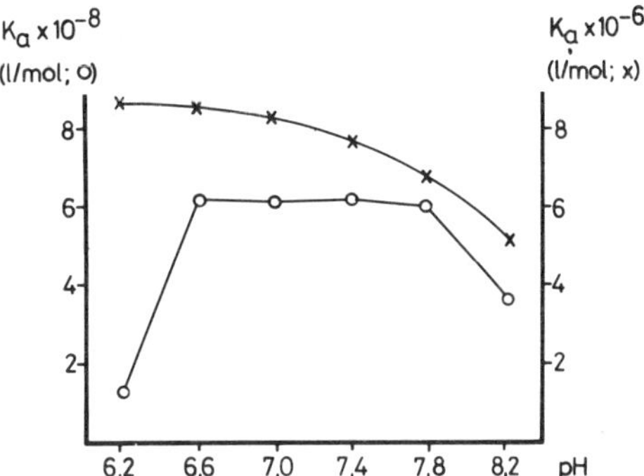

Fig. 2. Equilibrium association constants $K_a = 1/Kd$ as function of the pH for [^{3}H]-pirenzepine (x) and [^{3}H]dexetimide (o). *(An association constant is usually represented K_D, not K_a; see earlier footnote ** - Ed.)*

the filtration method. A further advantage is its speed. Moreover, the direct determination of the bound radioligand allows determination of low concentrations of receptor-bound ligand.

Assessment of non-specific binding in the assay

Non-specific binding of the radioligand and the drug can influence the detection limit and reproducibility of the assay. If the labelled ligand is non-specifically bound, its concentration can be altered to keep optimum conditions, i.e. a 50% receptor occupation and still a low limit of detection [2]. If the drug is non-specifically bound, this will raise the detection limit. When the ratio of specific to non-specific binding decreases for both ligands, the reproducibility will decrease as well.

In order to determine the amount of non-specific binding to membranes and filters, two concentrations of the receptor suspension are used for an inhibition experiment in which unlabelled dexetimide is added over a wide concentration range. This allows the choice of a suitable concentration of unlabelled ligand that inhibits specific binding of the radioligand to the receptor but not its non-specific binding. When the two inhibition curves are compared it can be seen that the non-specifically bound amount of radioligand is comparable in both experiments, is mainly localized on the filters, and is not affected by concentrations of unlabelled dexetimide up to 2 µM.

Binding experiments were carried out with Whatman GF/B and GF/C filters, which have about the same retention characteristics. We could not detect any difference between them in non-specific binding.

Another phenomenon which can disturb the radioreceptor assay is the non-specific binding of the radioligand to the incubation tubes. As expected, the extent (for ^{3}H-DEX) strongly depends on the material of the tubes and is diminished by addition of unlabelled ligand. This confirms that the loss of free radioligand is entirely due to adsorption. The relationship between the ^{3}H-DEX concentration and this binding has been determined.

About 5%, 20% and 50% of the added ^{3}H-DEX is non-specifically bound to polyethylene, polystyrene and glass tubes respectively, whereas at high concentrations saturation occurs. Therefore we chose polyethylene tubes for our experiments. Obviously, for each ligand it has to be ascertained which tubes give minimal adsorption, and with increasing affinity of the ligand the relative amount of non-specific binding increases.

Detection limit of radioreceptor assays

In the RRA the detection limit for a competitive drug is determined by its affinity for the receptor and the volume of the incubation medium. We prefer to define the limit of detection as the amount of competitive drug that causes a decrease of receptor-bound drug of 10%. This will lead to an error factor of ~10% in the reproducibility (accuracy).

The detection limit can be described thus:

$$D = 0.1 \times Ki \left(1 + \frac{F}{Kd}\right) \times V \cdot M_W$$

where [with, in parentheses (), the authors' representation if now altered; cf. earlier footnote * – *Editor*]:
Ki (K_i) and Kd (K_D) are respectively the equilibrium dissociation constants, in mol/l, of the competitive drug and the radioligand;
F is the concentration of free radioligand, in mol/l;
V is the incubation volume, in litres;
M_W is the mol. wt. of the competitive drug.

In this equation the limit of detection is the product of the free concentration of the drug, the incubation volume and the drug's mol. wt. The free concentration of the drug depends on the free concentration of the labelled ligand and its equilibrium dissociation constant. But as part of the drug is receptor-bound, the resulting limit of detection depends also on the receptor concentration, as can be seen in Fig. 3. This shows a range of drugs differing in affinity for the receptor. Thus, if the receptor concentration is decreased, a lower detection limit can be obtained; but on the other hand a minimum amount of receptor-bound radioligand has to be counted. We use a receptor concentration of 0.5 nM.

The influence of non-specific binding in our experiments was calculated, but for the quaternary anticholinergic drugs was trivial.

THE ASSAY AS DEVELOPED FOR QUATERNARY ANTICHOLINERGICS

Before consideration is given to the investigation of drug pre-isolation, the précis at the start of this article should be consulted for the assay conditions adopted – as amplified here in a few respects. Full methodological details are given elsewhere [3].

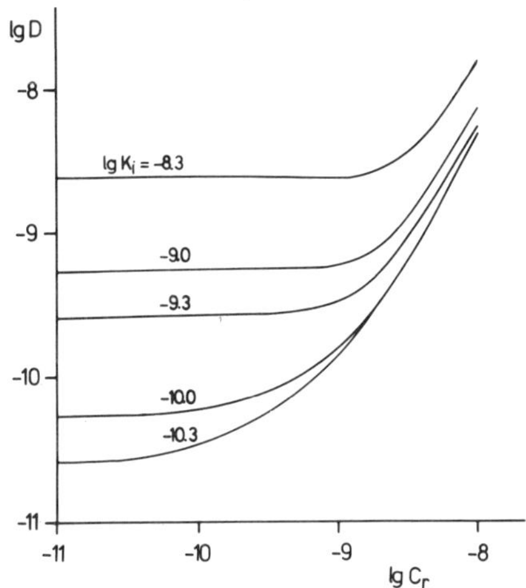

Fig. 3. Influence of the receptor concentration C_r (mol/l) on the detection limit, D (mol/l), for competitive ligands with increasing Ki values (50 fM to 5 nM). Ki (K_i in the Fig.) is the equilibrium dissociation constant for the competitive ligand.

Isolation of quaternary anticholinergics

The optimum conditions for the RRA are independent of the anticholinergic drug. The drug determines its own detection limit which can be predicted by its pharmacological potency. Direct addition of 0.1 vol. of plasma or urine to the assay medium inhibits the binding of ^{3}H-DEX by 20% or 10% respectively. Dilution of plasma or urine samples with buffer allows direct assay without prior extraction, yet at the expense of the detection limit. In view of the levels found, only urine levels can be monitored in this way after therapeutic dosing. For plasma an isolation step is necessary.

To isolate quaternary anticholinergic drugs, an ion-pair extraction seemed the most logical approach. Greving [6] determined the extraction constants of oxyphenonium with iodide, perchlorate and picrate from water into 1,2-dichloroethane. Picrate offered most promise for quantitative extractions at low concentrations. Experiments in buffer with oxyphenonium were successful, but when plasma was used losses were observed in re-dissolving the residue obtained after evaporating the dichloromethane extract. These losses may be ascribed to fatty endogenous compounds from the plasma. We then tried liquid-solid extraction using Amberlite XAD-2 columns. Again losses were observed when trying to re-dissolve the residue of the organic extract. [Cf. #F-3 in Vol.10, this series: de Zeeuw & Greving.–*Ed.*]

Because of these problems, re-extraction with tetrapentyl-ammonium (TPA) after ion-pair extraction with picrate was investigated. TPA has an extraction constant >100-fold higher with picrate than has oxyphenonium. So, when the amount of TPA is larger than the total amount of oxyphenonium in the organic layer, a nearly 100% re-extraction of oxyphenonium into the aqueous layer can be expected. This type of re-extraction may also be called a displacement extraction, as TPA displaces oxyphenonium in the picrate ion-pair. Unfortunately, however, the excess of TPA necessary for the extraction adversely affected the binding properties of the muscarinic receptors. When TPA was added to the RRA instead of oxyphenonium, the ^{3}H–DEX binding became substantially inhibited at concentrations >5 μM. A TPA concentration of 1 μM was considered sufficient for the range of oxyphenonium concentrations used in our experiments.

However, when inhibition curves with comparable oxyphenonium concentrations in buffer and in TPA extracts were determined, remarkable differences in total binding between the two curves were seen. The loss of 20–50% specific binding in the TPA extracts in comparison with the assay in buffer cannot be due to the TPA present. We found that the interference was caused by trace amounts of dichloroethane which remained in the aqueous phase after extraction. Despite the low solubility of dichloroethane in water (0.8% at 25°) it destroyed specific binding in a non-competitive manner. Organic solvents such as o-dichlorobenzene and chlorobutane were less destructive, but unsuitable for the ion-pair extraction. The only method for a complete removal of the dichloroethane which would not influence the reproducibility of the RRA seemed to be freeze-drying of the TPA extracts. Though it introduces an extra step in the overall procedure, freeze-drying offers extra advantages in that concentration of the sample is obtained too, thus lowering the detection limit. In addition, it provides a safe way to preserve and store plasma extracts. After freeze-drying for 48 h, no disturbances of specific binding were noticed.

Features of the assay, and conclusions

Summarizing our results, the RRA has a 10-fold higher sensitivity for oxyphenonium than has the chemical method (GC). This gain in sensitivity offers the possibility of determining plasma levels of oxyphenonium after inhalation. The RRA also offers an opportunity for determining the active enantiomer. Work on this subject and on synthesis of the enantiomers is in progress.

In principle, active metabolites may cross-react in an RRA. However, as metabolism of oxyphenonium has been found to be negligible, we have considered the possibility of cross-reactivity to be irrelevant.

RRA is simple in principle, but to obtain optimum conditions all the variables involved in the binding of a ligand to a receptor have to be investigated to ensure that small changes in these parameters do not cause changes in the amount of bound labelled ligand.

If direct assays on samples are precluded because endogenous compounds change the binding properties of the receptor or the free concentration of labelled or competitive ligand, a suitable isolation procedure that does not interfere with the assay has to be developed. Isolation and concentration of the anticholinergic drug may also be necessary to obtain the required sensitivity.

The assay is sufficiently reproducible when carried out with a re-suspended crude microsomal receptor preparation and a concentration of radioligand close to its equilibrium dissociation constant. The following C.V. values were obtained for a range of drug concentrations (pg in the assay; each 9 observations): 100, ±8%; 200, ±8%; 500, ±10%; 1000, ±14%. The effective linear range of the method lies between 100 and 5000 pg. For higher concentrations, a dilution of the sample is necessary. The overall recovery was 95-98% over the entire concentration range. Currently the assay is being used routinely for determining oxyphenonium in plasma and urine samples from patients and volunteers after administration of the drug via different routes.

Further advantages of the present method are that it offers stereoselectivity and can easily be extended to other quaternary and tertiary anticholinergics after a suitable extraction and concentration procedure (e.g. ipratropium, scopolamine).

Obviously if a RRA is used for drugs with negligible metabolism the data obtained, calculated concentrations, can be considered as a total biological activity as well as a proper measure of the concentration of a drug that can be used for pharmacokinetic studies. In the case of benzodiazepines with active metabolites the calculated concentration can be regarded only as a measure of total biological activity and is unsuitable for pharmacokinetic interpretation. More information can be obtained when a chromatographic method is used for the separation of parent compound and metabolites, in combination with an on- or off-line receptor assay to estimate the ratio of active to inactive enantiomer for the compounds present.

Acknowledgement

This project was supported by research grant 81023 of the Netherlands Asthma Foundation.

References

1. Greving, J.E., Jonkman, J.H.G. & de Zeeuw, R.A. (1977) *J. Chromatog. 148*, 389-395.
2. Ensing, K. & de Zeeuw, R.A. (1984) *Trends Anal. Chem. 3*, in press.
3. Ensing, K., Gerding, T.K., Kluivingh, F. & de Zeeuw, R.A. (1984) *J. Pharm. Pharmacol. 36*, in press.

4. Laduron, P.M. (1977) *Int. Rev. Neurobiol. 20*, 251–281.
5. Birdsall, N.J.M., Burgen, A.S.V., Hulme, E.C. & Wells, J.E. (1979) *Br. J. Pharmacol. 67*, 371–377.
6. Greving, J.E. (1981) *Bioanalysis and Pharmacokinetics of Oxyphenonium Bromide,* Ph.D. thesis, University of Groningen, The Netherlands.

#E-5

RECEPTOR BINDING AND HPLC ANALYSIS OF BENZODIAZEPINES IN A CLINICAL LABORATORY

N. Ratnaraj, V. Goldberg and P.T. Lascelles

Department of Chemical Pathology
The National Hospitals for Nervous Diseases
Queen Square, London WC1N 3BH

*The discovery that mammalian brain possesses specific recep-
tors for benzodiazepines has proved important not only for
research studies but also clinically, due to the possibility of
developing radioreceptor assays for their estimation in biological
fluids. Such assays should, theoretically, have the advantage
over purely chemical assays in that they reflect the total pharmacological 'activity' of both the parent compound and its active
metabolites at the receptor. The approach hinges on competition
between a radioactive ligand and the natural compound in the
sample. It also depends, however, on the availability of membrane receptors in a suitable form.*

*The preparation of membrane receptors in the fresh state,
although ideal, is basically a research procedure and is not
feasible in routine work; but the availability of stable freeze-
dried, detergent-treated receptors with high affinity provides an
acceptable alternative. Products now on the market as powders
that can readily be reconstituted in phosphate buffer represent a
significant advance. We have begun to study these assays so
as to evaluate them against HPLC techniques.*

Following their introduction into clinical practice 20 years
ago, benzodiazepines have become the most widely prescribed group
of drugs [1]. They possess a broad spectrum of activity as anxio-
lytics, hypnotics or sedatives, muscle relaxants and anticonvulsants
[2, 3]. There is increasing concern about aspects such as over-
prescription, the development of tolerance in some patients, and
withdrawal symptoms. Such aspects bear on the increasing prevalence
of therapeutic monitoring, although its usefulness is still contro-
versial [2, 4, *inter alia*].

The analysis of benzodiazepines represents a major challenge in the field of therapeutic drug monitoring for the following reasons.-

(1) There is the practical difficulty that benzodiazepines are present in serum in very low concentrations (typically 10-150 ng/ml) and many of the compounds are of similar structure. A number of GC methods have been published [cf. #A-2 by J.A.F. de Silva in Vol. 7, this series -*Ed.*] but these have several disadvantages.-
a) Some compounds, such as oxazepam and chlordiazepoxide, decompose under the conditions used in GC [2, 5, 6].
b) Some benzodiazepines, in particular the more polar compounds do not chromatograph well [2, 6, 7].
c) In some cases metabolites may decompose to give the parent compound, so that it is difficult to obtain accurate analyses of both [2, 5].
Some of these problems may be solved by the use of HPLC which is carried out near ambient temperature [2, 5], but several of the published procedures lack specific information concerning interference from co-prescribed drugs [2].

(2) There is considerable difficulty in finding a good correlation between serum levels of benzodiazepines and clinical response. Although a number of different techniques have been used including anxiety rating scales and choice reaction time [e.g. 8] these depend on the patient's cooperation and are essentially subjective. Recently, however, A. Richens' group have reported a correlation between serum levels of benzodiazepines and the peak velocity of saccadic eye-movement [4, 8]. This technique gives for the first time an objective measure of clinical response to benzodiazepines and has great potential for the future.

(3) Analysis of benzodiazepines by HPLC gives a full scan of parent drug and metabolites, but the results are often difficult to interpret clinically as it is not known which of the compounds detected are pharmacologically active. From this viewpoint the development of radioreceptor assays (RR assays) is an important technical advance, as RR assays measure total benzodiazepine *activity* in a sample, viz. only compounds with affinity for the benzodiazepine receptors.

The use of RR assays for monitoring serum levels of drugs and their active metabolites is a recent development. The first RR assay of this type, for phenothiazines, was published by Creese & Snyder [9] following the observation that neuroleptic drugs exert anti-schizophrenic therapeutic effects by blocking brain dopamine receptors [10]. In 1977 Squires & Braestrup [11] and Möhler & Okada [12] described highly specific receptors for benzodiazepines in rat brain, characterized by high-affinity [^{3}H]diazepam binding. The receptors were shown to be located in the synaptic membranes and to

be stereospecific [11, 12].   The receptors were found in greatest
density in the cortical, cerebellar and hippocampal regions in both
rat and human brain [11, 13].   The specific binding of [^{3}H]diazepam
was saturable [11, 12].   Non-specific binding, measured by carrying
out the assay with [^{3}H]diazepam in the presence of a large excess of
unlabelled diazepam was not saturable, but amounted to <10% of the
total receptor binding [11, 12].   The benzodiazepines with the most
potent pharmacological activity were found to have the highest affinity for the benzodiazepine receptors, and this supports the view
that these receptors constitute the site of action of benzodiazepines in the CNS [11 − 13].   Affinity for the receptor has been shown
to be associated with the presence of a carbonyl group at position 2
and a nitrogen atom at position 4 of the diazepine ring [3].

The first RR assays for benzodiazepines were described in 1979
and were based on the principle of competition between [^{3}H]diazepam
and unlabelled benzodiazepines in serum for binding sites on isolated
brain−membrane preparations [14-16].   More recently, [^{3}H]flunitraze-
pam has been used: it has a lower dissociation rate constant than
[^{3}H]diazepam, so that the reaction can be carried out at room temperature instead of at 0° [17, 18].

Ideally RR assays should be carried out with fresh preparations
of rat brain membranes, but in practice the techniques necessary to
obtain the preparations are too time-consuming for routine work: the
steps involve brain dissection, homogenization in ice−cold sucrose
or buffer, centrifugations, and washes with buffer followed by re-
suspension of the final pellet in buffer [11, 12]. In addition, it has
been found necessary to extract the benzodiazepines from serum
using an organic solvent before carrying out the assay, on account
of a serum component which inhibits receptor binding [15].   However,
in 1981 J. Lund [18] introduced dried, stable receptor preparations
from rat-brain membranes which had pharmacological specificity for
benzodiazepines closely similar to that of fresh membrane−bound rat
brain receptors.   Use of this type of preparation, now obtainable
from Amersham International plc (Amersham, U.K.), eliminates much
of the lengthy preparative procedure.

In our laboratory we have developed RR assays for diazepam and
clobazam and their active metabolites, and correlated the results
with those obtained by HPLC.

PROCEDURES

The RR assay is carried out using a modification of the method
developed by Lund [18, 19].   A portion (0.5 ml) of serum or standard
solution of benzodiazepine is extracted with 1 ml of ethyl acetate
and then 0.5 ml of the extract is evaporated to dryness at 35° under a
stream of nitrogen.   The residue is dissolved in 50 µl of 25 mM Na

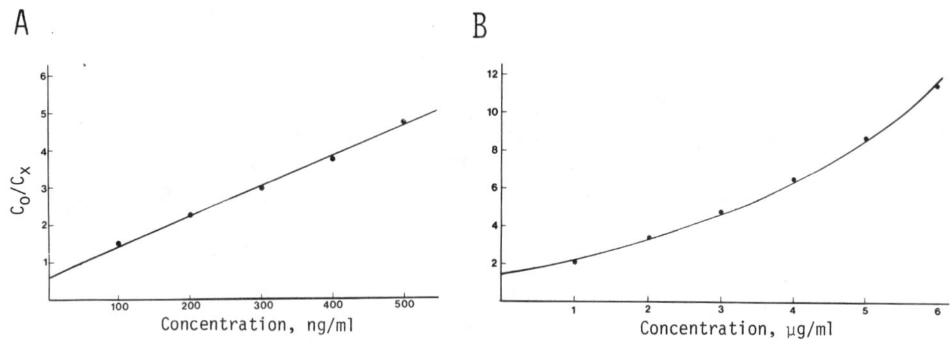

Fig. 1. RR assay calibration curves for clobazam   in serum over two concentration ranges: A, low; B, high.   Specific binding denoted C_O for the zero standard, C_X for the sample.

phosphate buffer (pH 7.4), then 400 µl of a prepared suspension of membranes (40 mg/ml in the phosphate buffer) is added, followed by 100 µl of [^{3}H]flunitrazepam to give a final concentration of 2 ng/ml. After 30 min the suspension is diluted with 10 ml of the buffer and filtered immediately through a glass-fibre filter (Whatman GF/C), using a Millipore XX 10025 filtering system with an XX 602205 pump. The filter is then triturated in 10 ml of Dimilume (Packard Insts., Reading) in a scintillation vial and the radioactivity read in a liquid scintillation counter.

The HPLC assay for diazepam and desmethyldiazepam is carried out using the method of Ratnaraj et al. [20] with a Spherisorb RP column (Spectra-Physics) and a methanol/water/acetic acid mobile phase. The method has been modified for the estimation of clobazam and desmethylclobazam [21]. The UV detector is operated at 254 nm for diazepam and 228 nm for clobazam with a sensitivity of 0.02 absorption units full-scale [20, 21].

RESULTS AND DISCUSSION

Over the concentration range studied the calibration curve for diazepam (or desmethyldiazepam) showed linearity with respect to the specific binding value for the zero standard (C_O) *versus* that for the sample (C_X): the ratio was 8 at 300 and 2 at 50 ng/ml (zero-concentration intercept = 1). Expressed inversely, the curve is linear when C_X is between 20% and 80% of C_O. Fig. 1A shows the corresponding calibration curve for clobazam (or desmethylclobazam); there is linearity between 10 and 800 ng/ml, but at very high concentrations the curve tends to follow a saturation kinetic pattern (Fig. 1B).

The correlation between RR assay and HPLC for diazepam and desmethyldiazepam is shown in Fig. 2, and that for clobazam and desmethylclobazam in Fig. 3. Good correlation was found in each case

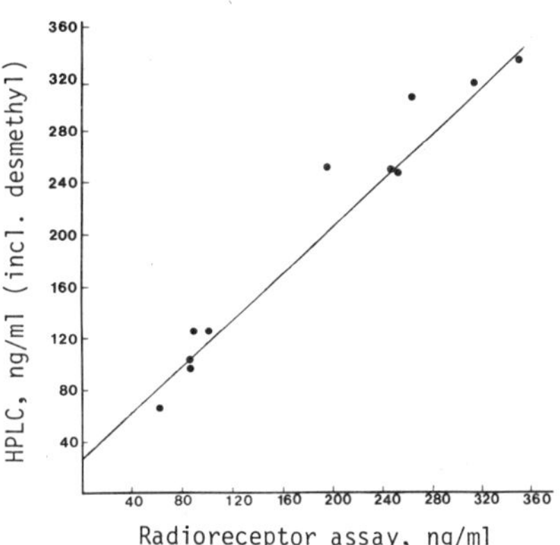

Fig. 2. Correlation between HPLC and radioreceptor (RR) assay values for serum concentrations of diazepam and desmethyldiazepam.

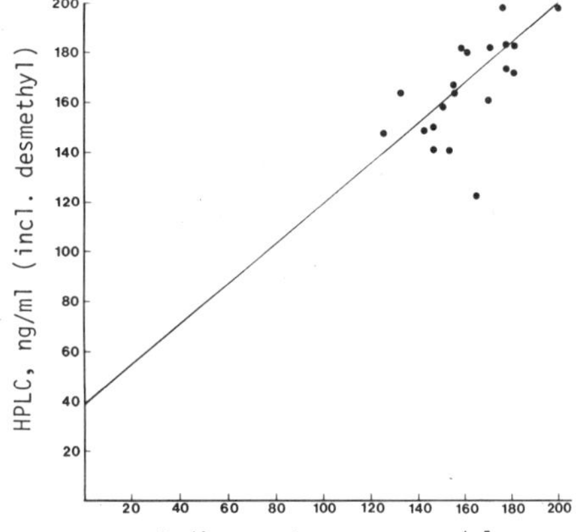

Fig. 3. Correlation between HPLC and RR assay values for serum concentrations of clobazam and desmethyl-clobazam.

between the concentration of parent drug and pharmacologically active metabolites as measured by RR assay and that derived by HPLC. The correlation coefficient was 0.94 for diazepam and desmethyldiazepam and 0.81 for clobazam and desmethylclobazam. This accords with reports of a good correlation between RR assay and well established chromatographic methods for benzodiazepines [2, 15, 16, 18].

Although the correlation between RR assay and chromatographic methods is acceptable, it is apparent that some differences in what is being measured do exist in biological terms. These differences when pursued may throw light on the mode of action of benzodiazepines [3]. It may be relevant that the benzodiazepines having the greatest effect on saccadic eye-movements [4, 8] are not the compounds which show the highest affinity for the receptor, despite the correlation with serum levels of the drugs.

A useful application for radioreceptor assays would, therefore, seem to be in the situation when new benzodiazepines are introduced into clinical practice. These assays would allow the relative contribution of the parent drug and its metabolites to the pharmacological activity of the whole system to be determined and thus facilitate the interpretation of results obtained in therapeutic monitoring.

References

1. Schütz. H. (1982) in *Benzodiazepines: A Handbook* (Schütz, H., ed.), Springer-Verlag, Berlin, pp. 5-6.
2. Lascelles, P.T., Ratnaraj, N. & Goldberg, V. (1983) in *Benzodiazepines Divided: A Multidisciplinary Review* (Trimble, M.R., ed.), Wiley, Chichester, 209-226.
3. Borea, P.A. (1983) *Arz. Forsch. 33*, 1086-1088.
4. Bittencourt, P.R.M. & Dhillon, S. (1981) in *Therapeutic Drug Monitoring* (Richens, A. & Marks, V., eds.), Churchill Livingston, London, pp. 255-271.
5. Lensmeyer, G.L., Rajani, C. & Evenson, M.A. (1982) *Clin. Chem. 28*, 2274-2278.
6. Greenblatt, D.J. & Shader, R.I. (1981) *as for* 4., pp. 272-280.
7. Klotz, U. (1981) *J. Chromatog. 222*, 501-506.
8. Bittencourt, P.R.M., Wade, P., Smith, A.T. & Richens, A. (1981) *Br. J. Clin. Pharmacol. 12*, 523-533.
9. Creese, I. & Snyder, S.H. (1977) *Nature 270*, 180-182.
10. Creese, I., Burt, D.R. & Snyder, S.H. (1976) *Science 192*, 481-483.
11. Squires, R.F. & Braestrup, C. (1977) *Nature 266*, 732-734.
12. Möhler, H. & Okada, T. (1977) *Science 198*, 849-851.
13. Braestrup, C., Albrechtsen, R. & Squires, R.F. (1977) *Nature 269*, 702-704.
14. Hunt, P. (1979) *Br. J. Clin. Pharmacol. 7 (Suppl. 1)*, 375-405.
15. Owen, F., Lofthouse, R. & Bourne, R.C. (1979) *Clin. Chim. Acta 93*, 305-310.
16. Skolnick, P., Goodwin, F.K. & Paul, S.M. (1979) *Arch. Gen. Psych. 36*, 78-80.
17. Duka, T.,Höllt, V. & Herz, A. (1979) *Brain Res. 179*, 147-156.
18. Lund, J. (1981) *Scand. J. Clin. Lab. Invest. 41*, 275-280.
19. Ratnaraj, N., Goldberg, V. & Lascelles, P.T. (1983) *Anal. Proc. 20*, 169-171.
20. Ratnaraj, N., Goldberg, V., Elyas, A. & Lascelles, P.T. (1981) *Analyst 106*, 1001-1004.
21. Ratnaraj, N., Goldberg, V. & Lascelles, P.T. (1984) *Analyst 109*, in press.

#NC(E)

NOTES and COMMENTS relating to

Ligand methods for drugs in forensic and other contexts

Comments related to particular contributions;

##E-1, #E-3 & #E-4, p. 317

#NC(E)-1

A Note on

ASSAY OF SECOVERINE AND ACTIVE METABOLITES: COMPARISON OF STABLE ISOTOPE DILUTION WITH RECEPTOR BINDING ASSAY

H. de Bree, P.H. van Amsterdam, D.J.K. van der Stel,
M.Th.M. Tulp, W.R. Vincent and L.W. de Zoeten

Duphar Research Laboratories
P.O. Box 2, Weesp, The Netherlands

Secoverine (formula below) is a new gastro-intestinal muscle relaxant whose pharmacokinetics and disposition are currently being investigated in different animal species and man. The drug is administered as a racemate of which only the (+) enantiomer has pharmacological activity. Human pharmacology revealed activity after either

1-cyclohexyl-4-[ethyl(p-methoxy-α-methylphenethyl)amino]-1-butanone hydrochloride

oral or parenteral administration. Plasma levels of the parent drug were hardly measurable (< 1 ng/ml) after oral administration, whereas a similar i.v. dose resulted in plasma levels of ~10 ng/ml. Accordingly we investigated the occurrence of active metabolites. As urinary metabolite profiles did not provide any clue in this respect, we started *in vitro* experiments with ^{14}C-labelled secoverine.

IN VITRO METABOLISM

The labelled compound was incubated with the 10,000 g supernatant from a rat liver homogenate. The composition of the incubation mixture was as described by Mazel [1], with 100 µg secoverine per mL. After incubation at 37°, metabolite patterns were ascertained using a reverse-phase (RP) HPLC system with radiochemical detection. Fig. 1 shows the metabolite pattern after incubation for 20 min.

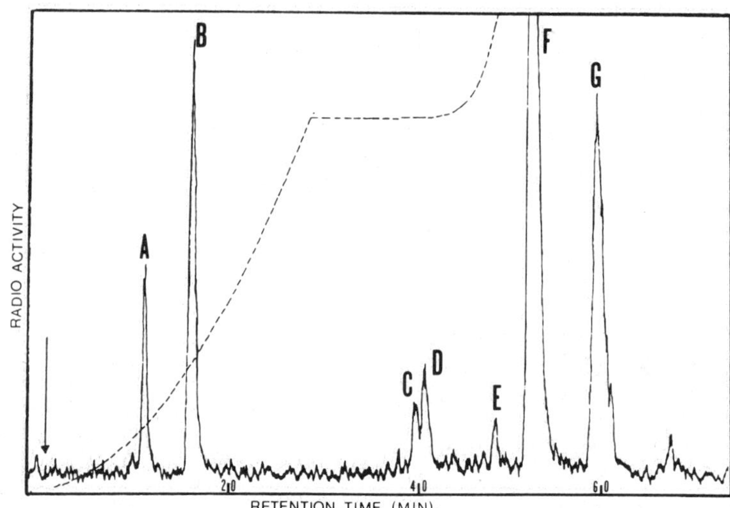

Fig. 1. RP-HPLC pattern of metabolites from secoverine incubated
for 20 min with rat-liver 10,000 g supernatant (cf. Table 1). Column:
7C8 Nucleosil. Mobile phase: gradient from water to methanol,
containing 5 g/l of ammonium bicarbonate and 1 g/l of sodium pentane
sulphonate. *Arrow* marks the starting point. Gradient profile
shown -----. Detection was with a solid-phase radiochemical flow-
cell. The sample was passed through a 1.2 μm filter before loading.

In order to detect and identify the pharmacologically active
metabolite(s), the experiment was scaled up to 10 mg of secoverine.
Subsequently all drug-related substances were isolated using pre-
concentration [2; cf. art. #B-1, this vol.] and HPLC with fraction
collection. The purified metabolites were subjected to a muscarinic
receptor affinity assay.

Receptor affinity

Affinity for cholinergic muscarinic receptors was determined by
receptor-binding investigation as described by Yamamura et al. [3].
After determination of the concentrations of the isolated metabolites
by ^{14}C-counting, each bulk solution was diluted so as to give appropriate concentrations for the binding assays, viz. 5 concentrations
ranging from 1 to 10,000 nM; each concentration was tested in triplicate. Table 1 gives the results, in terms of IC_{50} values and affinities relative to the parent drug, secoverine. Two metabolites,
E and F, clearly demonstrate a high affinity for the receptor. It is
surprising that the peak coinciding with the parent compound (G)
shows only a low affinity.

Table 1. IC$_{50}$ and activity, relative to secoverine racemate, of the compounds isolated from the incubation mixture. The IC$_{50}$ values are the concentration (nM) necessary for 50% displacement of ^{3}H-QNB.

Peak (cf. Fig. 1)	Amount (% of total radioactivity)	IC$_{50}$	Relative activity
A	6	>10,000	<0.001
B	14	>10,000	<0.001
C	2	>10,000	<0.001
D	4	900	0.004
E	4	37	0.13
*F: 1	40	14	0.34
2	10	8	0.6
G	20	200	0.024
Reference compounds			
Secoverine – racemate		5.5	(1.00)
(+) enantiomer		3.2	1.97
(−) enantiomer		260	0.021
O-Desmethylsecoverine		18	0.35
4-*trans*-(Cyclohexyl)hydroxy secoverine		7.8	0.71

* Peak F is composed of two components (see Fig. 2 legend) which were separated by TLC in the purification stage

Structure analysis

Fig. 2 shows the structures of the metabolites with a high affinity for the muscarinic receptor, elucidated by MS and proton magnetic

Fig. 2. Structures of the compounds isolated from the incubation mixture. Peak F contains besides O-desmethylsecoverine (80%) also 4-*trans*-(cyclohexyl)hydroxysecoverine (20%). A, B and D were not identified.

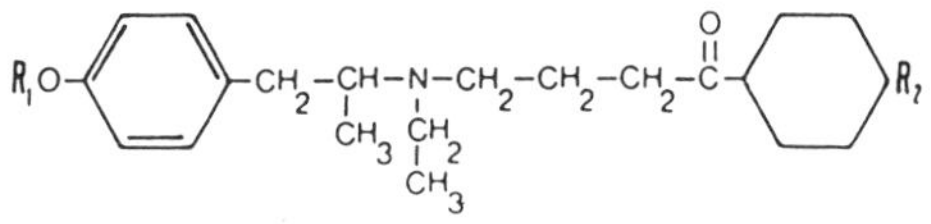

Metabolite	R$_1$	R$_2$
C	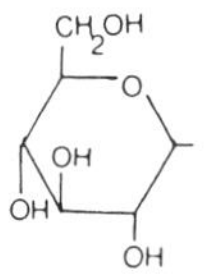 CH$_2$OH sugar ring (see structure)	H
E	H	OH
F	H and CH$_3$	H and OH
G	CH$_3$	H

resonance. The two compounds with the highest affinity in the receptor screen appear to be the *O*-demethylated parent drug (F1) and the 4-*trans*-cyclohexyl substance (F2), the latter having affinity comparable to the parent compound. Substance G was identified as secoverine, but in view of the low receptor affinity we must conclude that this substance is the inactive (−) enantiomer, apparently more slowly metabolized.

IN VIVO METABOLISM

As both the active metabolites F1 and F2 were virtually absent in human urine after oral or i.v. secoverine administration, human plasma was screened for the presence of these metabolites in order to explain the oral activity of secoverine. A sensitive stable-isotope GC–MS method (mass fragmentography) was developed to measure parent drug and both active metabolites.

Instrumental assay

The method is based on ion-pair extraction of the tertiary amine compounds into n-pentane, using heptane sulphonate. Clean-up is achieved by acid extraction and ion-pair re-extraction after neutralization. The hydroxyl groups of the metabolites are trimethyl-silylated with trimethylsilylimidazole (TSIM) and the product is subjected to GC-mass fragmentography with a moving-needle injector [4] and a support-coated open-tubular column [5]. As internal standards, [*ethyl* -d$_5$]secoverine and [*ethyl* -d$_3$]-4-*trans*-cyclohexylhydroxysecoverine were used.

Analysis of human plasma revealed the presence of 4-*trans*-cyclohexylhydroxysecoverine after oral administration, whereas i.v. administration gave mainly the parent drug. With neither route were there detectable levels of the *O*-desmethyl metabolite. Typical plasma level *vs*. time curves for oral and i.v. administration are presented in Fig. 3.

The instrumental assay, however, does not distinguish between enantiomers, and since only the (+) enantiomers are active it is desirable, for proper pharmacokinetic measurements, to have an assay for the active compounds only. Therefore a radioreceptor assay was developed based on the displacement of ^{3}H–QNB from cholinergic muscarinic receptors by active substances in the plasma sample.

Radioreceptor assay

After addition of butylated hydroxytoluene (BHT; antioxidant) the drug and its metabolites were extracted from 1–5 ml of the plasma by a Sep-pak-C$_{18}$ cartridge (Waters Associates Inc.). The bulk of endogenous material is washed from the cartridge with methanol/water (1:4 by vol.); secoverine and its metabolites are then eluted using

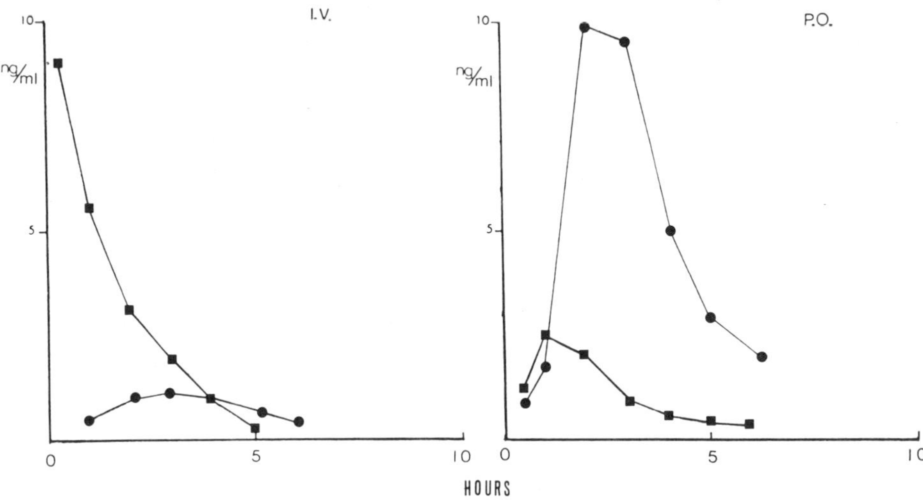

Fig. 3. Plasma level *vs.* time curves for secoverine and 4-*trans*-cyclohexylhydroxysecoverine after i.v. *(left)* and oral *(right)* administration of respectively 4 and 20 mg of drug to a human volunteer. Results obtained by instrumental assay.

Table 2.[*] Human plasma analyses: comparison of the instrumental with the radioreceptor assay (in subject BOS 2207), for secoverine (seco; determination limit 0.9 ng/ml) and the 4-*trans*-cyclohexylhydroxy metabolite (4-tr.OH; limit 0.6 ng/ml) individually in the case of the instrumental assay (GC-MF); the radioreceptor values (RRA) are expressed as equivalents of racemic secoverine in ng/ml.

| Sample time, h | Intravenous (0.05 mg/kg) | | | Oral (0.25 mg/kg) | | |
| | GC-MF | | RRA | GC-MF | | RRA |
	seco	4-tr.OH	total act.	seco	4-tr.OH	total act.
0	≤0.9	≤0.6	3.6	≤0.9	≤0.6	8.3
0.5	14.5	"	36.8	"	"	8.0
1	11.7	"	26.1	"	6.0	11.8
1.5	8.1	"	19.9	"	8.5	9.8
2	5.4	"	16.5	"	8.5	10.7
3	3.6	"	10.5	"	6.0	12.1
4	2.7	"	7.3	"	6.0	10.2
5	1.8	"	8.5	"	3.9	not analyzed
6	1.8	"	5.8	"	3.0	5.3
9	0.9	"	6.9	"	2.1	6.2

[*] discussed overleaf

pure methanol. The eluate is evaporated to dryness under reduced pressure and the residue is dissolved in phosphate buffer pH 7.4 containing 3% (v/v) of blank human plasma. An aliquot of the buffered residue is incubated with ^{3}H-QNB and an amount of freshly prepared receptor suspension that binds 25–30% of the ^{3}H-QNB. The receptor suspension was prepared from freeze-dried material according to Metcalfe [6]. After incubation for 60 min at 25° the suspension is filtered and the radioactivity in the residue is measured by liquid scintillation counting. Quantitation is done using a calibration graph made with the parent drug in blank plasma.

Comparative results

Table 2 (on previous p.) shows instrumental results for the drug and metabolite, and for total 'receptor activity' in the same plasma samples. Evidently the total receptor activity (expressed as secoverine racemate equivalents) slightly exceeds the sum of secoverine and 4- *trans*-hydroxysecoverine as measured instrumentally. The rather high baseline values observed by the radioreceptor assay are peculiar. Even if the i.v. results are corrected for these zero values, they remain higher than the instrumental values. Whether this is due to unknown active metabolites, stereoselective metabolism or other phenomena is not yet explained.

References

1. Mazel, P. (1972) in *Fundamentals of Drug Metabolism and Drug Disposition* (La Du, B.N., Mandel, H.G. & Way, E.L., eds.), Williams & Wilkins, Baltimore, pp. 527–536.
2. Ruijten, H.M., de Bree, H., Borst, A.J.M., de Lange, N., Scherpenisse, P.M., Vincent, W.R. & Post, L.C. (1984) *Drug Metab. Disp. 12,* 82–92.
3. Yamahura, H.I. & Snyder, S.H. (1974) *Proc. Nat. Acad. Sci. 71,* 1725–1729.
4. van den Berg, P.M.J. & Cox, Th.P.M. (1972) *Chromatographia 5,* 301–307.
5. German, A.L., Pfaffenberger, C.D., Thenot, J.P., Horning, M.G. & Horning, E.L. (1973) *Anal. Chem. 45,* 930–935.
6. Metcalfe, R.M. (1981) *Biochem. Pharmacol. 30,* 209–212.

#NC(E)-2

A Note on

SOME TRENDS IN LIGAND METHODS, INCLUDING FLUOROIMMUNOASSAY

J. Landon

Department of Chemical Pathology
St. Bartholomew's Hospital, London EC1A 7BE

Ligand assays depend upon the reversible, non-covalent binding between a specific binding protein and the ligand. The binding protein can be a receptor, a circulating binding protein (such as transcortin for the assay of cortisol) or an antiserum. Antisera, produced for immunoassays, now dominate because of the specificity and sensitivity that are possible and the ability to raise antisera against a vast range of small and large molecules.

Immunoassays need not have a labelled reactant as, for example, in those for proteins and glycoproteins which depend on the formation of a precipitin line or light scattering. However, use of ligand labelled with a radioisotope proved a major advance - enabling the assay of haptens as well as large molecules and markedly increasing potential sensitivity. More recently immunometric assays have been introduced, based on the use of labelled specific antibodies. These offer significant advantages, as compared with labelled ligand assays, for the user and will be employed increasingly.

Until recently, radioimmunoassays and immunoradiometric assays based on ^{125}I-labelled analytes and antibodies respectively have dominated. However, new labels are being introduced which avoid any health hazard, or emotive bias, have a virtually indefinite shelf-life and may not require a separation step. The use of fluorophores and phosphores offers several advantages as compared with such non-isotopic alternatives as enzymes. A number of fluoroimmunoassays and phosphoroimmunoassays have been developed in our laboratory.

AMPLIFICATION by Senior Editor, based on a review [1], of the foregoing Forum Abstract

Fluoroimmunoassay (FIA) entails use of antigen labelled with a fluorophore, e.g. fluorescein - whose dianion, if in alkaline solution, absorbs maximally at 490 nm and emits powerfully at 520 nm. When the species to be labelled lacks an amino group and hence is unreactive towards fluorescein isothiocyanate, one remedy may be to label a structurally related molecule, e.g. α,α-diphenylglycine where the analyte of interest is phenytoin [2]. Whilst there are alternative forms of non-isotopic immunoassay that, in common with FIA, have obvious advantages over RIA such as good stability of the labelled reagent, FIA has advantages over some enzyme-based methods, e.g. immediate measurement without actual removal of the bound fraction – which merely has to be absent from (or uniquely present in) the light path.

Even the latter constraint does not apply if the signal distinguishes free and bound label. Thus, fluorescein-labelled thyroxine gives little signal unless bound to antibody [3], and labelled gentamicin when bound to antibody gives little signal [4] or, conversely, in a polarization fluorimeter, gives an enhanced signal [5]. Polarization FIA is likewise feasible for the anti-cancer protein neocarzinostatin [6]. Whilst a separation step does commonly form part of FIA procedures, with possible benefit to sensitivity through diminishing interferences, the inconvenience of centrifugation can be avoided by the stratagem of pre-binding the antibody to magnetizable particles, as in a phenytoin assay [7].

Immunofluorimetric analysis (IFMA) entails labelling of the antibody rather than the antigen. An example of its application to small molecules is an assay for oestradiol-17β where the steroid was covalently pre-linked to Sepharose particles and the antibody carried an umbelliferone-type fluorophore [8].

Both FIA and IFMA call for a fluorimeter where possible troubles through even slight variations in light intensity are obviated, e.g. by 'ratio-recording'. Although sensitivity may not be as good as that achievable by RIA, even pg/ml levels in serum can be assayed for many analytes. The fluorimetric approach is notable for its speed, simplicity and measurement precision.

References

1. Smith, D.S., Al-Hakiem, M.H.H. & Landon, J. (1981) *Ann. Clin. Biochem. 18,* 253-274.
2. McGregor, A.R., Crookall-Greening, J.O., Landon, J. & Smith, J.S. (1978) *Clin. Chim. Acta 83,* 161-166.
3. Smith, D.S. (1977) *FEBS Lett. 77,* 25-27.
4. Shaw, E.J., Watson, R.A.A., Landon, J. & Smith, J.S. (1977) *J. Clin. Path. 30,* 526-531.
5. Watson, R.A.A., Landon, J., Shaw, E.J. & Smith, D.S. (1976) *Clin. Chim. Acta 73,* 51-55.
6. Maeda, H. (1978) *Clin. Chem. 24,* 2139-2144.
7. Kamel, R.S., Landon, J. & Smith, D.S. (1980) *Clin. Chem. 26,* 1281-1284.
8. Ekeke, G.I., Exley, D. & Abuknesha, R. (1979) *J. Steroid Biochem. 11,* 1597-1600.

#NC(E)-3

A Note on

PLASMA MORPHINE: COMPARISON OF RESULTS OBTAINED
BY HPLC AND RIA

G.W. Aherne, *A.R. Aitkenhead, E. Piall and N.K. Burton

Department of Biochemistry   *Department of Anaesthesia
University of Surrey   University of Leicester
Guildford GU2 5XH, U.K.   Leicester LE1 5WW, U.K.

Radioimmunoassay (RIA) has been used for several years to measure
morphine in biological fluids.   Recently some doubt has been cast
on the validity of the results obtained.   Over-estimation of unconjugated drug can occur with antisera having a relatively low % cross-
reaction with the main metabolite of morphine, morphine-3-glucuronide
[1].   In this study plasma morphine concentrations were measured
using two antisera and the results compared to those from HPLC.   The
plasma samples (25) were from terminally ill patients being treated
with oral doses of morphine sulphate (5-30 mg/dose) or diamorphine-
hydrochloride (10-60 mg/dose).

Two antisera were studied (Fig. 1), a goat antiserum raised
against a 6-succinyl morphine-BSA conjugate and a sheep antiserum
against a *N*-succinyl normorphine-BSA conjugate [2].   In the RIA
procedure (modified from that previously described [3]), 300 µl of
assay buffer (0.05 M phosphate pH 7.4, containing 6 g/l NaCl and
1 g/l gelatin), 100 µl of diluted antiserum, 100 µl of tritiated
dihydromorphine ($\simeq$100 pg) and 100 µl of standard or diluted sample
were incubated for 60 min.   Free and unbound radioactivity were
separated by use of dextran-coated charcoal.   The assay had an
inter-assay C.V. of 10% and a detection limit of <1 ng/ml.

The observed plasma morphine concentrations, in ng/ml, ranged
from 0 to 827 (mean 168) by HPLC, from 2.3 to 654 (mean 136) by RIA
using the sheep antiserum, and from 31 to 3250 (mean 777) using the
goat antiserum.   The non-comparability in results between the goat
antiserum and the sheep antiserum (or HPLC) is attributable to cross-
reactivity with morphine-3-glucuronide, 2.3% for goat antiserum and
only 0.26% for sheep antiserum. Comparison between HPLC values (y)
and sheep antiserum values (x) gave r = 0.977, y = -16 + 1.35x.

6-SUCCINYL MORPHINE - BSA

% CROSS REACTION	G/G/1
Morphine	100
Codeine	100
Diamorphine	100
Normorphine	< 1.0
Morphine-3-glucuronide	2.3

N-SUCCINYL NORMORPHINE - BSA

% CROSS REACTION	G/S/7
Morphine	100
Codeine	5.0
Diamorphine	2.5
Normorphine	16.0
Morphine-3-glucuronide	0.26

Fig. 1. The specificity of antisera raised against two BSA-conjugates of morphine.

RIA methods for measuring opiate alkaloids in biological fluids have many practical advantages over techniques such as GC and HPLC. However, results obtained with different antisera are not always comparable, reflecting relative cross-reaction with the glucuronide. [Examples of drug/metabolite cross-reactivity are given in Vol. 12, this series – Table 2 in #E.–*Ed.*]   Each antiserum should be carefully assessed before use in a particular application. Even where cross-reaction of an antiserum with the glucuronide is as low as 1-5% it appears necessary to incorporate a solvent-extraction step [4] to eliminate interference from this metabolite.

References

1. Aherne, G.W. (1983) *R. Soc. Med., Internat. Congr. & Symp. Series 58*, 21-28.
2. Morris, B.A., Robinson, J.D., Piall, E.M., Aherne, G.W. & Marks, V. (1974) *J. Endocrin. 64*, 6-7.
3. Aherne, G.W., Piall, E.M., Robinson, J.D., Morris, B.A. & Marks, V. (1976) in *Radioimmunoassay in Clinical Biochemistry* (Pasternak, C.A., ed.), Heyden, London, pp. 81-90.
4. Grabinski, P.Y., **Kaiko**, R.F., Walsh, T.D., Foley, K.M. & Houde, R.W. (1982) *J. Pharm. Sci. 72*, 27-30.

Comments relating to material in #E

Comments on #E-1, R.N. Smith - FORENSIC RIA

Question by Roberta J. Ward, partly related to mention by M.S. Moss of the statutory requirement to unequivocally identify the drug administered in racehorses (likewise in athletes): does this requirement apply in the forensic field also? *Reply by* R.N. Smith: since an RIA result by itself is of limited evidential value, positive RIA results are confirmed by alternative methods, although for cannabinoids in body fluids it is a second RIA that is used (with different labels, antisera and standards); however, the court also considers circumstantial evidence, e.g. possession of the drug, or medical evidence on the person's condition. A practice relevant to assay costs (R.N. Smith, *response to* J.S. Oliver) is that we save the EMIT-kit enzyme reagents routinely when we perform assays by RIA with the antisera; thereby we amass drug-protein conjugates that are now being used to raise antisera.

Comments on #E-3, S.H. Curry - LIGAND METHODS

Remarks on cross-reactivity, e.g. of metabolites.- D. Perrett advocated weighting, in published cross-reactivity comparisons, to take account of the relative amounts of parent drug and metabolites actually observed. R. Schmid and K. Ensing had misgivings about exploiting cross-reactivity to get an overall measure of 'biological' activity (in contrast with what chemical methods furnish); active and inactive enantiomers can indeed be distinguished, but the assay may not give a reliable picture of circulating species with clinical activity. The usefulness of the results depends on the context; it may be clinically advantageous to distinguish parent compound and an active metabolite, which can differ in action and side-effects as is observed with tricyclic antidepressants. Reiterating his misgivings about specificity, S.H. Curry (*in reply to* R.L. Williams) felt that with messy forensic samples (e.g. from a decomposed cadaver) there could be especial doubt. Whatever the analytical method (*agreeing with* I.S. Krull) one should obtain whatever confirmation is feasible.

Comments on #E-4, K. Ensing, & #E-5, R. Ratnaraj et al.
    - RADIORECEPTOR ASSAYS

Answering A. Gulaid: the enantiomer not disclosed by RRA in our study (revealed by GC) was quite inactive (K. Ensing); but with certain racemates one enantiomer might be, say, 20% active (H. de Bree). V. Goldberg (*reply to* R.M. Lee; cf. #E-5) appreciated that

the possible introduction of benzodiazepine antagonists could affect
RRA's, but the latter should generally be complemented by instrumental assays. *Answering* R.N. Smith: RRA's can be performed with
centrifugation instead of filtration; the purchased benzodiazepine
receptor preparation costs ~£1 per assay tube.

Remarks by R. Schmid.- Dr. Sieghart and I (at the Psychiatric
University Hospital, Vienna), in applying benzodiazepine binding
assay clinically, found some samples (~ 5-10%) seemingly had high
levels of benzodiazepine although, as confirmed by GC, none was in
fact present; evidently some endogenous substance can interfere with
benzodiazepine RRA. *Comment by* P.T. Lascelles: such anomalous observations warrant careful follow-up in the patients concerned. [In
#NC(E)-1, H. de Bree cites a comparable observation for secoverine.-
Ed. Also relevant: *Comments* at end of #D - B. Scales, tamoxifen.]

Some published observations (cited by Editor)

Human epidermal growth factor-urogastrone (ECF-URO; a peptide
of mol. wt. ~6000) has been assayed in patients, and a poor correlation found between RRA and RIA, whose respective merits and demerits
are discussed [1]. A book chapter [2] titled "A radioreceptor assay
for neuroleptic drugs" presents pioneer work together with valuable
practical guidance, e.g. on interferences from plastic containers
(a recurring theme in Vol. 10, this series), and with clinical applications of an RRA kit (Wellcome). Haloperidol gives rise to a
metabolite "more immunologically than pharmacologically active", such
that drug levels appear lower by RRA than by RIA. "The RRA will
measure the combined DA receptor blocking activity of all [neurolep-
tic] administered drugs and their active metabolites." The efficacy
of chlorpromazine may be due largely to its 7-hydroxy metabolite
(#E in Vol. 12 touches on chemical assay). Values relative to
haloperidol (= 100) by RRA are tabulated for various neuroleptics
(e.g. fluphenazine, 357; chlorpromazine, 18); for tricyclic antidepressants the RRA potencies were only ~1%. Drug binding to plasma
proteins is also considered.

Sodium cromoglycate may be specifically assayed in plasma by
RIA with use of a second antibody in separating bound and free radioligand [3]. A polarization fluoroimmunoassay [cf. NC(E)-2] can be
used to assay LSD in urine, although not serum [4].

1. Nexø, E., Lamberg, S.I. & Hollenberg, M.D. (1981) *Scand. J.
   Clin. Lab. Invest. 41, 577-582.*
2. Creese, I., Lader, S. & Rosenberg, B. (1981) in *Clinical
   Pharmacology in Psychiatry* (Usdin, E., ed.), Elsevier, N. York, pp. 79-104.
3. Brown, K., Gardner, J.J., Lockley, W.J.S., Preston, J.R. &
   Wilkinson, D.J. (1983) *Ann. Clin. Biochem. 20,* 31-36.
4. Hubbard, A.R., Miller, J.N., Law, B., Mason, P. & Moffat, A.C.
   (1983) *Anal. Proc. 20,* 606-608.

Section #F

VARIOUS ANALYTES IN BIOLOGICAL AND FORENSIC SAMPLES

#F-1

HPLC WITH ELECTROCHEMICAL DETECTION:
APPLICATIONS FOR THE NEUROSCIENCES

C.A. Marsden

Department of Physiology and Pharmacology
Medical School, Queen's Medical Centre
Nottingham NG7 2UH, U.K.

The original application of HPLC combined with electrochemical detection (LCEC) was for the assay of catecholamines. This article sketches current developments in LCEC assays available for measuring endogenous compounds in brain. These methods include single-run assays for catecholamines, indoleamines and their precursors and metabolites, using either reverse-phase (RP) or RP ion-pair separation. LCEC has also been used in the assay of (1) amino acids derivatized to make them electroactive, (2) neuropeptides containing one or both of the two readily oxidizable amino acids (tyrosine and tryptophan), (3) γ-aminobutyric acid (GABA), and (4) acetylcholine. The successful use of these methods depends on attention to three main areas: (a) suitable pre-injection sample preparation, (b) precise and reliable chromatography, and (c) suitable attention to the electrochemical electrode system.

HPLC with electrochemical (EC) detection has provided a major new research tool for the neuroscientist interested in the role of monoamines, e.g. noradrenaline (NA), dopamine (DA), 5-hydroxytryptamine (5HT) and their metabolites in peripheral and central neurones. The detection of these compounds by their oxidation at a carbon-based electrode following separation on an appropriate column has allowed the rapid, relatively cheap and very sensitive assay of amines in small samples of brain and plasma. However, while the major interest has centered on the monoamine assays there are many other compounds of interest to the neurochemist that are either electroactive or can be derivatized into electroactive compounds prior to detection. This group includes some amino acids, peptides containing one or more electroactive amino acids, GABA and acetylcholine (ACh). Many drugs used in general medicine and psychiatry are also electroactive (e.g.

certain neuroleptics, antidepressants and antihypertensives) and can
be measured by LCEC. Similarly, thiol-containing drugs are readily
oxidized (D. Perrett, #C-1, this vol.). Analytes amenable to LCEC
are listed in Table 1.

Three main requisites for successful LCEC are suitable pre-
injection sample extraction and preparation methods, consistent
chromatography, and detection at a carbon- or Pt-based electrode
with low residual current and high sensitivity. This article briefly
considers in general terms the electrodes available, and outlines
some specific applications of LCEC for measuring endogenous compounds
in brain samples. The basic principles have been discussed in
detail elsewhere [1, 2].

SELECTION OF ELECTRODES

Various carbon-based electrodes are now available. Maximum
sensitivity is obtained with carbon paste (e.g. BAS Inc.), though
these need more care in handling than the widely used glassy carbon
electrodes (e.g. BAS, Metrohm and EDT Research). Carbon paste electrodes provide high sensitivity combined with low residual current
and background noise. They are relatively easy to prepare and use
if a few basic guidelines are adopted: (a) they should never be
used with a mobile phase containing >15% methanol; (b) when preparing the electrodes care should be taken to use *fresh* carbon paste
and ensure that both the carbon-electrode surface and the plexiglass
electrode-housing surface are smooth and free from pits (the surface
of the electrode housing is best maintained by polishing with a
metal polish, e.g. Brasso); (c) the rim of the pit into which the
carbon paste is placed must not have rounded edges, otherwise the
flow of mobile phase over the electrode surface will be uneven. BAS
Inc., who provide carbon-paste electrodes, give full instructions
for their maintenance; but the extra points mentioned here will
further improve performance.

Glassy carbon electrodes are excellent for routine assays not
requiring maximum performance. The main problem with their use is
the difficulty in obtaining a baseline with low noise and so a good
signal-to-noise ratio at high sensitivity. Particular attention has
to be given to the mobile phase as this will considerably influence
the residual current detected by the electrode: for example, EDTA
addition to the mobile phase will remove any metals leached out of
the chromatographic system.

More recent additions to the range of EC detectors include the
Coulochem (ESA), which uses dual porous graphite electrodes. In our
laboratory this detector has been used successfully for measuring
certain neuropeptides (see later), and its application for measuring
catecholamines is discussed by Perrett (#C-1, this vol.). BAS also
provide a dual electrode system in which the glassy carbon electrodes

Table 1. Some compounds in brain showing EC activity, native *(above line)* or after derivatization *(below line)*.

DOPA	Tyrosine
Dopamine (DA)	Tryptophan
Noradrenaline (NA, NE)	Neuropeptides (see Table 3)
Adrenaline	
3-Methoxytyramine	Octopamine
3,4-Dihydroxyphenylacetic	Homovanillic acid (HVA)
acid (DOPAC)	3-Methoxy-4-hydroxyphenyl-
Normetanephrine (NM)	glycol (MHPG)
5-Hydroxytryptophan (5HTP)	Tryptamine
5-Hydroxytryptamine (5HT)	Tyramine
5-Hydroxyindoleacetic acid (5HIAA)	Melatonin
Catechol-related drugs	CNS drugs
e.g. α-Methyl DOPA & metabolites	e.g. some antidepressants
Isoprenaline	neuroleptics
	MAO inhibitors
	β-blockers
Ascorbic acid	
Uric acid	Glutathione

Native (left vertical label)

0	+0.5	+1.0 V

GABA   Amino acids (isoindole derivatives)
(trinitrophenylGABA ‒0.6 V)

Acetylcholine (ACh)
(Enzymatically produced H_2O_2)

Derivat^d· (left vertical label)

may be arranged either in parallel or in series, as discussed by R. Whelpton [#NC(C)-3, this vol.]. In parallel, peak purity is ascertainable from current ratios. In series, peak identification may be facilitated through having the first electrode set at a positive potential whereby analytes oxidizable at or below this potential are detected, and the second at a negative potential, detecting only compounds that are electrochemically reversible. Thus (Fig. 1), DA, DOPAC and NE (NA) show reversible oxidation reactions while the metabolite (of 5HT) 5HIAA and HVA are not reversible [3].

Alternative electrode materials include Pt, used successfully to detect hydrogen peroxide generated by the hydrolysis of ACh [4, & below].

AMINES

Most LCEC assays now use RP or ion‒pair RP chromatography with

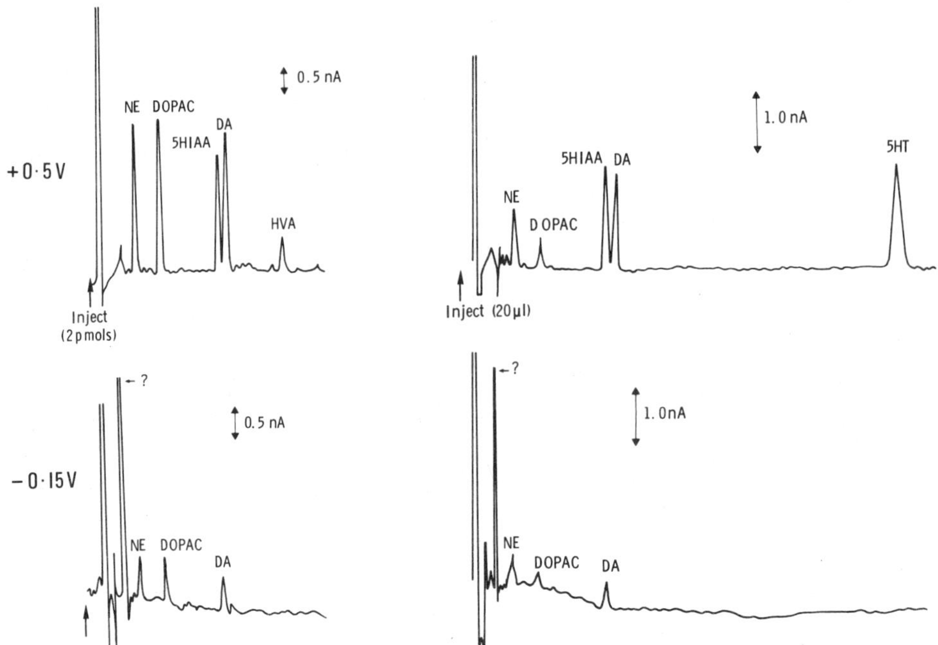

Fig. 1. HPLC with dual electrodes in series, the first set at
+0.5 V and the second at −0.15 V such that it reveals only
compounds that show reversible oxidation. HPLC in ion-pair RP
mode, with mobile phase (a) of Table 2. Abbreviations: see Table 1.
Left: Amine standards. Since 5HIAA does not show reversible
oxidation, the dual electrode clearly separates 5HIAA and DA.
Right: Rat brain sample. Note that 5HIAA and 5HT are both lacking
in the −0.15 V chromatogram.

detection at a glassy-carbon or carbon-paste electrode set at +0.65
or +0.90 V. The most common assays are summarized in Table 2 along
with relevant references. It must be stressed that while details of
solvents are given these should be treated only as a guide, since
the precise conditions need to be determined for individual columns
depending on their source and age. Details of the effects of pH,
buffer molarity and solvent composition on the separation of amines
and their metabolites are given elsewhere [4, 5].

(i) Pre-injection sample storage and preparation

Most of the early studies using LCEC to measure brain amines
involved an alumina extraction [1].[*] This is restrictive, as only
the catecholamines are extracted though alumina is still used for

____

[*] The Index entry 'Alumina' in Vol. 5 is pertinent, and 'Adsorbents'
in Vol. 12.– *Ed.*

Table 2. LCEC assays for brain amines and amine metabolites.
Note: The use of the newer 3 µm column materials with LCEC requires
some special attention: requisites include low dead volumes (5 µl
injections), increased mechanical damping and, in particular, a
rapid response from the recorder combined with decreased electronic
damping. Failure to establish fast response times can result in
peak merging, especially of early peaks.

	RP mode [refs. 6–9]	Ion–pair RP mode [refs. 3–6, 9–12]
Column	3 or 5 µm ODS (10 or 25 cm), e.g. Spherisorb, Ultrasphere, Biophase, µBondapak	3 or 5 µm ODS 2 (10 or 25 cm)
Mobile phase	0.1 M acetate/citrate buffer pH 4.6, with 10% (v/v) methanol	(a) 0.1 M NaH_2PO_4 buffer pH 3.6 & 0.1 mM EDTA, 0.05 mM Na octanyl sulphonic acid, 8–10% methanol (b) 0.1 M NaH_2PO_4, 1.0 mM EDTA, 1.5 mM octanyl sulphonic acid, 7–10% acetonitrile, pH 3.4
Electrode	Carbon paste [not with >15% (v/v) organic solvent] Glassy carbon Pyrolytic and Porous carbon	
Potential	+0.60 – +0.90 V Precise potential will depend on the analyte and on the electrode type	
Amines and metabolites *Abbreviations* – *Table 1*	DOPAC, 5HT/, tryptophan*, 5HIAA, HVA*, N-methyl-5HT	MHPG, NA (NE), DOPA, DHBA *(see below)*, DOPAC, NM, DA, tryptophan*, 5HIAA, 5HTP, HVA*, 3-methoxytyramine, (5HT/)
Run time	~14 min	12–40 min depending on whether 3 or 5 µm column packing
Pre–injection sample preparation	(a) Sonicate/homogenize in 10–20 vol. 0.2 M perchloric acid, spin, filter and inject† [12] (b) 0.5 M acetic acid, 0.5 M Na acetate, 0.4 N $NaClO_4$ (c) Sonicate/homogenize in 20 vol. mobile phase (b) adjusted to pH 4.0 [5]	
Internal standards	N-methyl-5HT	3,4-dihydroxybenzylamine (DHBA) N-methyl 5HT

/ Ion–pair RP separation difficult: long retention time (Fig. 1)
* Only detected using a high electrode potential, viz. +0.90 V
† Unsuitable for 5HT/5HIAA unless followed by immediate injection;
   rapid destruction, and their assay using automatic injection systems *needs care*; ensure full deproteinization, & add Na bisulphate

plasma catecholamine assays [13-16]. More recently, organic extraction methods have been adopted for assaying the latter [17]. With the advent of RP-HPLC and the consequent ability to measure amines and their metabolites in a single run, most pre-column separations now involve simple homogenization or, with small samples, sonication in perchloric acid or mobile phase (Table 2) followed by centrifugation and microfiltration prior to injection. Brain-region samples should be rapidly dissected and placed in sealed containers at $-80°$ for long-term storage (>2 weeks) or at $-20°$ for short-term storage. It is advisable to extract samples immediately prior to injection onto the column. We normally extract in batches of not more than 12 samples, since whilst catecholamines and their metabolites are relatively stable after extraction at $4°$, 5HT and 5HIAA are rapidly destroyed.

(ii) Internal standards

These are important as they (a) monitor the working efficiency of the chromatographic separation, (b) manifest the efficiency of the extraction procedure, and (c) minimize variations in values due to variations in detector response. *N*-methyl-5HT is a useful i.s. for RP separation, while 3,4-dihydroxybenzylamine is the compound normally used for ion-pair RP separation (Table 2). (The Index entry 'Internal standards' in Vol. 10 is pertinent.- *Ed.*)

(iii) Detection limits

Most EC detectors can measure 0.1 pmol noradrenaline, adrenaline, DOPAC, 5HT and 5HIAA on-column. Somewhat inferior (0.5 pmol) values are normal for the other amine metabolites (HVA, 3-methoxytyramine, MHPG) and for DA. This very high sensitivity allows the measurement not only of tissue levels but also of endogenous amine release *in vitro* from brain slices [e.g. 18] and *in vivo* using brain perfusate systems [e.g. 19, 20]. Recently LCEC has been applied to the measurement of endogenous DA release from post-mortem human striatal synaptosomes [21].

The ability of the newer LCEC assays to detect amines together with their precursors and metabolites allows the assay of the activity of most of the enzymes involved in the synthesis and degradation of both indoles and catecholamines, e.g. tyrosine hydroxylase, DOPA decarboxylase, dopamine- β -hydroxylase, phenethanolamine-*N*-methyltransferase, catechol-*O*-methyltransferase, monoamine oxidase, tryptophan hydroxylase and 5-hydroxytryptophan decarboxylase [22-24].

AMINO ACIDS

Only two amino acids, tyrosine and tryptophan, are readily electroactive (Table 3), but at higher potentials (~+0.85-0.95 V) than the amines [25]. Tryptophan can be measured together with 5HT,

Table 3. Peak oxidation potentials of tyrosine, tryptophan, cysteine and neuropeptides containing these amino acids at pH 4.6. The values (V) are the potential at which maximum current was generated.[*]

Substance	V	Substance	V
Tyrosine	0.84	Somatostatin	0.86–0.90
Tryptophan	0.88	Cholecysto-kinin (CCK-4)	0.88
Cysteine	0.90–1.00		
Neurotensin	0.78	Cholecysto-kinin (CCK-8)	0.88
Oxytocin	0.82		
Vasopressin	0.82	LH–releasing hormone	0.83
Caerulein	0.83		
Leu–enkephalin	0.84	α–MSH	0.80–0.86
Met–enkephalin	0.84	ACTH 1–24	0.80–0.88

Amino acids (1 µM) and neuropeptides (0.1 µM) were dissolved in 5 ml 0.15 M citrate/acetate pH 4.6 (at 7.4 V was ~0..5 V lower [18]. Oxidation was measured by differential pulse voltammetry (scan rate 10 mV /sec; step size 25 or 50 mV at sensitivity 200 or 500 nA).

[*] Peptides containing neither tyrosine nor tryptophan were not electroactive (e.g. substance P, thyrotrophin releasing hormone)

5HIAA, DOPAC and HVA using the RP separation described in the previous section (with the electrode set at +0.95 V), since it elutes between 5HT and 5HIAA. The problem with this method, however, is the high 'noise' caused by the oxidation of compounds in the mobile phase. This noise can be reduced by preventing metal contamination of the mobile phase through adding EDTA (to 1 mM) and replacing the metal filter on the solvent inlet with a sintered glass filter.

Recently, Joseph & Davies [26] have shown that the range of amino acids detected by LCEC can be radically extended by pre-column derivatization to iso-indoles, using o-phthalaldehyde and mercaptoethanol [27]. Another advantage of this approach is that the isoindole derivatives of the amino acids are well retained and resolved by RP chromatography, in contrast to the amino acids. The isoindoles are also fluorogenic though some (e.g. cysteine, kynurenine) show little or no fluorescence at the wavelengths used to monitor the other amino acids. Cysteine and kynurenine, however, are readily electroactive when derivatized to isoindoles. Fluorimetric and EC detection used in series with gradient RP chromatography (50 mM Na phosphate buffer pH 5.5 : methanol) provides a rapid separation and assay of a wide range of amino acids in a single run, combined with confirmation of their identities (Fig. 2). Isoindole derivatization has been successfully applied to the measurement of amino acids of interest to neuroscientists in plasma (Fig. 3), brain and cerebrospinal fluid (details in [26]).

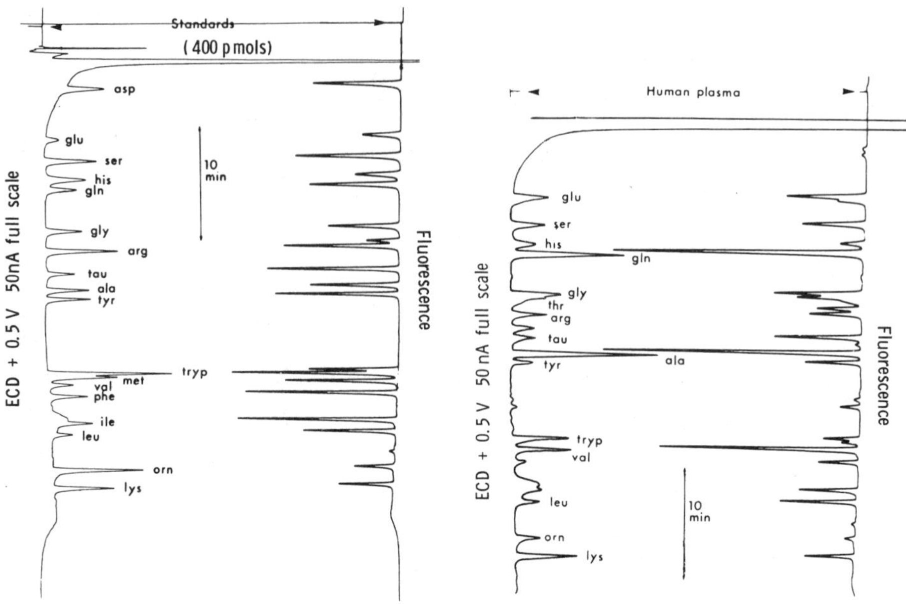

Fig. 2. Gradient elution of an
amino acid mixture (400 pmol of
each on the column), derivatized
with o-phthalaldehyde + mercaptoethanol [26]. *Figs. 2 & 3 kindly
supplied by Dr. M.H. Joseph.*

Fig. 3. Gradient elution of
derivatized amino acid in
human plasma. Injection =
50 µl plasma, methanol-deprot-
einized. 0.05 M Na phosphate
pH 5.5 with % methanol increasing.

NEUROPEPTIDES

With the interest in diverse intra-neuronal peptides as found
in brain in low concentration, assay has generally been by RIA
after production of a suitable antibody. There are, however, many
examples of families of peptides whose sequence differs by only one
amino acid, which poses difficulties in producing absolutely specific antibodies; HPLC is increasingly used to separate closely related neuropeptides prior to RIA [27]. An alternative approach would
be to combine HPLC with a sensitive on-line detection system.
Neither UV nor fluorimetric detection have adequate sensitivity to
measure the low levels found in brain. For electroactive neuropeptides (containing tyrosine and/or tryptophan; Table 3) in brain samples, LCEC offers promise of the requisite sensitivity, although
there are certain problems. These include the high working potential
required, and the need for gradient separation and for development
of a suitable pre-injection procedure for sample preparation. However, initial studies using isocratic separation have given encouragement [28, 29].

In our own studies [29] we have measured vasopressin and oxytocin levels in the neurointermediate lobe of rats homozygous for diabetes insipidus (Brattleboro rats, vasopressin-deficient) and of heterozygous controls (Long-Evans rats). RP-HPLC was carried out on a Hypersil 5 ODS column (250 X i.d. 5 mm) with a mobile phase containing 0.15 M Na_2HPO_4 and methanol (60:40 by vol.), pH 6.0, pumped at 1 ml/min. The glassy carbon electrode (BAS Inc.) was set at an applied potential of +1.0 V. Single neurointermediate lobes were extracted following sonication in 1.0 ml 0.1 M acetic acid / methanol (50:50) and centrifugation at 3000 rpm for 15 min. The supernatant was dried down under vacuum and the residue taken up in 1 ml of the mobile phase; 100 μl samples were injected onto the column. Vasopressin levels of 22.8 ±2.8 nmol/mg protein were measured in the sample from the control rats, while none was detected in the vasopressin-deficient rats. There was no significant difference in the oxytocin levels between the two strains of rat, and there was close agreement between the vasopressin values obtained by LCEC and those by RIA. The main problem with this assay has been the rapid loss of sensitivity of the BAS glassy-carbon electrode following injection of neuropeptide standards or brain samples. This is possibly due to adsorption of the peptides or their oxidation products onto the electrode surface.

More recently we have used the Coulochem dual-electrode detector in the peptide studies. While we have found no improvement with regard to catecholamine assays between this and the BAS detector, we have found increased sensitivity and, in particular, stability in the recorded peaks for peptides both with standards (Fig. 4) and with tissue extracts, using the Coulochem detector.

The combination of improved EC detection and increased column efficiency can be expected to promote interest in peptide chromatography together with on-line sensitive detection. There are many applications, but one particular area is the separation of sulphated and non-sulphated peptides, the former evidently being the active forms whilst both are found in the mammalian CNS. Only the non-sulphated forms of the electroactive peptides show EC reactivity, as sulphated tyrosine cannot be oxidized. Since desulphation of tyrosine is readily achieved either enzymatically (sulphatase) or by boiling with acid, EC detection may offer a means of measuring both the sulphated and non-sulphated levels of neuropeptides, by difference [30].

γ-AMINOBUTYRIC ACID (GABA)

GABA is an important inhibitory transmitter in mammalian brain which relates to neurological diseases such as Huntington's chorea and to the action of benzodiazepines. GABA exhibits no native electroactivity but readily undergoes pre-column derivatization to an electroactive compound, either using o-phthalaldehyde and mercapto-

Fig. 4. LCEC (RP) applied to
standards (10 pmol each):
L-enkephalin (ENK), met-
enkephalin (M-ENK), vasopressin
(VP) and angiotensin II (ANG II).
Replicate runs were performed
with two different chart speeds.
Dual-electrode detector
(Coulochem ESA) using porous
graphite electrodes: first
electrode set at +0.5 V; the
second at +0.8 V and recordings
made with that electrode.

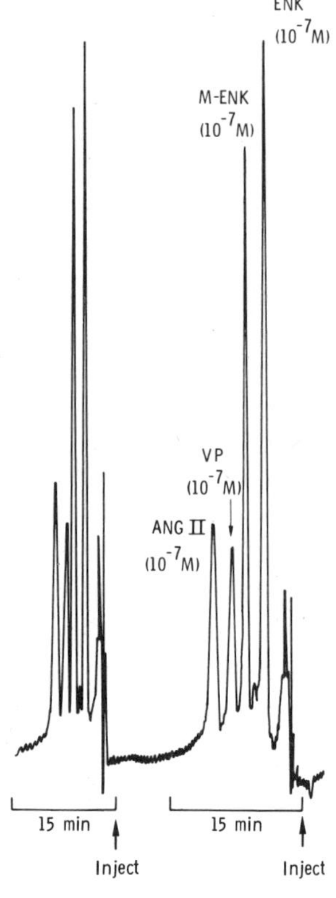

ethanol as described above [26], or using 2,4,6-trinitrobenzene
sulphonic acid (TNBS) [31]. The latter approach has been used to
measure brain levels of GABA and its release from brain slices. The
basis of the derivatization is the conversion of GABA to 2,4,6-tri-
nitro-γ-aminobutyric acid. The final sample solution is deoxygenated
by bubbling N_2 through it prior to analysis with reductive EC detection at a potential of −0.6 V following separation by RP chromatography. The extracts require a pre-column clean-up stage, to remove
impurities such as the products arising from reaction of the TNBS
with hydroxyl ions or trace impurities in the buffer.

ACETYLCHOLINE (ACh) AND CHOLINE

The assay of ACh has long posed problems, partly because of the
complexity of existing assays and partly because of their lack of
sensitivity. Potter and co-workers [4] have developed an LCEC assay
based on the ion-pair RP separation of ACh and choline, with ethyl
homocholine as i.s.; there is post-column enzymatic conversion of
ACh to choline and subsequent hydrolysis of choline to betaine and
hydrogen peroxide by choline oxidase. The hydrogen peroxide is

detected using a Pt electrode at +0.5 V set against a Ag/AgCl reference electrode. The detection limit for ACh is 2 pmol with hydrogen peroxide standards giving a linear response over a range of 1-2 pmol. The assay, applicable to ACh in brain samples, is described in detail elsewhere [4].

Acknowledgement

Studies from the author's laboratory have been financially supported by the Wellcome Trust.

References

1. Adams, R.N. & Marsden, C.A. (1982) in *Handbook in Psychopharmacology*, Vol. 15 (Iversen, L.L., Iversen, S.D. & Snyder, S.H., eds.), Plenum, New York, pp. 1-74.
2. Marsden, C.A. (1983) *Trends Pharmacol. Sci. 4*, 148-152.
3. Mayer, G.S. & Shoup, R.E. (1983) *J. Chromatog. 255*, 533-544.
4. Potter, P.E., Meek, J.L. & Neff, N.H. (1983) *J. Neurochem. 41*, 188-194.
5. Saller, C.F. & Salama, A.I. (1984) *J. Chromatog.*, in press.
6. Mefford, I.N. (1981) *J. Neurosci. Methods 3*, 207-224.
7. Reinhard, J.F., Moskowitz, M.A., Sved, A.F. & Fernstrom, J.D. (1980) *Life Sci. 27*, 905-911.
8. Sperk, G. (1982) *J. Neurochem. 38*, 840-843.
9. Marsden, C.A., Macdonald, I.A., Brazell, M.P. & Maidment, N. (1983) *Anal. Proc. 20*, 559-562.
10. Wagner, J., Vitali, P., Palfreyman, M.G., Zraika, M. & Huot, S. (1983) *J. Neurochem. 38*, 1241-1254.
11. Nielsen, J.A. & Johnston, C.A. (1982) *Life Sci. 31*, 2847-2856.
12. Mefford, I.N., Gilberg, M. & Barchas, J.D. (1980) *Anal. Biochem. 104*, 468-472.
13. Hjemdahl, P. (1984) *Acta Physiol. Scand. 120, Suppl. 527,* in press.
14. Hjemdahl, P. (1984) *Am. J. Physiol. 245*, in press.
15. Hjemdahl, P., Daleskog, M. & Kaham, T. (1979) *Life Sci. 25*, 131-138.
16. Allenmark, S. (1982) *J. Liq. Chromatog., Suppl. 1*, 1-41.
17. Smedes, F., Kraak, J.C. & Poppe, H. (1982) *J. Chromatog. 231*, 25-39.
18. Bennett, G.W., Sharp, T., Marsden, C.A. & Parker, T.L. (1983) *J. Neurosci. Methods 7*, 107-115.
19. Zetterström, T., Sharp, T., Marsden, C.A. & Ungerstedt, U. (1983) *J. Neurochem. 41*, 1769-1773.
20. Sharp, T., Maidment, N.T., Brazell, M.P., Zetterström, T., Ungerstedt, U., Bennett, G.W. & Marsden, C.A. (1984) *Neurosci.* in press.
21. Dowdall, M.J., Szafranski, J. & Marsden, C.A. (1983) *J. Neurochem. 41, Suppl. 1*, S139.
22. Borchart, R.T.. Vincek, W.C. & Grunewald, G.L. (1977) *Anal. Biochem. 82*, 149-157.

23. Borchart, R.T., Henazi, M.F. & Schowen, R.L. (1978) *J. Chromatog.* *152*, 255-259.
24. Blank, C.L., Wong, P., Bulawa, M. & Lin, P.Y. (1981) in *Function and Regulation of Monoamine Enzymes* (Usdin, E., Weiner, N. & Yondim, M.B.H., eds.), Macmillan, London, pp. 759-770.
25. Bennett, G.W., Brazell, M.P. & Marsden, C.A. (1981) *Life Sci.* *29*, 1001-1007.
26. Joseph, M.H. & Davies, P. (1983) *J. Chromatog.* *277*, 125-136.
27. Lindroth, P. & Mopper, K. (1979) *Anal. Chem.* *51*, 1667-1671.
28. Foyster, L.L., Marsden, C.A. & Bennett, G.W. (1982) *Regulatory Peptides* *4*, 361.
29. Bennett, G.W., Foyster, L.L., Greenwood, D.T. & Marsden, C.A. (1982) *Br. J. Pharmacol.* *77*, 442P.
30. Sauter, A. & Frick, W. (1983) *Anal. Biochem.* *133*, 307-313.
31. Caudill, W.L. & Wightman, R.M. (1982) *Anal. Chim. Acta* *141*, 269-273.

#F-2

GAS CHROMATOGRAPHY WITH ELECTRON-CAPTURE DETECTION FOR MEASUREMENT OF BIOACTIVE AMINES IN BIOLOGICAL SAMPLES

[1]Ian L. Martin, [2]Glen B. Baker and [2]Ronald T. Coutts

[1]MRC Neurochemical
Pharmacology Unit
Medical Research Council
Centre, Medical School
Hills Road
Cambridge CB2 2QH, U.K.

[2]Neurochemical Research Unit
Department of Psychiatry and
Faculty of Pharmacy and
Pharmaceutical Sciences
University of Alberta
Edmonton T6G 2GE, Canada

GC-ECD is used extensively for the analysis of a number of endogenous arylalkylamines and structurally related drugs in tissues and body fluids. In our laboratories, procedures have been developed for analysis of several bioactive amines after derivatization. With appropriate extraction conditions, several of the amines may be determined simultaneously. The methods described below offer sensitivity, speed and specificity, particularly when the GC analyses utilize capillary columns.

Investigations of the functional significance of the biogenic amines have largely depended on the development of methods for their quantification. It is only in the last decade that procedures with the appropriate sensitivity and specificity have become available to allow their confident examination in nervous tissue. Methodologies now used extensively include mass-spectrometry (MS; alone or combined with GC), radio-enzymatic procedures, HPLC with electrochemical detection [see accompanying arts. (and beware of this application of the term 'ECD' - *Ed.*)] or fluorescence detection, and GC-ECD. We will review here the use of the latter technique in its application to determining the catecholamines and 5-hydroxytryptamine (5-HT) and the non-phenolic trace amines with their phenolic analogues, as these provide representative examples of the techniques that have been developed.

In general, substances of this class are totally unsuitable for analysis by GC because of their lack of volatility. The first requirement was therefore to obtain derivatives which were both volatile and sufficiently stable for subsequent analysis. Together with this, the sensitivity requirement suggested that the detection mode be ECD, implying constraints in the choice of derivatives.

THE CATECHOLAMINES AND 5-HYDROXYTRYPTAMINE

The development of an effective assay procedure for these compounds in biological material (brain) was first reported in 1973 [1]. At that time a number of GC derivatization procedures were available which fulfilled the above-mentioned criteria; these involved trifluoroacetylation [2] or silylation of the hydroxyl groups, followed by formation of the heptafluorobutyryl amides [3, 4]. However, difficulties were encountered, not in preparing the derivatives but in the preliminary separation of these relatively hydrophilic amines from the tissue and their subsequent conversion to appropriate derivatives under anhydrous conditions without excessive losses. The use of ion-exchange resins at that time did not appear a practicable solution to the problem, and the poor solubility of these amines in most common organic solvents precluded a simple partitioning approach. The problem was eventually overcome with the use of the liquid ion-exchanger di-(2-ethylhexyl)phosphate (DEHPA), which had previously been employed by Temple & Gillespie for the recovery of some amines of biological interest [5].

The formation of the trifluoroacetyl derivatives of noradrenaline (NA), dopamine (DA) and 5-hydroxytryptamine (5-HT) was readily accomplished at room temperature using the appropriate anhydride with acetonitrile as the solvent The expected derivative structures were confirmed by MS: modifications were in each case at phenolic, hydroxyl, amino and pyrrole positions. However, because of the GC retention times of the catecholamine and 5-HT derivatives under the conditions used, it was decided to separate these compounds during the extraction procedure.

In the final protocol brain tissue was homogenized in acidified butan-1-ol and, after brief centrifugation, the amines were back-extracted into an aqueous phase by decreasing the polarity of the butanol with 2,2,4-trimethylpentane. The catecholamines were then adsorbed onto alumina at neutral pH while 5-HT was recovered from the alumina supernatant by multiple extractions into butanol. The organic layer was evaporated to dryness and the derivatives were formed with trifluoroacetic anhydride in acetonitrile medium. An aliquot of the reaction mixture could then be injected directly into the GC. The catecholamines were recovered from the alumina with perchloric acid, the pH adjusted to 7.8 with solid $NaHCO_3$, and the amines extracted into a 2.5% solution of DEHPA in chloroform. The catecholamines were then back-extracted into a small volume of 0.5 M formic acid, which was evaporated to dryness and the residue derivatized as for 5-HT.

The overall sensitivity of the assay procedure allowed the quantification of 5 ng of both NA and DA and 10 ng of 5-HT in a single piece of brain tissue. While this does not compare favourably with present-day methods, it marked a considerable advance at the time. The method did suffer from a number of disadvantages. The recovery of these relatively hydrophilic amines into DEHPA was concentration-dependent, and calibration curves were therefore curvilinear. The multiple butanol extractions for 5-HT were time-consuming though it has since been possible to recover this amine much more efficiently with the liquid ion-exchanger, as for the catecholamines. The second disadvantage was the direct injection of the reaction mixture onto the column, leading to large tailing solvent peaks. It was therefore necessary to limit the amount of trifluoroacetic anhydride used, consistent with maintaining complete derivatization. This latter problem was later largely overcome by introducing a precolumn, subsequent to which the more volatile anhydride could be bled off before it reached the analytical column [6]. The protocol was, however, far from ideal. Further work showed that both problems can be circumvented for many similar compounds of biological interest although, for the catecholamines, the difficulties have not been entirely resolved.

THE NON-PHENOLIC TRACE AMINES

This class of compound, notably 2-phenylethylamine (PEA) and tryptamine (T), may be recovered from aqueous homogenates by direct extraction into organic solvent under basic conditions. The compounds are, however, quite polar and direct extraction under these conditions tends to be inefficient and leads to little preferential purification of the amines of interest. Chattaway in 1931 observed that both phenolics and primary aliphatic amines may be efficiently acetylated directly in aqueous solution under slightly basic conditions with acetic anhydride [7]. The resulting acetylated derivatives, having a considerably reduced polarity compared to their parents, can be efficiently extracted into organic solvents. In our hands the extraction of PEA with ethyl acetate, subsequent to acetylation, results in 95% overall recovery of the compound in this 2-step procedure. The amines may then be further derivatized with one of the perfluoroacylating agents to yield compounds which are suitable for ECD. This procedure therefore overcomes the first hurdle, the efficient recovery and derivatization of such compounds under anhydrous conditions.

The second major difficulty arose from the removal of excess derivatizing agent from the preparation prior to GC. Evaporation of the sample to dryness and re-constitution in an appropriate solvent was found to be unreliable because of the volatility of the derivatives produced, i.e. the amine derivative was partially lost during the evaporation stage. [Table 3 in #A-6 by E. Reid, Vol. 7 of this series, documents this problem for two drugs.- *Ed.*] It was commonly believed that such perfluoroacylated derivatives were extremely water-sensitive, and under certain conditions this is in fact the case; but we made use of the fact that the acetylated perfluoroacyl derivatives are also very lipophilic. Purification was achieved by partitioning the reaction mixture between a small volume of cyclohexane and a much greater volume of saturated sodium tetraborate. The unreacted perfluoroacyl anhydride was presumably hydrolyzed rapidly to the acid and so partitioned readily into the aqueous phase, while the lipophilic amine derivative remained in the organic solvent which was dried down for the GC step. Thus in a single step we could remove the derivatizing reagent and produce a degree of concentration of the derivative with little if any loss.

The modified procedure allowed not only an excellent recovery of the amine of interest but also a remarkable improvement in the GC analysis of the resulting derivatives for PEA [8]. The methodology was later extended for the determination of tryptamine [9, 10].

THE PHENOLIC TRACE AMINES

Although a procedure such as is described above for the non-phenolic analogues can likewise be applied to compounds containing phenolic substituents, a disadvantage immediately becomes apparent. Acetylation of primary amino functions in aqueous solution allows subsequent derivatization of the acetylated amine with the perfluoroacyl anhydrides; but in the case of the phenolics the position is blocked to further derivatization. In order to increase sensitivity and attain acceptable GC retention times, addition of more fluorinated substituents to the molecule would be advantageous. It has now proved possible to selectively hydrolyze acetylated phenolics, leaving the acetylated amino derivative intact. Subsequently both the phenolic and the acetylated amino functions can be perfluoroacylated to yield products with good sensitivity for ECD.

These various procedures have now been modified and combined into one to provide for simultaneous analysis of PEA, *m*- and *p*-tyramine (*m*- and *p*-TA), 3-methoxytyramine (3-MTA), normetanephrine (NME), tryptamine (T) and 5-HT in a single sample [11, 12].

Multi-analyte assay procedure as performed for brain

The brain tissue is homogenized in ice-cold 0.1 M perchloric acid containing 10 mg % EDTA. One or more of the following amines can be added as internal standards: 5-methyltryptamine (for T and 5-HT only), benzylamine, 3-phenylpropylamine, 2-(4-chlorophenyl)-ethylamine or tranylcypromine. After centrifugation, the supernatant is adjusted to pH 7.8 (NaHCO$_3$) and shaken with DEHPA (2.5% in chloroform). Following separation of the phases, the lower layer is retained and back-extracted with 0.5 M HCl. This acid phase is neutralized with solid sodium bicarbonate and reacted with acetic anhydride (1 vol.:10 vol. aqueous phase). After cessation of effervescence, the sample is shaken with ethyl acetate to extract the acetylated amines. The ethyl acetate layer is then divided into two portions: **A** for analysis of T and 5-HT, and **B** for analysis of PEA, *m*-TA and *p*-TA, 3-MTA and NME.

Portion A is shaken briefly with 0.1 vol. of dist. water (to remove residual sodium bicarbonate) and, after a brief centrifugation, the organic layer is retained and evaporated to dryness under a stream of nitrogen. To the residue is added 25 µl of ethyl acetate and 75 µl of pentafluoropropionic anhydride (PFPA). The mixture is reacted at 60° for 30 min, then partitioned between cyclohexane (300 µl) and satd. sodium tetraborate buffer (3.0 ml). A portion of the organic phase is injected onto a GC-ECD equipped with a 10 m OV-101 fused silica capillary column or a 2 m packed column containing 3% OV-17 on Gas-Chrom Q.

Portion B is shaken with 0.1 vol. of 10 M ammonia solution, to hydrolyze the acetylated phenolic groups. After adding 6 M HCl to the aqueous phase (0.1 vol.), the organic phase is retained and taken to dryness under a stream of nitrogen. The residue is reacted with 75 µl of tetrafluoroacetic anhydride in the presence of 25 µl of ethyl acetate. The reaction is allowed to proceed for 30 min, and then the mixture is partitioned between cyclohexane (300 µl) and satd. sodium tetraborate buffer (3.0 ml). An aliquot of the cyclohexane layer is used for GC-ECD analysis using the same capillary column as described above for T and 5-HT. The derivatives formed under these conditions are *N*-acetyl,*N*-trifluoroacetyl-PEA [12], *N*-acetyl-*N,O*-di(trifluoroacetyl)-*m*-TA, -*p*-TRA,-3-MTA and -NME [13, 14], and spirocyclic derivatives of T and 5-HT [15]. PEA can be analyzed using either procedure, but the method employed for portion **B** has proved more satisfactory for PEA in urine.

Values thereby obtained are: (a) control rat brain (ng/g), PEA, 1.1 [8]; *p*-TA, 1.8 [13]; 3-MTA, 16; NME, 8 [16]; T, <1; and 5-HT, 581 [10]; and (b) human urine (µg unconjugated amine/24 h), PEA, 22; *p*-TA, 101; *p*-TA, 506; 3-MTA, 48; NME, 28; T, 96; and 5-HT, 139 [14]. These levels are in good agreement with values obtained using MS techniques [17-21]. Urinary PEA levels are somewhat higher than those reported using integrated-ion-current MS [20] and GC-MS [21]; but levels of the other amines agree well with MS values. Control (normal) values for *m*-TA and T in brain could not be obtained because of background interference, but both can be quantitated in urine samples and in brains of rats treated with monoamine oxidase inhibitors.

CONCLUDING COMMENTS

It is clear, then, that such techniques can be used for the quantification of a number of arylalkylamines. The initial problem of the inefficient recovery of many of these compounds from aqueous homogenates has been largely circumvented, and techniques have been developed for the removal of excess derivatizing agents. Advances in instrumentation (e.g. GC capillary columns) have greatly facilitated these studies, and the development of some novel derivatizing agents may be expected to result in further increases in the sensitivity of similar methodologies.

Acknowledgements

Funding for parts of the above work was provided by the Medical Research Council of Canada, the Alberta Mental Health Research Fund, and the Alberta Heritage Foundation for Medical Research.

References

1. Martin, I.L. & Ansell, G.B. (1973) *Biochem. Pharmacol. 22*, 521-533.
2. Kawai, S. & Tamura, Z. (1968) *Chem. Pharm. Bull. Tokyo 16*, 699-701.
3. Horning, M.H., Moss, A.M., Boucher, E.A. & Horning, E.C. (1968) *Anal. Lett. 1*, 311-321.
4. Anggard, E. & Sedvall, G. (1969) *Anal. Chem. 41*, 1250-1256.
5. Temple, D.M. & Gillespie, R. (1969) *Nature 209*, 714-715.
6. Martin, I.L. (1974) *J. Chromatog. 96*, 232-234.
7. Chattaway, F.D. (1931) *J. Chem. Soc.*, 2495-2496.
8. Martin, I.L. & Baker, G.B. (1981) *Biochem. Pharmacol. 26*, 1513-1516.
9. Baker, G.B., Calverley, D.G., Dewhurst, W.G. & Martin, I.L. (1979) *Br. J. Pharmacol. 67*, 469P-470P.
10. Calverley, D.G., Baker, G.V., McKim,H.R. & Dewhurst, W.G. (1980) *Can. J. Neurol. Sci. 7*, 237.
11. Baker, G.B., Coutts, R.T. & Martin, I.L. (1981) *Prog. Neurobiol. 17*, 1-24.
12. Baker, G.B., Coutts, R.T. & Martin, I.L. (1984) in *Neurobiology of the Trace Amines* (Boulton, A.A., Baker, G.B., Dewhurst, W.G. & Sandler, M., eds.), Humana Press, Clifton, NJ, pp. 57-68.
13. Baker, G.B., LeGatt, D.F. & Coutts, R.T. (1982) *J. Neurosci. Methods 5*, 181-188.
14. Coutts, R.T., Baker, G.B., LeGatt, D.R., McIntosh, G.J., Hopkinson, G. & Dewhurst, W.G. (1981) *Prog. Neuropsychopharmacol. 5*, 565-568.
15. Blau, K., King, G.S. & Sandler, M. (1977) *Biomed. Mass Spectrom. 4*, 232-236.
16. LeGatt, D.F., Baker, G.B. & Coutts, R.T. (1981) *Res. Comm. Chem. Path. Pharmacol. 33*, 61-68/
17. Durden, D.A., Philips, S.R. & Boulton, A.A. (1973) *Can. J. Biochem. 51*, 995-1002.
18. Philips, S.R., Durden, D.A. & Boulton, A.A. (1974) *Can. J. Biochem. 52*, 336-373.
19. Philips, S.R., Durden, D.A. & Boulton, A.A. (1974) *Can. J. Biochem. 52*, 447-451.
20. Slingsby, J.M. & Boulton, A.A. (1976) *J. Chromatog. 123*, 51-56.
21. Karoum, F., Masrallah, H., Potkin, S., Chuang, L., Moyer-Schwing, J., Philips, U. & Wyatt, R.J. (1979) *J. Neurochem. 33*, 201-212.

#F-3

ANALYSIS OF THE METABOLITES OF BUMETANIDE
IN URINE BY HPLC

[1]M.R. Howlett, [1]W.H.R. Auld and [2*]G.G. Skellern

[1]Department of Clinical
Biochemistry
North Ayrshire District
General Hospital
Kilmarnock KA2 OPE, U.K.

[2]Drug Metabolism Research
Unit
Department of Pharmacy
University of Strathclyde
Glasgow G1 1XW, U.K.

Requirement   *Sensitive and specific assay for 4 oxidation products of bumetanide in urine.*

End-step   *RP-HPLC (ODS-silica radial compression cartridge, 10 x 0.8 cm). Fluorescence detection (excitation, 337 nm; emission, 400 nm). Eluent: acetonitrile/tetrahydrofuran/phosphate buffer pH 3.5 (15:10:75 by vol.). Internal standard (i.s.): frusemide.*

Sample preparation   *Urine (0.5 ml) adjusted to pH 4.2, and extracted once with diethyl ether (3 ml); urine taken to pH 2 and re-extracted with ether; combined extracts dried down, and residue taken up in eluent (100 µl) for injection in toto.*

Comments   *Eluent pH and nature of solvent components affect resolution and quantum efficiency of fluorescence. Disadvantage of the RP system is that bumetanide has a high k' value; necessary to wash and re-equilibrate column periodically.*

The pharmacokinetics and pharmacodynamics of the loop diuretic bumetanide (3-*n*-butylamino-4-phenoxy-5-sulphamoylbenzoic acid, [I])

[I]
R_1, R_2 = H
R_3 = CH_3

$-NH.CH_2.CH_2.CH_2.CH_3$
     α   β   γ   δ

Substituent positions in butyl group (see over)

---

* Author to whom any correspondence should be addressed

have been studied in detail in normal subjects [1] and in patients
with renal and liver disease [2], with the aid of a simple HPLC
method [3] using fluorescence detection. Bumetanide is normally
given in small doses (1-5 mg), and the amounts excreted unchanged
are 12% and 69% in patients with renal and hepatic disease respectively [2]. After oral administration of ^{14}C-bumetanide (2 mg) to a
group of volunteers, 50% was excreted unchanged in the urine within
24 h [4]. Oxidation products of the butyl group, and conjugates,
accounted for a further 24% of the urinary ^{14}C-radioactivity. The
relative amounts of the oxidation products excreted were 0.5% δ-hyd-
roxybutyl-, 4.0% γ-hydroxybutyl- and 2.1% β-hydroxybutyl-metabolite
([II], [III] and [IV] respectively), also 0.2% aminobutyric metabolite
[VI; COOH replacing CH_3] and 0.4% desbutylbutenanide, VII. Though metabolite IV has not been found in rat and dog [5], the γ,δ-dihydroxy-
butyl metabolite [V] is formed.

In order to study the fate of bumetanide in patients with renal
disease it was necessary to develop a specific and sensitive assay
for quantitatively determining bumetanide metabolites in biological
fluids, since small amounts of the drug are administered and at
least 6 metabolites may be formed.

CHROMATOGRAPHIC CONDITIONS

Both RP and adsorption HPLC were considered as possible methods
for the separation of bumetanide from its metabolites. However,
for either mode it would appear that the choice of a particular
organic solvent is critical not only for a good chromatographic
separation but also for fluorescence sensitivity. For example,
eluents containing hexane, although chromatographically suitable for
the separation of the compounds on a 5 μm silica column, quenched
their fluorescence, increasing the limit of detection from 4 ng to
1 μg column load. In addition, when the apparent pH of the eluent
exceeded 3.5 in the RP mode the quantum efficiency of fluorescence
of the compounds was considerably reduced (Fig. 1A). For a given
concentration of metabolite the order of detector response was VII >V
>III >VI (or 1.0:0.76:0.64:0.47).

The existing RP-HPLC method [3] which utilizes an ODS-silica
column and a methanol-water-acetic acid (70:30:1 by vol.) pH 2.9
eluent was not suitable for the measurement of the metabolites since
they eluted near the void volume. The eluent was modified initially
by increasing the water content to increase its polarity, and by
including acetonitrile. Acetic acid was replaced by phosphoric acid.
With the acetonitrile/methanol/phosphoric acid (0.01 M) eluent, the
available metabolites, III, V, VI and VII were separated from bumetanide (Fig. 2A). Unfortunately with this eluent bumetanide has a
very high k' value (~70); endogenous materials in urine extracts
co-eluted with the diol (V) and desbutyl (VII) metabolites; there
was, too, a loss of resolution with time, in that the peaks in each

pair of compounds coalesced. The effect of the pH of the eluent on
the separation was examined in an attempt to improve the chromatographic separation. Metabolites III and IV were particularly sensitive to changes in the pH of the eluent: their k' values markedly
decreased on increasing the pH by 1 unit from 3.5 to 4.5 (Fig. 1B).
Although a more efficient separation of metabolites was obtained,
the quantum efficiency of fluorescence for the compounds was considerably reduced at pH values above 4.

Two approaches were attempted in order to clean up the urine
extract and selectively to remove bumetanide. Firstly, prior to
metabolite extraction from the urine at pH 2 with diethyl ether, a
liquid-solid extraction step with ODS-silica in cartridges was tried
but, unfortunately, this did not provide a cleaner extract and
retrieved only 50% of the bumetanide. Secondly, a preliminary
liquid-liquid extraction step was included by adjusting the urine

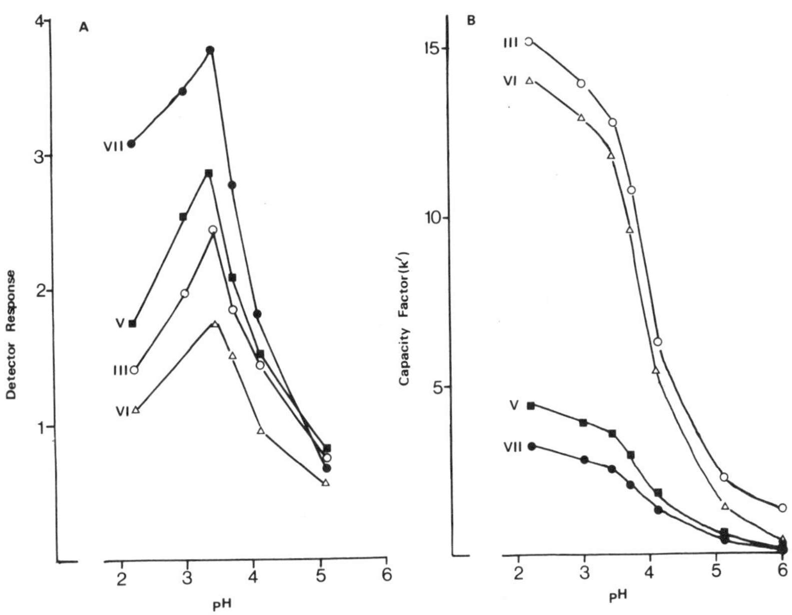

Fig. 1. Behaviour of some bumetanide metabolites (those for which
reference samples were available; see text for designations, e.g. **V**),
in relation to pH of eluent (methanol/0.01 M phosphate buffer, 30:70
by vol.). Influence on: A, fluorescence; B, capacity factors.
Column: 5 μm ODS-silica, 250 x 4 mm. Flow-rate: 1.5 ml/min. Fluorescence detection: ex. 337, em. 400 nm.

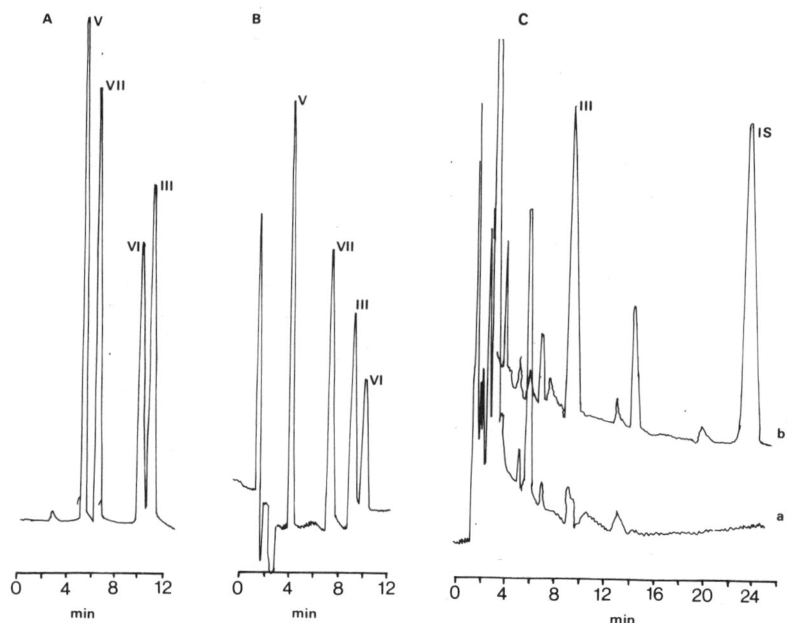

Fig. 2.   HPLC chromatograms for bumetanide metabolites (see text for designations, e.g. V).
A. Metabolite standards. Eluent: acetonitrile/methanol/phosphoric acid (0.1 M) pH 3.5 (22:15:63 by vol.).
B. Metabolite standards. Eluent: see first p. of article.
C. Control (a) and test (b) urine extracts.
For column (5 μm ODS-silica) and detection, see first p. of article.
Flow-rate: 2 ml/min.

sample to pH 6 with a 0.1 M citrate-phosphate buffer and extracting with diethyl ether. This did not, however, remove the interfering endogenous material, and although the losses were negligible for III, V and VI a significant amount of VII was removed.

To overcome the problem of endogenous material interfering and of loss of resolution with time, the composition of the eluent was also examined. Replacing the methanol in the original eluent with tetrahydrofuran resulted in a better separation of metabolites (Fig. 2B), although bumetanide was still strongly retained on the column (2 h). The effect of the addition of ion-pairing reagents to the eluent was also examined (Table 1). It was necessary, however, to adjust the pH of the eluent to 4, since the presence of ion-pairing reagents did not affect the k' values of the compound at pH values less than 4.

Table 1. The effect of adding ion-pairing reagents to the eluent
on the k' values of the metabolites. The column was pre-loaded
with the reagent in solution, viz. tetrabutylammonium phosphate
(TBAP; 70 mg) or di-n-octylamine (DOA; 50 mg). The eluent was
acetonitrile/tetrahydrofuran/[phosphoric acid (0.01 M)/KH$_2$PO$_4$
(0.01 M) pH 4] (12:10:78 by vol.). The values are capacity
factors, k'.

Eluent	V	III	VII	VI	Frusemide (i.s.)
No addition	1.86	5.64	3.79	6.64	7.57
+ 5 mM TBAP	3.23	9.23	6.38	10.0	11.0
+ 0.04 mM DOA	3.15	8.85	6.38	10.38	13.77

The resolution of metabolites III and VI and frusemide (internal
standard) decreased when tetrabutyl ammonium phosphate was added.
When dioctylamine was added to the eluent, k' increased for all the
compounds with concomitant peak broadening. Thus there was no advantage in adding ion-pairing reagents at pH 4, since they did not
improve the separation of the metabolites, and they increased the
total analysis time. Increasing the pH of the eluent above 4 resulted in a reduction in fluorescence.

Gradient elution with an RP column should overcome the problem
of the high k' value of bumetanide. Alternatively, with an adsorption system bumetanide has a low k'; however, with the eluents which
contained either hexane or chloroform there was a loss in detection
sensitivity due to quenching.

The internal standard,   4-benzyl-3-n-butylamino-5-sulphamoyl-
benzoic acid, used in the original RP-HPLC method [3] for assaying
bumetanide in biological fluids was unsuitable since it was strongly
retained on the column with a k' value greater than that of bumetanide. Frusemide was chosen as an i.s. since its physico-chemical
properties are similar to those of bumetanide and its metabolites,
in that it fluoresces and is co-extracted with the bumetanide metabolites, but has a lower k' value than bumetanide. Frusemide is,
however, not wholly ideal as an i.s. since its k' value lengthens
the time of analysis. Whilst frusemide solutions are not stable
even if refrigerated, deep-freeze storage is satisfactory.

FEATURES OF THE ASSAY

With concentrations of 10 to 200 ng/ml of metabolites in urine,
there was a rectilinear relationship between peak-height ratio of
metabolite : i.s. concentration (r = 0.99). The overall extraction
yields for metabolites III, V, VI and VII and frusemide from urine

were 87%, 65%, 87%, 95% and 90% respectively, when extracted initially at pH 4.2 and then at pH 2 with diethyl ether. The C.V.s for these metabolites were respectively 4.2%, 4.95%, 3.62% and 11.4% at 60 ng/ml.

Fig. 2C(b) exemplifies chromatograms obtained from urine extracts from a subject who had received 1 mg of bumetanide orally. A cleaner control extract was obtained if the patient fasted prior to taking the drug, as was the case for the Fig. 2C pattern. The test urine extract contained measurable amounts of metabolite III.

Our use of radial compression columns (5 μm ODS-silica) rather than rigid columns was beneficial to resolution; the shorter length and high efficiency of the radial columns permitted a higher eluent flow rate without excessive column back-pressure.

Acknowledgements

The authors thank Dr. P.W. Feit (Leo Pharmaceutical Products, Denmark) and Dr. S.J. Kolis (Hoffman-La Roche, Nutley, N.J., U.S.A.) for providing the authentic bumetanide metabolites.

References

1. Marcantonio, L.A., Auld, W.H.R., Skellern, G.G., Howes, C.A., Murdoch, W.R. & Purohit, R. (1982) *J. Pharmacok. Biopharm. 10*, 393-409.
2. Marcantonio, L.A., Auld, W.H.R., Murdoch, W.R., Purohit, R., Skellern, G.G. & Howes, C.A. (1983) *Br. J. Clin. Pharmac. 15*, 245-252.
3. Marcantonio, L.A., Auld, W.H.R. & Skellern, G.G. (1980) *J. Chromatog. 183*, 118-123.
4. Halladay, S.C., Sipes, I.G. & Carter, D.E. (1977) *Clin. Pharmacol. Ther. 22*, 179-217.
5. Kolis, S.J., William, T.H. & Schwartz, M.A. (1976) *Drug Met. Disp. 4*, 169-176.

#F-4

ASSAY OF COMPOUNDS OF FORENSIC INTEREST

R.L. Williams

Metropolitan Police Forensic Science Laboratory
109 Lambeth Road, London SE1 7LP, U.K.

The forensic scientist is concerned with three groups of materials:-
(i)   those whose possession is controlled by law;
*(ii) those which have been administered to a person with the intention of endangering life; ***
(iii) those whose ingestion affects the ability of motorists.

These present different analytical problems. For the first group it suffices to show that the analytical specimen contains one out of some 120 substances, and a semi-quantitative estimate may be adequate. The other two groups are more demanding analytically and usually require quantitation. Thus the scientist who may have to deal with the first group needs only one or two of the analytical methods available. For the other two he may have to choose from a range of techniques those most appropriate to his particular problem.

Although the standard methods are thin layer chromatography (TLC), high performance liquid chromatography (HPLC), gas chromatography (GC), gas chromatography-mass spectrometry (GC-MS) and radioimmunoassay (RIA), developments have taken place which have improved their performance. For example, multiple wavelength detection and electrochemical detection in HPLC and positive and negative chemical ionization (CI) in MS have been of considerable benefit to the forensic analyst. These are discussed, and also another technique, isotachophoresis, which seems to hold promise in some specialized areas.

The term 'assay' carries a connotation of quantitative analysis as exemplified by the definition in Chambers 20th Century Dictionary "to determine the proportion of a metal or other component in". However, the alternative description therein, "to test", accords much more closely with the activities of forensic scientists.

*The complementary survey by J.S. Oliver, #A-2, concerns cadaver samples

For the first of the three groups of materials listed above, it suffices to show the presence of one or more of some 120 substances [1], usually with an estimate of the amount present. For the others quantitation is generally necessary, but the precision needed varies considerably depending on the nature of the case. Thus, the only law in which concentrations are specified is that dealing with drunken driving, maximum permissible levels of alcohol in the driver being set at 35 µg/100 ml of breath or 80 mg/100 ml of blood or 107 mg/100 ml of urine [2]. In this instance high levels of accuracy and precision are essential; on the other hand, a toxicological assay on viscera from a badly decomposed cadaver would not be expected to achieve this certainty.

ANALYSES FOR CONTROLLED DRUGS

The procedures used in forensic laboratories follow the same general pattern but with local variations. That of the Metropolitan Police Laboratory is shown in Scheme 1 (and see below). Approximate concentrations can be estimated from the TLC spot sizes. Greater precision is obtainable in one of the following ways:
- HPLC with UV detection - the preferred method of quantitation;
- UV spectrometry, if it is known that only one drug is present;
- GC, resorted to when low levels of material are encountered, e.g. residual amphetamine in a hypodermic syringe, or in special circumstances, e.g. cocaine/cinnamoylcocaine mixtures.
These are standard procedures well described elsewhere [e.g. 3] and covered in recent comprehensive literature surveys [4, 5].*

Increasingly, separation for quantitative determination is being coupled with positional measurement in order to achieve the identification of the analyte more speedily with little extra expenditure of effort and without resort to the more demanding methods of analysis. Thus, Moffat [6, 7] and de Zeeuw [8], *inter alia*, have shown that accurate chromatographic retention data, e.g. the TLC R_f values preferably in two non-correlated separatory systems, are good identifiers, particularly if used in conjunction with a locating spray reagent which has some specificity [3].

Both authors suggested ways of deciding the effectiveness of the systems used, based on statistical evaluations similar to that proposed by Parker [9, 10] who considered the general problem of comparing two sets of attributes in order to identify an object. Thus, Smalldon & Moffat [11] defined a discriminating power DP_k given by the formula $DP_k = 1 - 2M/N(N-1)$ where N is the number of compounds examined in each system and M the total number of matching pairs. de Zeeuw [8] used an analogous equation from which the probability of an identification could be defined given a specified R_f search window.

---

* For packed-column GC in our laboratory, most analytes are derivatized.

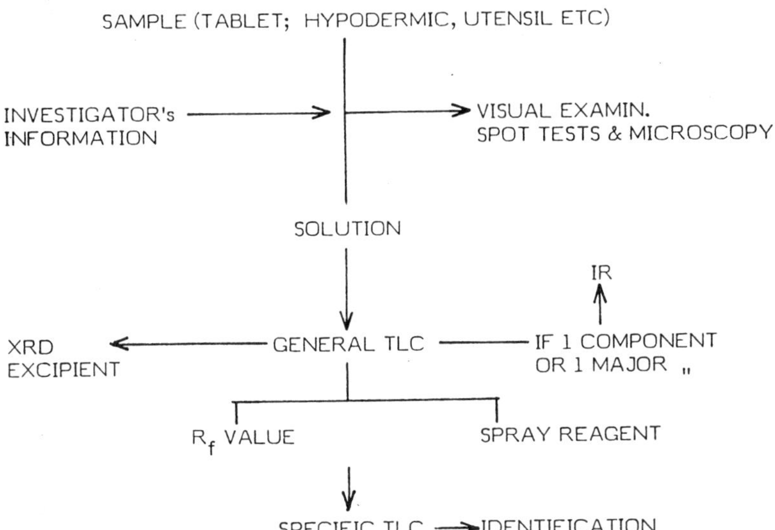

Scheme 1. Identification approach in possession-of-drugs cases.
IR denotes infra-red spectroscopy; XRD, X-ray diffraction, in the
context of e.g. mannose (excipient) or sulphate (drug-salt anion).
For subsequent quantitation (HPLC/UV; UV; GC), see text.

The same philosophy has been applied to values for GC retention
volume, V_r [12]. Capillary columns can enable these values to be
measured with much greater precision because of the sharpness of
the peaks associated with the high column resolution. Thus, Alm et
al. [13] have described split injection analyses carried out on twin
silica capillary columns with immobilized stationary phases of SE-54
and OV-215 and, respectively, FID and AFID. With each system it was
possible to achieve a relative retention-time window of 0.01, which
is sufficiently narrow for drug identification. They tested the validity of the method over a period of 6 months by recording mass spectra
in parallel with the GC runs, and obtained no false positives for 80
compounds in more than 2000 casework samples. Quantitation using
internal standards was also reported as satisfactory.

It is not surprising that, besides exploitation of retention
behaviour with different stationary phases [14], the relative responses to FID and to AFID have been considered as means of charact-
erizing the eluting substance. Baker [15] showed that amongst 70
commonly occurring drugs 59% had relative V_r's within 2% of another
of them; if it were also required that the FID and AFID relative
responses differ by at least 20% then only 16% were similar. Despite
the attractiveness of this approach, difficulty in maintaining reproducible response ratios has evidently detracted from further development (e.g. L. Stromberg, pers. comm.; & work done in this Laboratory).

Table 1. Characterization of drugs by chromatographic retentions and detection features [18]. HPLC system A: μBondapak C_{18}; 0.025 M NaH_2PO_4 in methanol/water (2:3 by vol.), pH adjusted to 7.0. HPLC system B: μ Porasil; methanol/2 M NH_4OH/M NH_4NO_3 (27:2:1 by vol.). Retention time denoted t_r in HPLC, and likewise in GC (= V_r x flow rate). For GC response index (FID $vs.$ AFID) see text, and for GC general conditions see ref. [15]; analytes were not derivatized.

First parameter	Second parameter	No. of compounds	Identifiable with 1st param.	Identifiable with both param.
t_r, system A	A_{254}/A_{230}	78	9%	95%
GC t_r	Response index	71	41%	85%
t_r, system A	GC t_r	51	12%	100%
t_r, system A	t_r, system B	35	23%	83%

However, more success has been obtained in the application of this concept to HPLC where detectors can often be used sequentially thereby avoiding the problems of stream splitting and consequent sensitivity to flow rate which occurs with GC detectors. However, some of the earlier applications were concerned with quantitative analysis rather than identification. For example, Smith & Vaughan [16] used serial UV detectors set at 220 and 254 nm for the analysis of cannabinoids. Cannabinol (CBN) and cannabigerolic acid (CBGA) were not separated in the analysis, but satisfactory quantitation was achievable from peak-height measurements at the two wavelengths without having to calibrate for CBGA, although a pure sample of the latter was required. The formal treatment of the quantitative determination of unresolved or partially resolved compounds was given in a subsequent paper [17].

The use of UV absorption ratios in combination with retention data has been reported by Baker et al.[18], who examined 101 drugs. Their results as summarized in Table 1 confirm that a high degree of identification could be obtained from the HPLC data alone, being surpassed only by a combination of HPLC and GC parameters. A more specialized extension of this work has arisen from the proposal to control barbiturates in the U.K. White [19] performed measurements at 220, 240 and 254 nm respectively and showed that the combined use of the absorbance ratios A_{220}/A_{254} and A_{240}/A_{254} and retention data enabled 27 out of 29 barbiturates to be positively identified.

Such a process needs more than a single ordinary UV detector. One remedy lies in the development of photodiode array detectors, now commercially available with computer facilities included [20], whereby one can, without stopping the solvent flow, ascertain the UV spectrum of a single component or, if more than one is present, the change in spectrum within the profile of an eluting peak, and achieve quantitation. Because such instruments are a recent development and are relatively expensive, their full potential has yet to be realized. Moreover, forensic specimens seldom contain completely unknown substances, so there is no great need for a full UV spectrum, which in any case is not a strong structural diagnostic.

In a possible alternative approach, selected absorbance ratios can be determined with a single detector [21, 22], through interruption of deuterium-lamp radiation at 2 cycles/sec. by a rotating chopper containing appropriate narrow band-pass filters. The two outputs, processed by a small microcomputer, yield either individual absorbance profiles or the ratio of any pair of substances as the chromatography progresses. Detection of unresolved components is exemplified in Fig. 1, where the A_{279}/A_{254} vs. time plot for a single compound has a flat-topped profile whereas that of an unresolved mixture does not. Application of the system to the identification of barbiturates [23] in a blind trial on several current formulations gave correct results with no need for any pre-analytical extraction.

Other HPLC detectors are possible candidates to provide an identifying parameter, e.g. the electrochemical detector since molecular structure governs the oxidation- or reduction-potential needed for a response. However, use other than qualitative is as yet unpromising.

TOXICOLOGICAL ANALYSIS

As in the foregoing drug-analysis area, variants of a basic protocol are used in the Forensic Science Service, that of the Metropolitan Police Laboratory for general screening being as in Scheme 2. For confirmation, and also for small samples of blood (e.g. Road Traffic Act cases) or urine, use is made of GC-MS (EI, $+^{ve}$ CI or $-^{ve}$ CI) or possibly RIA (specific for insulin, diazepam, temazepam). Specialized techniques are used too for quantitation, viz. GC, HPLC and RIA, whereas the general screening of Scheme 2 is more conventional. The main problem in quantitation has been to find suitable internal standards. Thus, in the analysis of paraquat in blood plasma by HPLC the detection relies on the conversion of the substance by oxidation with alkaline hexacyanoferrate (III) to a fluorescent product. The ethyl or butyl homologues of paraquat are suitable in this case as internal standards (I. Jane, unpublished work).

SPECIALIZED TECHNIQUES

These are used only when the standard procedures fail to give an unambiguous result and generally arise when the sample for examination is very small or when the concentration of unknown substance is too low to be detected by the normal tests. A modification of a

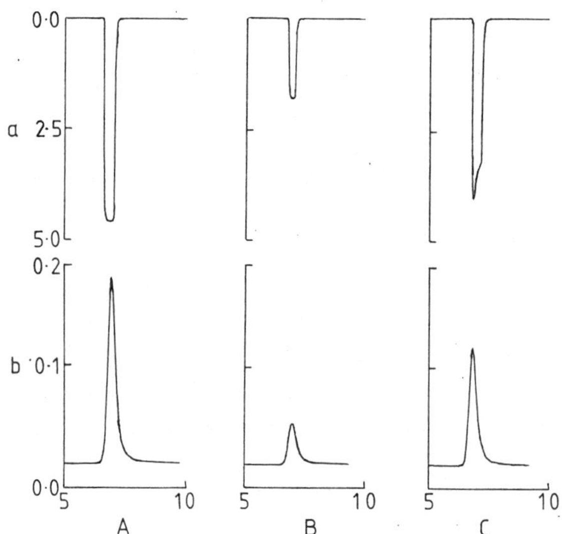

Fig. 1. Peak absorption monitored on-line. (a) Absorbance ratio 279:254 nm; (b) absorbance at 279 nm.
(A), 2 µg timolol; (B), 2 µg alprenolol; (C), 1 µg of each, mixed.

standard method can sometimes be developed to solve the problem, as with the paraquat example given above.

HPLC DETECTION MODES OTHER THAN UV ABSORBANCE

When HPLC was first taken up for forensic analyses, the UV detector was the only one in use. It is still the main detector because of its ease of operation particularly in quantitation, but the somewhat low sensitivity and the absence of a fully general response has led to the introduction of alternatives. A recent literature survey [24] gave a world distribution for usage as follows, the usage in the Metropolitan Police Laboratory being given *in italics* : UV, 70.7%, *61.5%;*   fluorescence, 15.0%, *15.4%;*   refractive index, 5.5%, *7.7%;* electrochemical, 4.4%, *11.5%;* others, 4.4%, *3.9%* (conductive). It is interesting that forensic applications follow the world trend.

Fluorescence detectors are well established, relying mainly on natural fluorescence as in LSD assay, but with a few instances of derivatization. Examples of the latter, besides paraquat as above, include morphine in urine by on-column oxidation [25]; oestriol, oestrone and oestradiol by pre-column reaction with EDTN or ACE*(B.B. Wheals, unpublished work); warfarin with coumachlor as internal standard by post-column pH change to alkalinity (R. Fysh, unpublished investigations); and quinine by post-column pH reduction to strong acidity (I. Jane, R. Fysh & A. McKinnon, unpublished work). An inter-

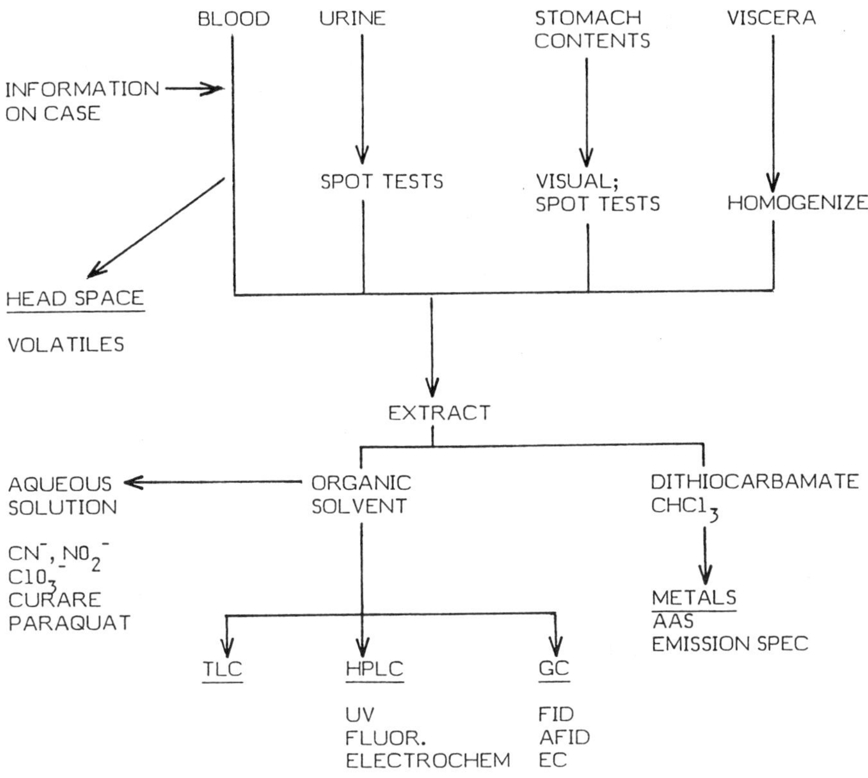

Scheme 2. General analytical approach in toxicology cases. The
extraction to furnish a solvent phase and an aqueous phase is
typically performed with ethanol or chloroform.

esting variant on this theme is the use of UV radiation itself [26]
for selective detection.   Thus cannabinol on irradiation forms a
fluorescent product; LSD after irradiation at the correct wavelength
yields non-fluorescing lumi-LSD.   Hence it is possible to gain in
sensitivity in both instances by comparing chromatograms before and
after UV treatment.

It is perhaps surprising that greater application of derivatization has not been made to other compounds of forensic interest,
since the possibilities are a challenge to the ingenuity of the
organic chemist in finding suitable labels [25].   However, the development of electrochemical detectors, which eliminate the need for
pre- or post-column treatments, has reduced the interest in fluores-

---

* 1-ethoxy-4-(dichloro-*S*-triazinyl)-naphthalene; 5-dichloro- *S*-triazinyl)-
acenaphthene

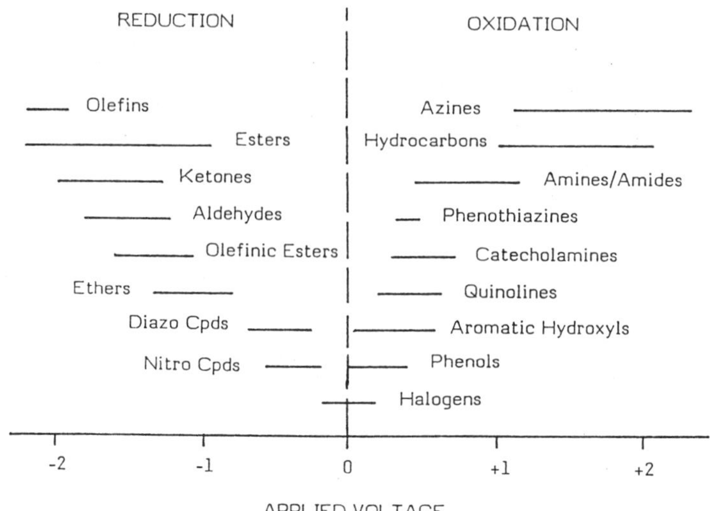

Fig. 2. Electrochemical detection in HPLC: compound types in relation to oxidative and reductive modes.

cent methods. Although several varieties of detector are available (cf. other arts. in this vol.), only those based on conductimetry and voltammetry (electrochemical) have been employed extensively. The latter has been the more important, and the conductivity detector has been used only for simple anions [28]. Of the two possible ranges of operation of the electrochemical detector (Fig. 2), the reductive mode is very sensitive to adventitious oxygen and necessitates a good experimental technique from the operator. Its use has therefore been restricted to materials containing $-NO_2$ that arise from explosives cases [29, 30]. The oxidative mode presents no such difficulties [cf. art. by P.T. Kissinger in Vol. 7, this series-*Ed.*] and the detector is often coupled in series with one of the UV type.

Except possibly for amphetamine and barbiturates, most of the drugs of interest have structural features which are susceptible to oxidation, as borne out by our own experience. Materials analyzed include phenothiazines such as chlorpromazine; tricyclic antidepressants, e.g. amitriptyline; diphenhydramine; dextropropoxyphene; LSD; morphine in blood [31; cf. #NC(E)-3]; oxytocin; leu-enkephalin, angiotensin II, lys-vasopressin [32]; uric acid, nitrite and thiocyanate in saliva [26].

Detection sensitivity is good, being generally in the 1-20 ng range but sometimes better. Additionally there is the above-mentioned advantage that the half-wave potential necessary for the oxidation to proceed gives an indication of the nature of an unknown. Moreover when interfering substances are present the correct choice of the potential can often minimize or eliminate their signals.

GC-MS

Forensic use of MS falls into two clearly defined areas. It is a specialized technique in its own right, directly applicable to problems as in polymer characterization via pyrolysis MS, or serves as a very sophisticated detector for GC, as now considered. The GC-MS combination is now standard equipment in the Home Office and our own laboratories. The medium-resolution machines as acquired by forensic laboratories, only within the past decade because of cost, are quadrupole (QMS) or magnetic sector (MMS);   the development of cheaper instruments will make GC-MS accessible to many more laboratories [33]. Our laboratory has one MS of each type. The MMS is used mainly for pyrolysis work, but also on those occasions when it is necessary to examine metastable ions or to obtain an accurate mass number so as to establish the identity of a new substance.

The QMS is coupled to a 12.5 m fused capillary column with OV bonded phase and is used for the bulk of the drugs and toxicology work.   Standard procedures are followed. Thus the total ion current serves for maximum sensitivity as a GC detector and quantitation carried out by multiple ion monitoring of 2 peaks in the analyte specimen and 2 in that of the internal standard. Quantitation is becoming commoner where there are interferences with FID or AFID.

Applications to the first group of samples considered earlier include fragments of tablets; tablets which were unusual from a visual examination (shape, colour, etc.); syringe washings; white powders. The criterion for resort to GC-MS would be ambiguous results by routine procedures, or too small a sample or too low sensitivity. White powders are therefore examined because they often consist of mixtures. There were 103 such cases in 1982. The second group, viz. toxicological samples of which there were 265 cases in 1982, exploit the high sensitivity and identification capabilities of GC-MS to their fullest extent.   Apart from straightforward analyses, many now result from the need to show that the level of a particular substance was less than a specified amount, relevant to defences put forward in court that the behaviour of the accused or the victim resulted from the taking of a drug.

The third group of samples arises from the analysis of blood or urine samples (106 cases in 1982) taken from motorists under the Road Traffic Acts. The situation here is changing because of the switch to breath alcohol levels as a measure of intoxication. Nevertheless there is still a not inconsiderable number of blood samples averaging less than 2 ml in volume from motorists whose blood alcohol is below the legal limit but whose incompetence to drive might have resulted from a combination of drugs and alcohol. GC-MS [33] and RIA [35, & #E-1, this vol.] are the preferred methods of dealing with this problem. Barbiturates are the major class of drug found (31.5% in 1982), but 17 other types were identified in 1982 (opiates, 11.2%;

benzodiazepines, 9.0%; amphetamine, 7.9%; cocaine, 6.7%;   dipipanone, 5.6%; methadone, 4.5%; cyclizine, 3.4%; pemoline, 3.4%; propoxyphene, 3.4%; paracetamol, 3.4%; and, 32% in all: meprobamate, glutethimide, chlomethiazole, methaqualone, pethidine, pentazocine and moramide). It is notable that barbiturates are especially amenable to negative-ion chemical ionization (CI) MS [36, 37], and this mode together with its positive ion CI counterpart is requisite in ~5% of the cases examined by MS in the Metropolitan Police Laboratory. However, the possibilities of enhanced sensitivity of detection with negative ion CI by concentrating most of the ion current in one ion such as pentafluorobenzoyl [38] have yet to be exploited.

Other developments which seem likely to have an impact particularly when present-day equipment is being replaced are the coupling of microbore HPLC columns to the MS, and the use of fast atom bombardment [39] as a means of ionization of difficultly volatile substances such as polypeptides which are of increasing importance because of their function as CNS transmitters.

RADIOIMMUNOASSAY (RIA)

As R.N. Smith deals with forensic ligand assays in #E-1, suffice it to say that amongst the possible versions of immunoassay only RIA has achieved any significant usage so far, and that even so it can be justified only when there are sufficient cases to ensure that the technique is in frequent use. The well-known wish for antisera specific to the analyte can rarely be met in practice, but the property of cross-reactivity with similar compounds has some advantages. Thus it is possible to detect and identify groups of compounds having a common molecular structure such as benzodiazepines at very low concentrations, but another analytical method is then needed to confirm which particular member of that group is present.

If there were sufficient need, monoclonal antibodies might be used to generate reagents with notably higher specificity. There is also scope for alternative modes of analyte or antibody labelling, eliminating the need for the rather costly β- or γ-counting equipment; an example is enzyme labelling ('ELISA').

IMPAIRED DRIVING ATTRIBUTABLE TO ETHANOL

In respect of ethanol (drugs having been dealt with above), for blood and urine samples quantitation is based on variants of a standard GC method [40] which is also used for non-motoring samples. Though results for blood are quoted on the basis of S.D. ±2 mg at the legal limit of 80 mg%, the outcome of many hundreds of test analyses shows that the precision approaches ±1 mg at this level. Following recent legislation, blood or urine samples are being partly replaced by breath samples [41]; the precision and accuracy of the machines for measuring these are comparable to that achievable by GC [42].

ISOTACHOPHORESIS

Isotachophoresis is the analysis of a mixture of anions or cations by electrophoresis under constant current. Under these conditions the ions are separated quite sharply in the order of their mobilities and may be detected either by changes in conductivity, UV absorption or thermally through the alteration in ohmic heating. The principles have been discussed recently in a review [43].

Sample handling is minimal; sample size is small, e.g. 2 µl of urine can be injected directly into the instrument; quantitation is simple and sensitivity good, viz. 0.5-1 mM minimum detectable concentration. Since many drugs and metabolites can be converted to an ionized form, the technique has obvious attractions for forensic analysis. However at present it appears still to be at the evaluation stage.

In the Metropolitan Police Laboratory it is being used in a study of whether or not alcohol in a urine sample had resulted from ingestion by the donor or had originated from microbial processes. The standard procedure is to plate samples from the urine and assay any growth of yeasts or other microorganisms which can generate ethanol in their metabolism [44]. This is satisfactory only if viable organisms are present. It gives no information on whether there had been a non-viable population which had at some stage affected the alcohol level.

Studies are therefore being carried out on the growth of various bacteria in urine both with and without glucose present. For example. the 'mixed acid' fermentation of glucose by *E. coli* produces either as products or intermediates lactate, acetate and pyruvate. Fig. 3 shows the growth of *E. coli* in the absence or presence of glucose; only in the latter instance is ethanol produced. Fig. 4 shows the corresponding curves for levels of acetate, lactate and pyruvate; the last two are not formed in the absence of glucose. Hence if a urine sample contains alcohol and also substantial quantities of acetate, but shows no signs of microbiological activity from ethanol-producing organisms, it would be very inadvisable to assert that all the alcohol had been derived from ingestion [45]. Studies are now proceeding on other organisms such as proteus, staphylococcus and yeasts and on blood samples.

References

1. Government publication (1971) *Misuse of Drugs Act 1971* (Chap. 38), Her Majesty's Stationery Office, London, Schedule 2, pp. 29–32.
2. Government publication (1981) *Transport Act 1981* (Chap 56), Her Majesty's Stationery Office, London, p. 77.
3. Clarke, E.G.C. (1969, Vol. 1; 1975, Vol. 2) *Isolation and Identification of Drugs*, Pharmaceutical Press, London.

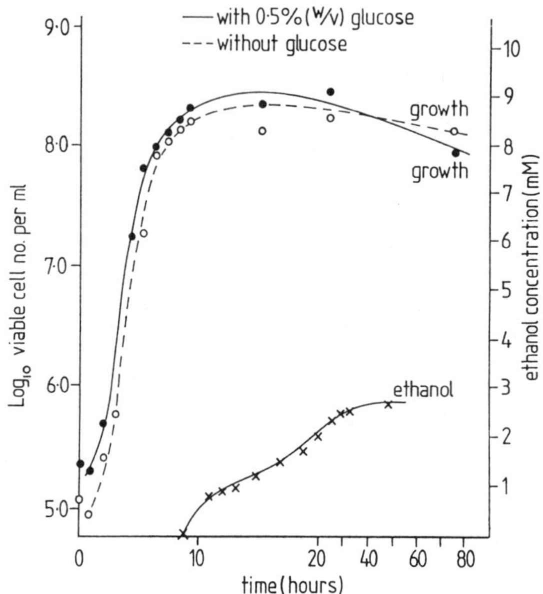

Fig. 3. Growth of *E. coli* in urine at 37° and associated
production of ethanol (none detected in the absence of glucose).

4. Gough, T.A. & Baker, P.B. (1982) *J. Chromatog. Sci. 20*, 289–
    329.
5. Baker, P.B. & Phillips, G.F. (1983) *Analyst 108*, 777–807.
6. Moffat, A.C., Smalldon, K.W. & Brown, C. (1974) *J. Chromatog.
    90*, 1–7.
7. Moffat, A.C. & Smalldon, K.W. (1974) *J. Chromatog. 90*, 9–17.
8. de Zeeuw, R.A., Schepers, P., Greving, J.E. & Franke, J-P. (1978)
    in *Instrumental Applications in Forensic Drug Analysis*
    (Klein, M., Kruegel, A.V. & Sobel, S.P., eds.), U.S. Govt.
    Printing Office, Washington D.C., pp. 167–169.
9. Parker, J.B. (1966) *J. Forensic Sci. Soc. 6*, 33–39.
10. Parker, J.B. (1967) *J. Forensic Sci. Soc. 7*, 134–144.
11. Smalldon, K.W. & Moffat, A.C. (1973) *J. Forensic Sci. Soc. 13*,
    291–295.
12. Moffat, A.C., Stead, A.H. & Smalldon, K.W. (1974) *J. Chromatog.
    90*, 19–33.
13. Alm, S., Jonson, S., Karlsson, H. & Sundholm, E.G. (1983) *J.
    Chromatog. 254*, 179–186.
14. Eklund, A., Jonsson, J. & Schuberth, J. (1983) *J. Anal. Toxicol.
    7*, 24–28.

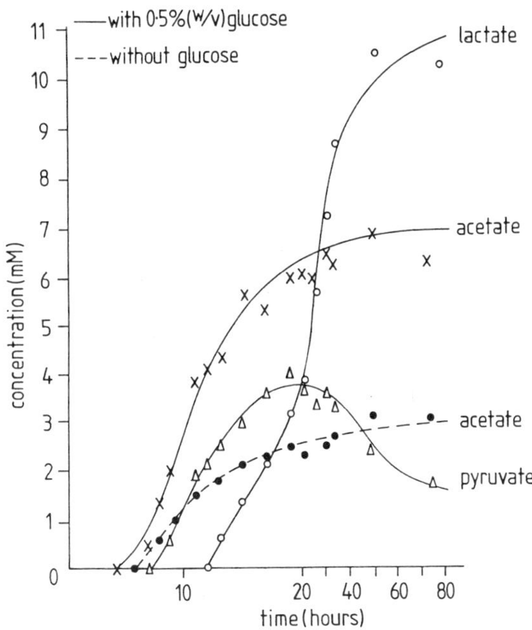

Fig. 4. Metabolites produced by *E. coli* during growth in urine at 37°.

15. Baker, J.K. (1977) *Anal. Chem.* 49, 906-908.
16. Smith, R.N. & Vaughan, C.G. (1976) *J. Chromatog.* 129, 347-354.
17. Smith, R.N. & Zetlein, M. (1977) *J. Chromatog.* 130, 314-317.
18. Baker, J.K., Skelton, R.E. & Cheng-Yu Ma (1979) *J. Chromatog.* 168, 417-427.
19. White, P.C. (1982) *J. Chromatog.* 200, 271-276.
20. Borman, S.A. (1983) *Anal. Chem.* 55, 836A-842A.
21. Catterick, T. (1983) *J. Chromatog.* 259, 59-67.
22. Russell, J. (1983) *J. Chromatog.* 280, 370-375.
23. White, P.C. & Catterick, T. (1983) *J. Chromatog.* 280, 376-381.
24. Majors, R.E., Booth, H.G. & Lochmuller, C.H. (1982) *Anal. Chem.* 54, 323R-363R.
25. Jane, I. & Taylor, J.F. (1975) *J. Chromatog.* 109, 37-42.
26. Twitchett, P.J., Williams, P.L. & Moffat, A.C. (1978) *J. Chromatog.* 149, 683-691.
27. Seitz, W.R. (1980) *Crit. Rev. Anal. Chem.* 8, 367-405.
28. Wheals, B.B. (1983) *J. Chromatog.* 262, 71-76.
29. Lloyd, J.B.F. (1983) *J. Chromatog.* 257, 227-236.
30. Lloyd, J.B.F. (1983) *J. Chromatog.* 261, 391-406.

31. White, M. (1979) *J. Chromatog. 178*, 229–240.
32. White, M. (1983) *J. Chromatog. 262*, 420–425.
33. Borman, S.A. (1983) *Anal. Chem. 55*, 726A–730A.
34. Whitehouse, M.J. & Jones, L.V. (1979) in *Recent Developments in Mass Spectrometry in Biochemistry and Medicine* (Frigerio, A. & McCornish, M., eds.), Vol. 6, Elsevier, Amsterdam, pp. 303–316.
35. Smith, R.N. (1981) in *Alcohol, Drugs and Traffic Safety*, Proc. 8th Internat. Conf., 1980 (Goldberg, L., ed.) Almquist & Wiksell, Stockholm, pp. 469–477.
36. Brandenberger, H. & Ryhage, R. (1978) in *Blood Drugs and Other Analytical Challenges* [Vol. 7, this series] (Reid, E., ed.) Horwood, Chichester, pp. 173–184.
37. Whitehouse, M.J. & Jones, L.V. (1981) *Biomed. Mass Spec. 8*, 231–236.
38. Hunt, D.F. & Crow, F.W. (1978) *Anal. Chem. 50*, 1781–1784.
39. Barber, M.. Bordoli, R.D., Sedgwick, R.D., Tyler, A.N., Garner, G.V., Gordon, D.B., Tetler, L.W. & Hider, R.C. (1982) *Biomed. Mass Spec. 9*, 265–268.
40. Curry, A.S., Walker, G.W. & Simpson, G.S. (1966) *Analyst 91*, 742–743.
41. as for ref. 2, p. 74.
42. Emerson, V.J., Hollyhead, R., Isaacs, M.D.J., Fuller, N.A. & Hunt, D.J. (1980) *J. Forensic Sci. Soc. 20*, 3–70; also Isaacs, M.D.J., et al. (1982) IBSN 0 11 340783 1, HMSO, London.
43. Holloway, C.J. & Trautschold, I. (1982) *Fresenius Z. Anal. Chem. 311*, 81–93 [In English].
44. Corry, J.E.L. (1981) as for 35., pp. 600–613.
45. Harper, D.R. & Martin, P.J. (1984) 9th Internat. Conf. on *Alcohol, Drugs and Traffic Safety* –to appear in Proceedings.

#F-5

DETECTION OF SOLVENT ABUSE BY DIRECT
MASS SPECTROMETRY ON EXPIRED AIR

John D. Ramsey

Toxicology Unit, Chemical Pathology Laboratory
St. George's Hospital, London SW17 1QT

Solvent abuse currently causes >60 deaths each year in Great Britain and could be detrimental to the health and development of teenage users. Epidemiological surveys to detect possible health effects can be considerably strengthened by toxicological analysis but this can be difficult by conventional techniques for ethical reasons. Breath has unique advantages as a specimen in such circumstances, being readily obtainable by non-invasive means.

Measurements on breath can be effected by use of a transportable mass spectrometer (MS; 'PETRA'), as already shown with samples from volunteers. Enrichment of organics over the permanent gases in the sample is achieved by a membrane separator. Benefit to sensitivity or identification is obtainable by collecting onto a solid adsorbent and desorbing into a GC or MS system.

We have surveyed the problem of solvent abuse, or more generally volatile substance abuse (VSA) [1, 2], and are engaged in an extended study with monitoring of relevant media reports and court proceedings. Volatiles from glues (especially toluene) are associated with the deaths, as are halogenated compounds (mainly 1,1,1-trichloroethane, tetrachloroethylene and trichloroethylene) and butane from fuel gases. Altogether ~20 compounds are found in cases which come to official attention either in reports of deaths [2] or from clinical surveys [1]. No systematic survey of a normal population has yet been undertaken. A complicating factor is the fact that the population at risk is below the age of consent and that diverse products and substances are involved. Almost any product that contains a solvent or a volatile component can be and has been misused. Ingenious ways are often used by the sniffers to separate off the volatile components from the fluid. Glues are usually put in small plastic

bags, and the headspace breathed and rebreathed. Aerosol deodorants
have been bubbled through water to remove the non-respirable alumi-
minium salts from the propellants. Most deaths are attributed to
asphyxia and occur at home.

The compounds may be sequestered in body fat for long periods
and so put the CNS at risk; they are slowly excreted, largely unmetabolized, in breath. It is normally impracticable and perhaps
unethical to use analytical techniques which require a blood specimen. Few of the volatiles are metabolized to analytically useful
substances which can be detected in urine. A notable exception is
toluene which is partly excreted as hippuric acid; but this occurs
naturally. Glue sniffers can be distinguished by the hippuric acid/
creatinine ratio, but only for a few hours after the last episode [3].

Breath has unique advantages as a clinical specimen, requiring
little or no pre-treatment and apt for field or bedside testing [4].
Some measure of cooperation is required from the subject to obtain
a valid sample for quantitative assay, and respiratory disease can
affect results. Alveolar air is needed, undiluted with tidal air.
Established applications besides checking drivers for alcohol ingestion [R.L. Williams, #F-4, this vol.] include malabsorption inves-
tigations (labelled fat $\rightarrow CO_2$ [5]; colonic CHO $\rightarrow H_2$ by bacterial action
[6]) and GC/MS characterization of normal breath [7] [review: 8]. Checking
for tobacco-withdrawal compliance in the clinic may entail breath CO
measurement [9], and breath serves to monitor exposure to combustion
products and methylene chloride in industry [10].

Various analytical principles are employed ranging from disposable glass indicator tubes to MS. Dedicated electrochemical analyzers for alcohol, CO and H_2 are in routine use. The current evidential
alcohol analyzers are non-dispersive IR devices.   Such analyzers
are mostly applicable to only one compound. GC is used to measure
breath H_2 [6], anaesthetics [11] and many solvents [12]. The major
drawback for some purposes is the relatively long analysis time, sometimes several min, which may necessitate specimen storage – a problem
in itself [13].

DIRECT MASS SPECTROMETRY

The development of 'PETRA' (Personnel and Environmental TRace
gas Analyzer) by VG Gas Analysis Ltd., Middlewich, Cheshire, U.K.) has
established direct MS in the industrial environment to monitor solvent exposure [14]. It serves for a range of compounds, essentially
simultaneously; it has a fast response time and is very sensitive
and discriminating. For personnel monitoring, short analysis times
make frequent testing of many subjects possible.

The main difficulty with MS is interfacing the high vacuum of
the machine with the sampling system, without adversely affecting

response time or sample integrity. For the major inorganic constituents of respired air a simple molecular leak is suitable since sensitivity is not a problem and response time is of the essence [15].
However, for trace analysis of organic components some form of
enrichment is required.   The choice of device is limited partly
because we have no control over the 'carrier gas', unlike GC-MS. A
simple permeable-barrier molecular separator with a methyl silicone
membrane can achieve enrichments of organics of ~2 orders of magnitude over the permanent gases. Water vapour is a problem because it
is also enriched, but it causes little MS interference because of
its low mass. CO_2 is enriched 50-fold but this is put to good use.
Multi-stage separators (Llewellyn-Arnold) with a water scrubbing
stage have been built [16]. Diffusion through thin silicone membranes
is surprisingly fast, <100 msec for most compounds. Transmission
time depends on solubility in the membrane and is therefore affected
by temperature as well as the pressure differential.

A simple single-stage device is used in PETRA, consisting of a
methyl silicone rubber membrane supported on a nickel grid at a
source pressure of 5×10^{-5} mbar by an ion pump. The quadrupole MS
(mass range 0-200 M/Z) is controlled by an Apple microcomputer. As
shown in Fig. 1, the heated sampling system consists of a loop round
which gas can be circulated (past the membrane) by a pump. Solenoid
valves allow sample to be acquired and discarded at will. A calibration reservoir of 1 l capacity can also be valved in as required,
and a septum inlet allows standards to be injected.

The response time of the total system is ~1 sec but does vary
with the polarity of the analyte: 0.5 sec for halogenated hydrocar-
bons to 10 sec for isopropanol.   The detector is a high-gain (10^5)
secondary emission multiplier (SEM) giving detection limits of from
1 ppb for carbon disulphide to 1000 ppb for dimethyl sulphate at a
signal-to-noise ratio of 2:1 [1 billion = 1000 million; for carbon
disulphide, 1 ppb = 3.4 mg/l of air].   Detection limits are subject
to several influences.- *MS factors:* electron multiplier noise; background spectrum; ionization efficiency. *Physical & chemical factors:*
volatility; memory effects; chemical reactivity; membrane permeabi-
lity.   Depletion of the sample in the calibration loop is so slow
that the signal decreases at only ~2% per h, making it particularly
easy to check short-term stability. The response to toluene in air
is linear from background to 100 ppm.

The PETRA assembly consists of an analyzer module housing the
quadrupole and sampling system, and a control unit containing the
quadrupole electronics, oscilloscope and the computer with disk drive.
In addition an 8-channel oscilloscope and chart recorder are used to
output analogue data from breath analysis, as desired; up to 8 peaks
can be monitored. The assembly is transportable in a small estate
car and can be operational within 1 h of unloading.

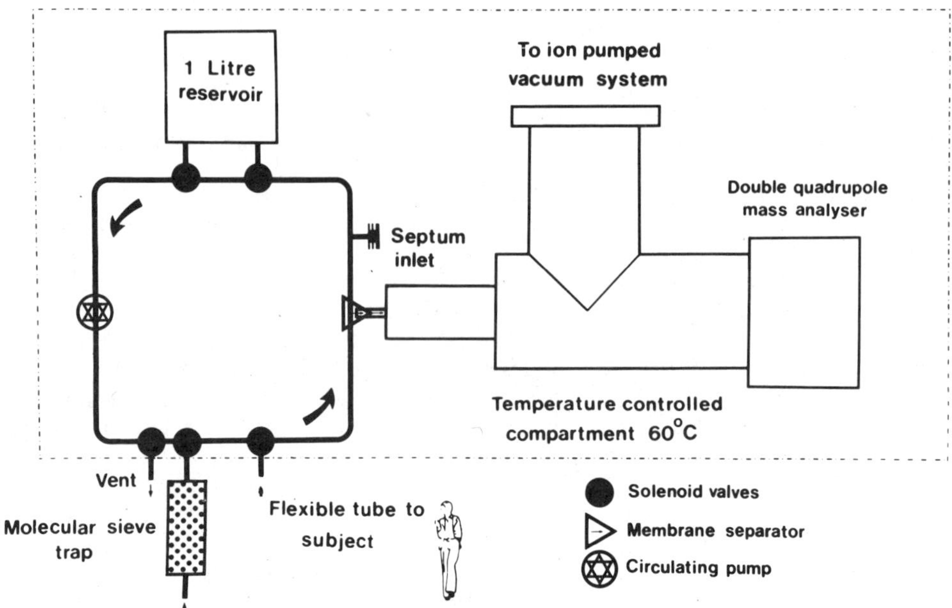

Fig. 1.   Loop system interfaced to the MS (see text).

Breath is sampled with a Haldane–Priestley tube, viz. an open tube 50 x 2.2 cm fitted with a disposable mouthpiece. A 7.5 mm o.d. teflon tube is fitted into the tube 5 cm from the mouthpiece and connected to the inlet of the sampling system.   The sample is drawn down this tube at a flow-rate of 5 l/min by the circulating pump. Calibration is achieved by a static technique, in which liquid is added to a heated vessel of known volume. A circulation pump ensures that the contents are homogeneous and also purges the system after use.   The apparatus consists of a 1.8 l reservoir connected to a 200 ml reservoir so that a 10-fold dilution can be made using valves. Vapour is transferred to PETRA via a syringe and the septum port.

USE OF THE SYSTEM

The apparatus has been used to follow the elimination of solvents in breath from volunteers after exposure in a controlled environment chamber.   As shown in Fig. 2, the alveolar excretion of 1,1,1-trichloroethane could be followed for 7 days after exposure to a level 40% of the Threshold Limit Value (TLV).   Toluene was detectable in breath 6 h after exposure for 10 min to 1,900 mg/m^3 (5 times the TLV);   it has been detected in glue sniffers 4 days after the last episode.   It was possible, with monitoring of every breath, to follow the excretion of dichlorodifluoromethane (Halon 12) for 20 min after one oral inhalation (from a metering aerosol can sprayer) of vapour from 15 μl of liquid. Similarly, with 5 ml of butane the

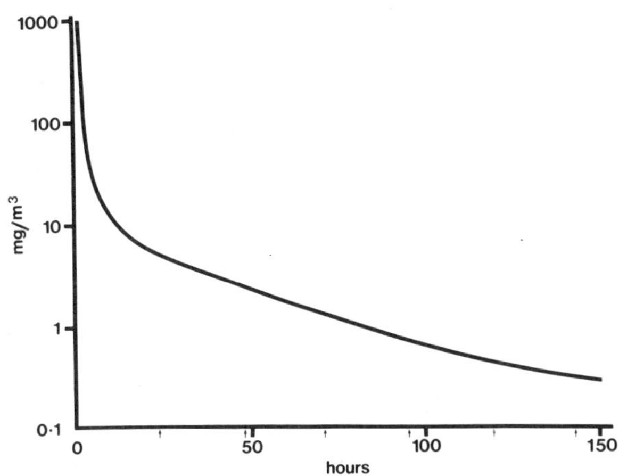

Fig. 2. Elimination in breath, followed by direct MS, of 1,1,1-trichloroethane after chamber exposure for 4 h to 750 mg/m^3 (125 ppm).

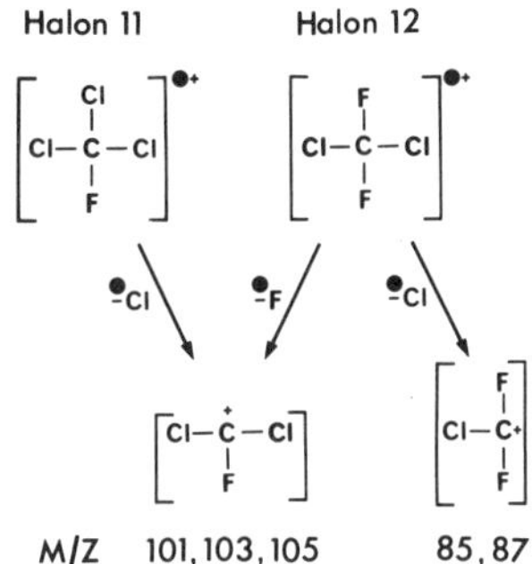

Fig. 3. Fragmentation behaviour of Halons 11 & 12 (trichloro- & di-chlorodi-fluoromethane); latter gives a characteristic fragment (one Cl) besides the two-Cl fragment common to 11 and 12.

wash-out after one breath can be monitored for 30 min. One fragment can be selected (Fig. 3) to screen for the presence of more than one compound and then to select another fragment for differentiation.

The system will be used to screen a large population of subjects, mostly not solvent-exposed. Breath CO_2 can be used as a standard to which quantitative results may be normalized in a manner similar to creatinine for urine analysis [6]. It also serves to ensure that the subject is exhaling cooperatively and not trying to deceive the operator. Using the 7 remaining channels we can screen (Fig. 3) for 15 or more compounds by judicious choice of fragments. The computer system records on disk the maximum outputs of all the channels. Having established that a subject is positive, the breath can be retained in the loop and a complete background-subtracted spectrum recorded. Tentative identification is by comparison with standard spectra, but at high gain interference from normal breath constituents could be a problem.

Thermal desorption analysis gives us a useful means of confirming our identification. A litre or more of breath can be loaded onto a tube packed with Tenax or a mixed-bed adsorbent. The volatiles may then be desorbed onto a GC or into the MS. Retention data for 200 likely compounds have been published [17]. The tube can be attached (Fig. 1, 'trap') to the PETRA sample loop and heated to desorb the volatiles; a mass spectrum can then be recorded, and the volatiles re-trapped by cooling the tube. It can then be removed and analyzed by GC.

No problems have been encountered with normal breath constituents, although 100-200 have been identified [7, 8]. Similarly compounds from the oral cavity have not proved troublesome [18]. Urban air pollutants [19] can cause difficulties because they are absorbed by the subject's lungs, and his breath then contains less than the ambient air, giving a negative peak on the SIM analogue output. This usually matters only in the laboratory or factory that uses solvents.

Acknowledgements

Volunteers were exposed to solvent vapours at the HSE's laboratories with Dr. K. Wilson's help. A DHSS grant covered buying the PETRA.

References

1. Francis, J.,Murray, V.S.G., Ruprah, M., Flanagan, R.J. & Ramsey, J.D. (1982) *Human Toxicol. 1*, 271-280.
2. Anderson, H.R., Dick, B., Macnair, R.S., Palmer, J.C. & Ramsey, J.D. (1982) *Human Toxicol. 1*, 207-221.
3. Ramsey, J.D. & Flanagan, R.J. (1982) *Human Toxicol. 1*, 299-301.
4. Dubowski, K.M. (1974) *Clin. Chem. 20*, 966-972.
5. Schwabe, A.D. & Hepner, G.W. (1979) *Gastroenterol. 76*, 216-218.
6. Niu, H., Schoeller, D.A. & Klein, P.D. (1979) *J. Lab. Clin. Med. 94*, 755-763.
7. Krotoszynski, B., Gabriel, G., O'Neill, H. & Claudio, M.P.A. (1977) *J. Chromatog. Sci. 15*, 239-244.
8. Manolis, A. (1983) *Clin.Chem. 29*, 5-15.
9. Jarvis, M.J., Russell, M.A.H. & Saloojee, Y. (1980) *Br. Med. J. 281*, 484-485.
10. Putz, V.R., Johnson, B.L. & Setzer, J.V. (198) *J. Environ. Path. Toxicol. 2*, 97-112.
11. Corbett, T.H. & Ball, G.L. (1973) *Anesthesiology 39*, 342-345.
12. Stewart, R.D., Hake, C.L. & Peterson, J.E. (1974) *Arch. Environ. Health 29*, 6-13.
13. Pasquini, D.A. (1978) *Am. Ind. Hyg. Assn. J. 39*, 55-62.
14. Willson, H.K. & Ottley, T.W. (1981) *Biomed. Mass Spectrom. 8*, 606-610.
15. Fowler, K.T. (1969) *Phys. Med. Biol. 14*, 185-199.
16. Green, D.E. (1970) *Intra-Sci. Chem. Repts. 4*, 211-221.
17. Ramsey, J.D. & Flanagan,R.J. (1982) *J. Chromatog. 240*, 423-444.
18. Kostelc, J.G., Zelson, P.R., Preti, G. & Tonzetich, J. (1981) *Clin. Chem. 27*, 842-845.
19. Lamb, S.I., Petrowski, C., Kaplan, I.R. & Simoneit, B.R.T. (1980) *J. Air Pollut. Control Assn. 30*, 1098-1115.

#NC(F)

NOTES and COMMENTS relating to

Various analytes in biological and forensic samples

Comments related to particular contributions:

#F-2, p. 371 (#F-1 material is touched on in #C – p. 196)
#F-4 & #F-5, p. 372
#NC(F)-1 to -3, p. 373

#NC(F)-1

A Note on

EXPLOSIVES AND OTHER NITRO COMPOUNDS DETERMINED BY LIQUID CHROMATOGRAPHY WITH PHOTOLYSIS-ELECTROCHEMICAL DETECTION

I.S. Krull, X-D. Ding, C. Selavka, [†]K. Bratin and [†]G. Forcier

Institute of Chemical
 Analysis
Northeastern University
360 Huntingdon Avenue
Boston, MA 02115, U.S.A.

[†]Analytical Research
 Department
Pfizer Central Research
Pfizer & Co., Inc.
Groton, CT 06340, U.S.A.

Post-column photolysis has long been used in HPLC to confer UV or fluorescence detectability, but very little has been described for electrochemical detection (LCEC) [1-3]. A commercial photoconductivity detector has a built-in high-intensity UV source whereby the column effluent can be irradiated [4]; the ionic species thus generated are detected non-specifically by conductivity. Photohydrolytic methods have already been described for trace analysis of *N*-nitroso compounds. However, reports are lacking on photolytic studies with LCEC using commercially available flow-through cells of thin-layer type [5].

With suitable photolytic reactions[†], several classes of organic or inorganic compounds can efficiently form inorganic anions, e.g. nitrite from a range of compounds now studied by LCEC. These include nitrate esters ($R-O-NO_2$), organic nitrites (esters; $R-O-NO$), aromatic *C*-nitro derivatives, and *N*-nitramines ($R-N-NO_2$). Depending on the particular analyte, nitrite (NO_2^-) and/or nitrate (NO_3^-) may arise on-line in the HPLC effluent, with a low-pressure Hg arc lamp giving ~190-350 nm output [6]. The photolysis module (also usable directly without HPLC) is shown in Fig. 1. Wound round the light source is a narrow-bore flow tube of Teflon, the i.d., length and transparency of which are all critical for successful generation of nitrite and/ or nitrate from nitro analytes. Our optimization, especially of detection limits, has also involved HPLC flow rates, pH, salt concentrations, and specific electrolyte salts in the mobile phase; we have had to minimize band-broadening effects in the irradiation step.

[†] *Note by Eds.*- The term '(photochemical) derivatization' as sometimes used in the MS. seemed somewhat inappropriate and has been avoided.

Fig. 1. Schematic diagram of post-column irradiation of the HPLC effluent followed by oxidative EC detection of the nitrite generated from nitro compounds [6]. Flow-tube i.d. 763 µm (o.d. 1.6 mm). For nitroglycerin, mobile phase methanol/0.1 M NaCl (1:1), 0.6 ml/min.

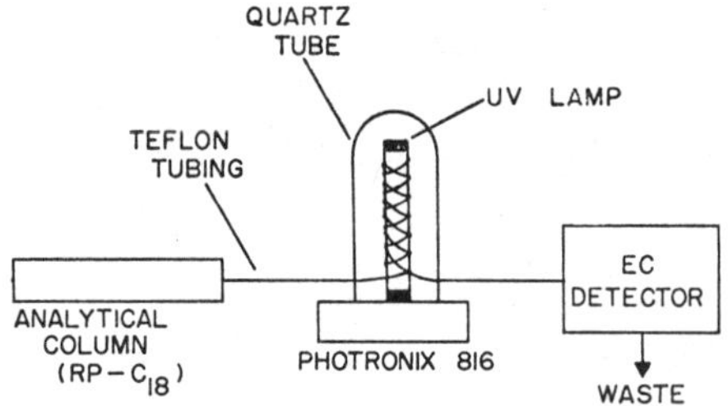

For the photolytically generated nitrite, optimal oxidative and reductive EC potentials have been established. Oxidatively, +0.8 to +1.0 V appears ideal [7]. Rather than a single working electrode, two may be used, in series or [#NC(C)-3, this vol.] in parallel, both oxidative, both reductive, or one oxidative and the other reductive; one can serve for measurement and the other to verify nitrite identification. Analysis and speciation has been thus achieved for (e.g.) isosorbide dinitrate and for certain nitro explosives. Characteristic dual-electrode response ratios include 4.69 ±0.24 (S.D.) for nitroglycerin, 2.04 ±0.15 for TETRYL, 2.26 ±0.18 for TNT and 2.94 ±0.22 for RDX; the potentials used were +0.8 V and +0.9 V.

Limits of detection for explosives have ranged from 6 to ~50 ppb depending on the particular analyte, and could be improved with 200 rather than 20 µl injection volumes. There is linearity over 2-3 orders of magnitude. Already the approach has been applied to post-blast extracts, confirming TLC results but with much better overall specificity. (Extractant: acetonitrile; sensitivity to be ascertained.)

Acknowledgements

Part-support came from a grant by Pfizer's Analytical Research Department and from NIH Biomedical Science Support Grant No. RRO7143, to Northeastern University.

References

1. Krull, I.S. & Lankmayr, E.P. (1982) *Am. Lab., May issue,* 18-32.
2. Frei, R.W. & Lawrence, J.F., eds. (1981) *Chemical Derivatization in Analytical Chemistry*, Vol. 1: *Chromatography*, Plenum, New York, pp. 316-340.
3. Lefevre, M.J., Frei, R.W., Scholten, A.H.M.T. & Brinkman, U.A.Th. (1982) *Chromatographia 15*, 459-467.
4. Popovich, D.J., Dixon, J.B. & Ehrlich, B.J. (1979) *J. Chromatog. Sci. 17*, 643-650.
5. Snider, B.G. & Johnson, D.C. (1979) *Anal. Chim. Acta 106*, 1-13.
6. Krull, I.S.. Selavka, C., Ding, X-D., Bratin, K. & Forcier, G. (1984) *Proc. Internat. Symp. (FBI Academy, March 1983) on Analysis and Detection of Explosives*, U.S. Govt. Printing Office, Washington, D.C., to be published.
7. Krull, I.S., Ding, X-D., Selavka, C., Bratin, K. & Forcier, G. (1984) *J. Forensic Sci. 29*, in press.

#NC(F)-2

A Note on

NITROGLYCERIN ASSAY IN PLASMA BY THERMAL ENERGY ANALYZER

[1]Stephen H. Curry, [2]Gary Algozzini and [3]Wing Yu

[1]University of Florida (Box J-4), Gainesville, FL 32610

[2]Pharmacy Department, L.W. Blake Memorial Hospital
Bradenton, FL 33529

[3]Thermo-Electron Corporation, Waltham, MA 02254, U.S.A.

Measurement of nitroglycerin (NG[*]) in plasma is notoriously difficult. Considerable success has followed the use in various laboratories of GC with electron-capture detection (ECD), but this method is insufficiently sensitive for detection of NG in low-dose pharmacokinetic studies [1-6]. Greater sensitivity has been obtained by use of GC-MS, but this approach is expensive and time-consuming [7]. We have recently obtained good results by use of a thermal energy analyzer (TEA) coupled to an HPLC system [8, 9].

In assessing the NG content of plasma samples, care is needed at several stages including the sample-collection and storage stages. The time between the blood leaving the body and plasma becoming frozen for storage must be minimized. This is achieved by chilling the blood-collection tubes before use, and cooling the tubes again immediately after placing blood in them (ice-bath, $0°-1°$). Centrifugation must be for the shortest possible time, in a refrigerated centrifuge if possible. Once the plasma is separated it should be frozen for storage immediately. Storage at $-4°$ or below is satisfactory, including overnight in dry-ice packages if transfer between locations is needed. Frozen samples once thawed for assay should not be re-frozen for repeat assay at later times.

We extract NG into pentane or hexane, and nitric ester NG metabolites into ethyl acetate. The extracts are filtered through an 0.5 μm filter and concentrated to small volume. Our HPLC system has a 250 x 4.5 mm CN-silica (5 μm) column, with iso-octane/dichloromethane/methanol (75:20:5 by vol.) as the mobile phase at 1.5 ml/min and a Model 502 TEA Analyzer (Thermo-Electron).

---

[*] Editor's abbreviation; GTN is an alternative.

Very few analysts have presented sample NG chromatograms. Most traces published are of poor quality. Our results show clear NG or NG metabolite peaks rising from peak-free, near-horizontal baselines. The C.V. on replicate assays of spiked plasma is in the range 7.7% (0.2 ng/ml) to 10.9 (1.0 ng/ml). Comparison of true and found data shows good agreement. The signal is linear.

We tested our system using samples from a patient previously treated with a transdermal NG preparation (TTS), transferred to i.v. NG during haemodialysis. The baseline concentration (on the TTS) was 0.2 ng/ml. At 20 min from the start of i.v. infusion the NG level had risen to 2.1 ng/ml, and it remained at ~2.5 ng/ml during 3 h of dialysis except for an initial drop when the NG input line was switched from a human vein to a 'vein' of the artificial kidney.

We conclude that the TEA is the most stable detection system available at a reasonable price for NG assay at present. In our hands, the chromatographic trace is superior to that from GC-ECD. There is sufficient sensitivity for the study of low-dose NG dosage forms.

References

1. Rosseel, M.T. & Bogaert, M.G. (1972) *J. Chromatog.* 64, 364-370.
2. Yap, P.S.K., McNiff, E.F. & Fung, H.L. (1978) *J. Pharm. Sci.* 67, 582-584.
3. Armstrong, P.W., Armstrong, J.A. & Marks, G.S. (1979) *Circulation* 59, 585-588.
4. Wei, J.W. & Reid, P.R. (1979) *Circulation* 59, 588-592.
5. Spanggord, R.J. & Keck, R.G. (1980) *J. Pharm. Sci.* 69, 444-446.
6. Taylor, I.W., Ioannides, C., Turner, J.C., Koenigsburger, R.U. & Parke, D.V. (1981) *J. Pharm. Pharmacol.* 33, 244-246.
7. Gerardin, A., Gaudry, D. & Wantiez, D. (1982) *Biomed. Mass Spectrom.* 9, 333-335.
8. Yu, W.C. & Goff, E.U. (1983) *Anal. Chem.* 55, 29-32.
9. Yu, W.C. & Goff, E.U. (1983) *Biopharm. Drug Dispos.* 4, 311-319.

#NC(F)-3

A Note on

A COMPARISON OF HPLC-THERMAL ENERGY ANALYSIS AND GC-ELECTRON CAPTURE DETECTION FOR DETERMINING ISOSORBIDE DINITRATE AND ITS MONONITRATE METABOLITES IN HUMAN PLASMA

A.J. Woodward, P.A. Lewis & J. Maddock

Analytical Division, Simbec Research Ltd.
Merthyr Tydfil, Mid Glamorgan CF48 4DR, U.K.

Isosorbide dinitrate (ISDN) is widely used for prophylactic treatment of angina pectoris and for the treatment of respiratory congestive heart failure. Sustained or controlled release preparations in a variety of dosage forms are becoming available. Thus there is considerable interest in the pharmacokinetics of the compound and the bioavailability of these new dosage forms. Further, the 5-mononitrate metabolite (IS5MN) has also been shown to be a potent anti-anginal substance, and thus more interest is being placed on determining this and the 2-mononitrate metabolite of ISDN (IS2MN).

Most published methods [1-4] for plasma ISDN utilize GC-electron capture detection (ECD) techniques employing instrumentation which is generally available and relatively inexpensive. Although adequate sensitivity for the parent compound can be achieved with only 200 µl plasma, this approach suffers disadvantages. The non-specificity of the detector results in chromatography prone to interference from extraneous impurity peaks, necessitating multiple extractions and leading to relatively long run times per sample. The internal standards used are not chemically related and are difficult to obtain, and some methods use different internal standards for different concentration ranges.

Recently a technique employing HPLC-thermal energy analysis (TEA) has been published [5]. Our aim was to compare the two techniques for determining ISDN and its metabolites. The first of our two separate investigations was an intra-laboratory study employing the HPLC-TEA method slightly modified from the published procedure and the GC-ECD technique of Fung et al. [4], for the analysis of 18 human plasma samples spiked with ISDN over the range 1-25 ng/ml. The following between-method correlations were obtained: GC-ECD *vs.* HPLC-

TEA, $y = 0.93x + 0.52$, with $r = 0.93$;   spiked-in value *vs.* GC-ECD, $y + 0.95x + 0.40$, with $r = 0.92$;   spiked-in value *vs.*   HPLC-TEA, $y = 0.98x + 0.46$, with $r = 0.99$.

The second investigation was an inter-laboratory study where 60 plasma samples from human volunteers receiving single oral doses of ISDN were analyzed for ISDN and its mononitrate metabolites by HPLC-TEA and a GC-ECD procedure. Data on 59 of the 60 samples was produced by both collaborating laboratories (Table 1).

Table 1. Inter-laboratory study results[*](rec. = recovery; sens. = sensitivity, $pg \times 10^2/ml$: GC-ECD & *(italicized) HPLC-TEA*.

Para-meter	ISDN	IS2MN	IS5MN	CORRELATION (Cor.) OF DATA:		
					ISDN	IS2MN	IS5MN
Rec.	85, *98%*	66, *76%*	76, *78%*	Pearson, raw	0.88	0.91	0.85
Sens.	2-5, *1*	20, *5*	100, *10*	Cor. log	0.95	0.84	0.85
C.V.	18.0, *9.9%*	9.0, *13.4%*	3.0, *9.9%*	Spearman cor.	0.97	0.94	0.90

[*]No. of 'Not Detectable' samples: ISDN: 5, 7; IS2MN: 19, 7; IS5MN: 16, 6.

The data show fairly good agreement between the two methods, although better correlation with the spiked-in values is obtained by HPLC-TEA, indicating superior accuracy. HPLC-TEA is more sensitive than GC-ECD, particularly for the mononitrates, due both to thermal instability and to interfering peaks on GC. The specificity of the TEA contributes to the higher sensitivity and results in cleaner chromatograms, allowing simpler sample preparation procedures and a very much more rapid throughput.

The HPLC-TEA system can also be applied to the determination of sub-ng/ml concentrations of other organic nitrates, e.g. nitroglycerin, pentaerythritol tetranitrate, and their metabolites, and is widely used for the analysis of *N*-nitroso compounds. HPLC-TEA is surely a chromatography detector system worthy of greater recognition.

References

1. Rosseel, M.T. & Bogaert, M.G. (1973) *J. Pharm. Sci.* 62, 754-758.
2. as for 1. (1979) 68, 659-660.
3. Doyle, E., Chasseaud, L.F. & Taylor, T. (1980) *Biopharm. Drug. Disp. 1,* 141-147.
4. Fung, H.L., McNiff, E.F., Ruggirello, D., Darke, A., Thadari, U. & Parker, J.D. (1981) *Br. J. Pharmacol. 11,* 579-589.
5. Maddock, J., Lewis, P.A., Woodward, A.J., Massey, P.R. & Kennedy, S. (1983) *J. Chromatog.* 272, 129-136.

Comments on material in #F

Comments on #F-2, I.L. Martin et al. - GC–ECD FOR BIOACTIVE AMINES

Replies to U.A.Th. Brinkman, O. Gyllenhaal.- For the perfluoro derivatives of the amines the detection limits are ~1 ng/g of biological material (~1 pg on-column). In hydrolyzing the acetyl derivative of pyramine, there seems little risk that the liberated phenol group will become oxidized, since the pH is ~10 and the organic phase has a volume 10 times that of the aqueous ammonia phase.

Supplementary refs. relevant to bioactive amines, with Editor's warning that the field abounds in temperamental assay methods.—

Muskiet, F.A.J., Thomasson, C.G., Gerding, A.M., Fremouw-Ottevangers, D.C., Nagel, G.T. & Wolthers, B.G. (1979) *Clin. Chem. 25,* 453-460.-Urinary catecholamines and their O-methyl metabolites were amenable to GC-MS (mass fragmentography) assay as suitable derivatives. 'Persilylation' problems are discussed, and the need with PFP derivatives to preserve them during GC by having a small amount of the PFP anhydride present in the ethyl acetate solution of the dried-down derivatives. HPLC-EC following cation-exchange clean-up was rather time-consuming, and the detector was unstable.

Hegstrand, L.R. & Eichelman, B. (1981) *J. Chromatog. 222,* 107-111.- Catecholamines in brain by HPLC with EC detection(cf.#F-1).

Keller, R., Oke, A., Mefford, I. & Adam, R.N. (1976) *Life Sci. 19,* 995-1004 - *Ditto;* includes tips on using carbon paste electrodes.

Baldessarini, R.J. & Fischer, J.E. (1978) *Biochem. Pharmacol. 27,* 621-626.- A 'Commentary' giving useful background on trace amines (including octopamine) and alternative neurotransmitters in the CNS. (For background on receptors, see Vol. 13, this series.)

Davies, C.L. & Molyneux, S.G. (1982) *J. Chromatog. 231,* 41-45.- Ion-pair RP-HPLC with EC detection applied to plasma catecholamines; includes tips on overnight sample storage, on obviating impairment of the bonded-silica packing by the ion-pairing agent, and on use of glassy-carbon electrodes (can be troublesome).

Cross, A.J. & Joseph, M.H. (1981) *Life Sci. 28,* 499-505.- Acidic (DOPAC, HVA, 5HIAA) and neutral (MHPG) metabolites of monoamines in brain, estimated in a pH 3 extract by HPLC with both EC detection and, essentially confirmatory (title misleading!), fluorimetry.- Cf. ref. [26] in #F-1. The following also relates to #F-1:

Lyness, W.H. (1982) *Life Sci. 31,* 1435-1443.- DA and metabolites (5HT, 5HIAA, tryptophan) in brain by HPLC with EC detection. Cf.:

Ishikawa, K., Shibanoki, S. & McGaugh, J.L. (1983) *Biochem. Pharmacol. 32,* 1473-1478.- RP-HPLC/EC determination of butanol-extracted morphine [cf. NC(E)-3] and the foregoing metabolites.

Comments on #F-4, R.L. Williams – FORENSIC ASSAYS
        & #F-5, J. Ramsey – SOLVENTS IN BREATH, BY MS

Question by E.P. Lankmayr.– Concerning the advocacy of Fast
Atom Bombardment MS as a forensic tool, would it indeed be useful
for drugs insofar as its advantages lie in analyzing compounds of
high mol. wt. (over 1000)? *Reply by* R.L. Williams. – I agree about
the especial applicability to 'high' mol. wt. compounds. The discovery that many neurotransmitters are polypeptides suggests very
strongly that these or synthetic analogues may become available in
chemotherapy in the future; since they are often mood–affecting, it
it also highly probable that they will be abused and become items of
concern to forensic analysts.   Meanwhile there are problems with
more commonplace compounds that are labile.   Thus, the salts of
various alkaloids, e.g. morphine and the ergot group, appear to
yield reasonable FAB spectra from which the anion can be identified.
The spectra have the advantages of simplicity associated with chemical ionization, and both negative and positive ions can be produced.
However, more work needs to be done to establish the technique fully.
But R. Schmid *remarked (rejoinder: costly!):* with MS/MS you could
get a lot more information from your sample.

Answer to A. Gulaid.– We have difficulty in deciding the accuracy of our estimates in forensic toxicology, especially when the
pathologist furnishes us with the sample when he has done everything
that *he* wants to do. But our estimate is not taken as sole evidence
and other information may be available. B.S. Thomas *asked:*   Can a
toxic agent perhaps be identified by its biological effects when normal
identification is not feasible? – Usually the effects are slow in
materializing and the toxic agent has long disappeared.   This is a
very difficult situation.

J. Ramsey, *replying to* K. Ensing.– The possibility of endogenous interference, e.g. by acetone if the person has diabetes mel-
litus, is small; acetone is not frequently used for sniffing, and
in any case the amount would be small compared with that inhaled to
obtain central effects. *Reply to* R.L. Williams.– No euphoria or
other effects were experienced when I inhaled toluene, but the level
was only 5 times the TLV, giving a very small dose in terms of solvent abuse.

General remarks by R.P. Maickel.– Forensic analysts and indus-
trial/pharmaceutical analysts seem to represent two different groups
and research (academic) pharmacologists and toxicologists represent
a third group. *Forensic analysts* accumulate various bits of information so that they can identify (and often assay) a drug, chemical,
toxin, etc. *Pharmaceutical analysts* already have knowledge of the
drug and/or metabolites and so can concentrate on more sophisticated
procedures.   Both these groups need to 'cross-transfer' information
so that each can apply advances made by the other in their own field.

Comments related to #NC(F)-1 to -3 - NITROGLYCERIN etc.

R.L. Williams, *remark to* I.S. Krull: J. Lloyd (at the Forensic
Lab. in Birmingham) has been able to pick up 10 pg quantities reductively with an EC detector, but the experimental technique was very
demanding. *A nitroglycerin ref. noted by Senior Editor:*
Penton, Z. (1983) *Am. Lab. 15(2),* 86-88.- A procedure entailing
capillary GC-ECD with non-vapourizing on-column injection allows the
detection of 0.2 ng nitroglycerin/ml plasma, after extracting with
pentane and drying down. Isosorbide dinitrate serves as i.s.

A forensic extraction method noted by Editor (cf. #A-2*):*
Kristinsson, J. (1982) *Acta Pharm. Toxicol. 50,* 318-320.- For the
extraction of antidepressant drugs from post-mortem samples of blood,
brain (treated initially with dil. sulphuric acid), liver or urine,
the sample deposited onto kieselguhr with Na carbonate present was
hexane-extracted after keeping at least 30 min at room temp. After
back-extraction into formic acid containing 1% methanol, and drying
down, GC was performed with a nitrogen detector.
Trial of pre-digesting liver with β-glucuronidase in a drug assay:
Holzebecher, M., Perry, R.A. & *[pers. comm. also]* Ellenberger,
H.A. (1982) *J. For. Sci. 27,* 715-717.- Following an unvalidated laboratory practice, this treatment was applied to an aqueous homogenate
of liver from a fatal case of metoprolol overdosage, but it led to
little increase in the drug value (obtained by GC-ECD after benzene
extraction at alkaline pH, then derivatization).
Cf. possible efficacy of protease digestion: #C-1 in Vol. 10; #A-2.

Some drug-assay refs. in therapeutic or related contexts

Krylov, A.I. & Khlebnikova, N.S. (1982) *J. Chromatog. 252,*
319-324.- If urinary pemoline   (2-amino-5-phenyl-2-oxazolin-4-one;
may be assayed in athletes) is to be assayed by GC, rather than HPLC
(refs. cited), *N*-methylation as hitherto performed with diazomethane
can be conveniently achieved by reaction, in the injection port, with
trimethylanilinium hydroxide (TMAH).
Trautmann, K.H. & Haefelfinger, P. (1981) *J. High Resol.
Chromatog. Chromatog. Comm. 4,* 54-59.- A cephalosporin ('Ro 13-
9904') in plasma (deproteinized with ethanol), urine or bile could be
assayed by RP-HPLC with a quaternary ammonium counter-ion.
Masoud, A.N. & Krupski, D.M. (1980) *J. Anal. Toxicol. 4,*
305-310.- RP-HPLC (232 nm detection) can be used to assay cocaine in
plasma, down to 20 ng/ml, after successive extractions: into diethyl
ether, back-extraction into dil. acetic acid, then re-extraction at
alkaline pH into hexane and dry down.
Pacifici, G.M., Placidi, G.F., Fornaro, P. & Goneni, R. (1983)
Int. J. Clin. Pharm. Res. 3, 331-337.- Pinazepam and Phase I metabolites (including desmethyl) were assayed in plasma (solvent-extr-
acted) and urine by GC.

Tanner, R.J.N., Martin, L.E. & Oxford, J. (1983) *Anal. Proc. 20*, 38-41.- The GC-MS assay of salbutamol in plasma is considered from the viewpoint of automation.

Stable isotopes in clinical pharmacology - cf. #A-3; *ref. noted by Co-Ed.*

Browne, T.R., van Langenhove, A., Costello, C.E., Biemann, K. & Greenblatt, D.J. (1984) *Ther. Drug Monitoring 6*, 3-9.- Review of the use of stable isotopes in pharmacokinetic studies on antiepileptic drugs in the young. Includes consideration of drug assay in small volumes of biological fluid.

Flow-injection analysis: relevant to drug-level determination?

At the Forum, A.G. Fogg (no publication text) outlined the scope of flow-injection analysis with voltammetric detection. With very simple equipment, assay down to ~1 µM has been achieved for various analytes, not only inorganic but also, bromimetrically on-line, drugs such as isoniazid [papers from the Loughborough laboratory in *Analyst*]. A simple use of flow injection is to present an electroactive material to an electrode.

Discussion points.- The technique seems inapplicable to biological drug-containing samples unless initial clean-up is performed (*reply to* R.D. McDowall). Interfering EC signals are hard to prevent, even if precautions such as deoxygenation are taken; pressure pulses limit use at high sensitivity (D. Perrett, R. Schmid). *Comment by* I.S. Krull.- Unless there is a regular need for high analytical throughput (say 250 samples/h), automated HPLC with its high specificity would be preferable to flow-injection analysis.

Can saliva or sweat be used for detecting drugs/steroids?

At the Forum, M.S. Moss (Newmarket) outlined approaches, e.g. RIA, for screening racehorses. *In answer to* J.W. Paxton, he felt saliva was a possible test specimen, since non-ionized drugs such as steroids should cross into saliva; but there are problems in collecting sufficient saliva after a race. Moreover, the steroids are highly protein-bound in plasma, and only the free fraction passes into saliva. He knew of no published information on the transfer of drugs into sweat. (R. A. de Zeeuw in #D-2, Vol. 10, has surveyed drug transfer into saliva.- *Ed.*)

Morphine pharmacokinetics: RIA compared with GC [cf. #NC(E)-3]

Stanski, D.R., Paalzow, L. & Edlund, P.O. (1982) *J. Pharm. Sci. 71*, 314-317: RIA over-estimated plasma morphine by 27%, but with little effect on pharmacokinetic parameters.

Section #G

ANALYTICAL QUALITY ASSURANCE

#G-1

THE ROLE OF EXTERNAL QUALITY ASSESSMENT SCHEMES IN THE DETERMINATION OF PLASMA DRUG LEVELS

John Williams

Department of Pharmacology and Therapeutics
Welsh National School of Medicine
Heath Park, Cardiff CF4 4XN, U.K.

The advent and obvious advantages of therapeutic drug monitoring in the treatment of several disease states has resulted in a vast increase in the number of laboratory requests made for the measurement of plasma levels of drugs concerned. Although stringent internal quality control helps ensure the precision of the laboratory's results, participation in external quality assessment schemes establishes the degree of accuracy of the assay procedure employed. This article describes a typical scheme and outlines benefits from the long-term feedback of information made available to scheme coordinators.

Early studies [1, 2] clearly showed that the accuracy with which antiepileptic drugs were being assayed was subject to an excessive inter-laboratory variability. Thus, for phenytoin in a pooled plasma sample distributed by Pippenger et al. [2] the following results were obtained for all 109 of the laboratories reporting and *(italics)* for 5 reference laboratories.- Mean, µg/ml: 13.1,   *12.8*;   C.V.: 57.3%, *15.7%*; range: 0.0-70.0, *10.7-16.0*. Corresponding results for ethosuximide were as follows for 71 laboratories/*5 reference laboratories.-* Mean: 14.9, *1.4* ; C.V.: 504.7%, *156.4%*; range: 0.00-633.0, *0.0-5.0*. For all four antiepileptics tested, the reference laboratories exhibited significantly lower C.V. values.

These findings were the stimulus to the setting up of external quality control schemes, for antiepileptics in the first instance and later for antidepressants, theophylline and digoxin. Prof. Richens had been the initiator whilst at St. Bartholomew's Hospital, London (in the early 1970's), and with his move to Cardiff and the transfer of the original scheme to our laboratory, the term Heathcontrol has now been applied.   Besides national recognition, the schemes have attracted participation from 21 overseas countries (U.S.A. included).

Monthly assays are requested but, for convenience, distribution is 3-monthly. Human serum screened for Australia antigen is spiked with the respective scheme analytes at various levels, dispensed and freeze-dried. Each laboratory reconstitutes and assays the sample and reports the estimated value to the scheme coordinator by a set date. The return cards incorporate method coding that enables the performance of different methods to be compared and their relative popularity evaluated. One conclusion (Table 1) is that the EMIT method is the most popular for antiepileptic drugs. Fig. 1 shows a typical print-out as sent monthly to each laboratory, with the method it uses (x) distinguished from others (o) in the histograms. The laboratories rejected ('outliers') are those whose estimate exceeds the initial group by >3 S.D. units. Once removed the group mean is recalculated and all subsequent calculations are based on this 'trimmed' mean value. The method for removing outliers is discussed elsewhere [3].

PERFORMANCE APPRAISAL

The PERFORMANCE INDEX values shown in Fig. 1 are an attempt to categorize individual laboratory performance rather than to provide any useful statistical parameter. The 1-10 scale is derived by dividing the 3 S.D. units either side of the trimmed mean into 10 zones each of 0.3 S.D. units; thus a laboratory with a S.D. of < 0.3 for a particular analyte will be indexed as 1. The index for overall per-

Table 1.   Percentage of participants in Heathcontrol (to Sept. 1982) using particular methods to assay anti-epileptic drugs. Spec. = spectrophotometric. 'Wallace' is colorimetric. 'n' = no. of labs.

Drug	GC, no deriv.	GC, deriv.	EMIT	RIA	TLC	HPLC	Spec.	Wallace	(n)
Phenytoin	16	21	47	2	–	12	1	1	(207)
Phenobarbitone	17	20	47	0.5	–	13	1.5	1	(201)
Carbamazepine	17	9	55	–	–	17	1	1	(190)
Primidone	26	17	43	–	–	11	1.5	1.5	(122)
Ethosuximide	57	5	33	–	–	5	–	–	(107)
Valproic acid	52	5	38	–	–	5	–	–	(170)
Carbamazepine-10,11-epoxide	–	–	–	14	–	86	–	–	(7)
Clonazepam	35	35	–	–	–	30	–	–	(20)

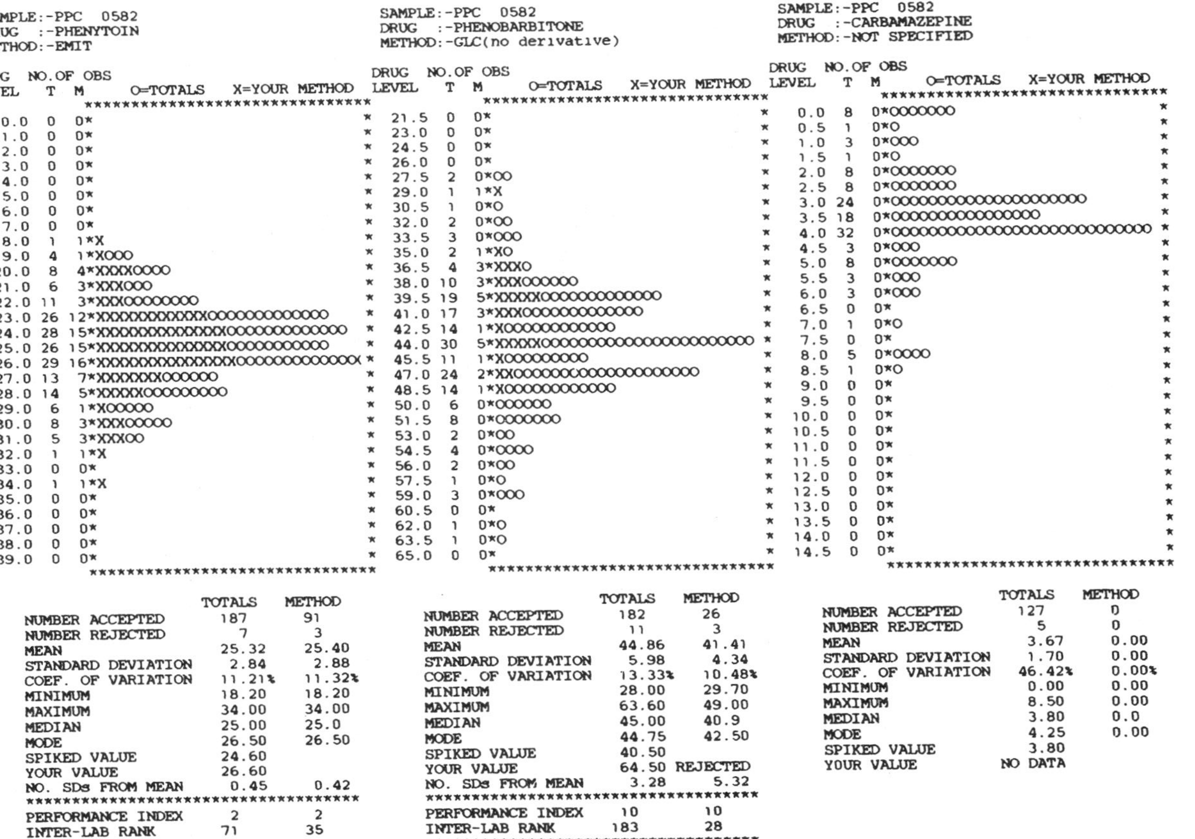

Fig. 1. Computer print-out as sent to a participating laboratory, 'No. 43'; headed HEATHCONTROL May 1981 OVERALL PERFORMANCE INDEX 6.

formance, taking all drugs into consideration, is calculated by summating the (S.D.) for each analyte, dividing this value by the no. of analytes assayed, and then obtaining the square root of this value. The rating is as above (0.3 S.D. = 1).

Another important feature of the laboratory print-out is that a frequency histogram for the participant's method is superimposed on the overall histogram. Thereby any bias that a particular method may exert on the group as a whole can be identified. Moreover, if a participant's data are not returned the laboratory still receives a print-out of the overall group performance and so can evaluate its own, on a monthly basis.

Laboratories also receive an annual performance summary which enables supervisors to assess their laboratory performance data over an extended time and concentration range. Fig. 2 shows excerpts from two summaries (a) for all methods, and (b) for the particular method used for assaying the drug in that laboratory. Fig. 3 shows a trend analysis – an attempt to identify where laboratory estimates are showing consistent deviations from the mean values obtained by the group as a whole. For obvious reasons trend analysis applies only if more than 3 estimates for the drug are returned during the year. It provides a plot of the laboratory's drug estimates against their respective group means. Ideally there should be a line with a slope of 1 passing through the origin. To identify where laboratories are departing from expected results, the absolute mean % differences of the estimates from the expected results are calculated, and the average % difference used to classify the results into poor, satisfactory, good, very good and excellent. The delimiters for these categories have been arbitrarily set at 25%, 12.5%, 10%, 7.5% and 5%.

Where the difference exceeds 25%, a statistically significant trend is looked for by fitting a line through the laboratory estimates by weighted linear regression. This fitted line appears on all trend analysis graphs (Fig. 3) irrespective of the accuracy of the laboratory data. By applying a variance ratio test, laboratories can be identified where there is no significant relationship between their estimates and the group means. Where a significant regression can be fitted the slope is tested against the expected value of 1 and, if there is no significant difference from 1, the intercept is tested against zero to look for parallel shifts in the line. Such a shift would indicate a possible deterioration in accuracy as a direct result of assay performance.

DRUG	NUMBER OF MEASUREMENTS	OVERALL SD'S FROM MEAN	OVERALL PERFORMANCE INDEX	LAB RANKING	
PPC-PHENYTOIN	10	0.96	4	88 OUT OF 217	
PPC-PHENOBARBITONE	10	1.12	4	111 OUT OF 217	
PPC-CARBAMAZEPINE	10	1.96	7	162 OUT OF 193	
H8-PHENYTOIN	12	0.89	3	82 OUT OF 225	
H8-PHENOBARBITONE	12	1.15	4	132 OUT OF 219	(a) All methods
H8-PRIMIDONE	12	0.83	3	44 OUT OF 142	
H8-CARBAMAZEPINE	12	1.01	4	105 OUT OF 199	
H8-ETHOSUXIMIDE	12	0.40	2	5 OUT OF 117	
H8-VALPROIC ACID	12	0.56	2	20 OUT OF 172	
H8-10,11 EPOXIDE	0			NO DATA	
H8-CLONAZEPAM	0			NO DATA	
ALL DRUGS	102	1.04	4	64 OUT OF 230	
PPC-PHENYTOIN	0			NO DATA	
PPC-PHENOBARBITONE	10	0.97	4	18 OUT OF 40	
PPC-CARBAMAZEPINE	10	1.90	7	31 OUT OF 37	
H8-PHENYTOIN	0			NO DATA	(b) Yours: GC, no deriv.
H8-PHENOBARBITONE	12	1.09	4	20 OUT OF 42	
H8-PRIMIDONE	12	0.78	3	10 OUT OF 37	
H8-CARBAMAZEPINE	12	1.15	4	23 OUT OF 39	
H8-ETHOSUXIMIDE	12	0.40	2	3 OUT OF 71	
H8-VALPROIC ACID	12	0.57	2	20 OUT OF 119	
H8-10,11 EPOXIDE	0			NO DATA	
H8-CLONAZEPAM	0			NO DATA	
ALL DRUGS	80	1.05	4	57 OUT OF 129	

Fig. 2. Excerpts from representative annual performance sheets *[2 sheets amalgamated for the purpose of the Fig.- Ed.],* as sent to 'Lab. 161'. (a) Summary of all methods (1981 performance); (b) results from the laboratory concerned.

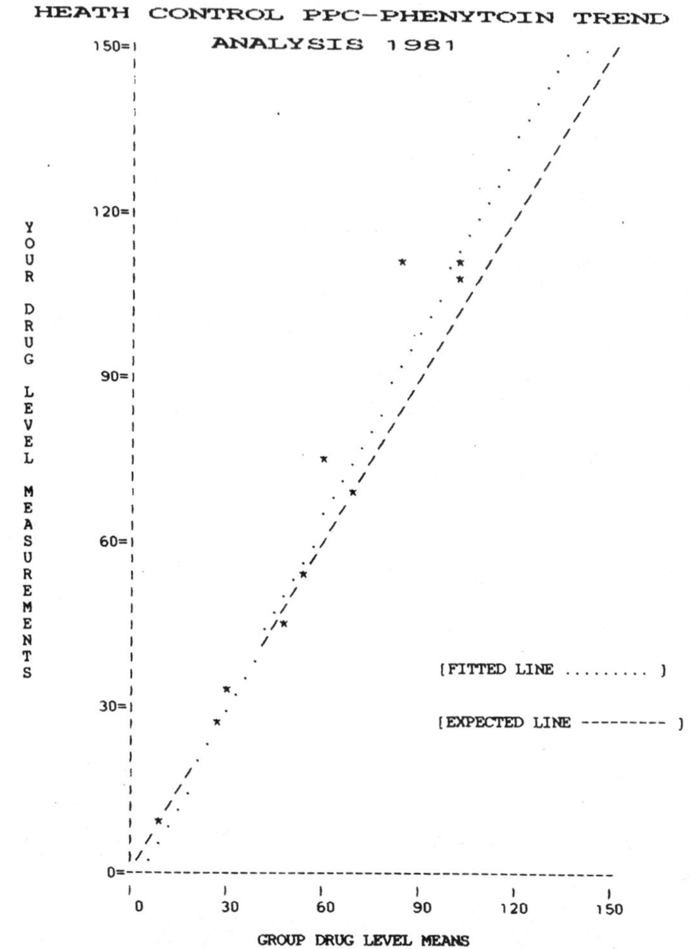

Fig. 3. Annual performance summary: a representative trend
analysis print-out, as sent to 'Lab. 161'. For amplification,
see text.

COMPARABILITY OF DIFFERENT ASSAY METHODS

Central analysis of data can not only indicate the accuracy of a laboratory's assay procedure, but can illuminate possible sources of error by comparing different assay techniques. Two aspects of data collected over 26 months were used to compare methods. Firstly the S.D.'s of the sample means for a particular method have been compared with those of other methods, on the assumption that methods with greater precision will demonstrate narrower distributions in their data and hence smaller S.D. values. Secondly, the accuracy of a method has been assessed from the closeness with which the method means approached the true drug level (as spiked into the serum).

This analysis revealed that no method could be identified which performed better or worse over a limited portion of the concentration range. However, several significant differences were observed which indicated that certain methods produced significantly lower S.D.'s over the whole range of concentrations and were therefore of greater precision. Table 2 gives the results for 'mid-therapeutic' concentrations. Thus, RIA appeared to be the best method for phenytoin, whereas spectrophotometric assay was significantly worse for phenobarbitone than any of the other methods, confirming earlier results [4]. TLC, although not the most popular method for measuring carbamazepine, exhibited significantly less variability than other methods. For ethosuximide HPLC appears to be the method of choice. GC with derivatization was always worse than the no-derivatization procedure although not always significantly so. In general no one method was consistently good for all drugs in the Scheme.

Table 2. C.V.'s (%) for drug-level measurements calculated at the mid-point of the therapeutic range for different methods of drug measurement (this mid-point being given as μmol/l *and* μg/*ml*).

Drug (& mid-point)	GC, no deriv.	GC, deriv.	EMIT	RIA	TLC	HPLC	Spec.	Wallace
Phenytoin (60, *15.1*)	9.8	10.0	10.8	7.2	–	11.0	–	–
Phenobarbitone (120, *27.9*)	11.2	12.9	10.0	–	–	11.4	18.6	–
Primidone (40, *8.7*)	11.2	14.5	9.4	–	–	14.0	–	–
Carbamazepine (40, *9.5*)	12.8	14.0	11.1	–	6.8	10.6	10.6	7.7
Ethosuximide (500, *70.6*)	11.8	13.0	14.1	–	–	8.3	–	–
Valproic acid (500, *71.4*)	12.2	14.9	13.3	–	–	–	–	–

RELEVANCE TO NON-THERAPEUTIC INVESTIGATIONS

External quality assessment schemes, although hitherto confined to laboratory measurement of drugs in therapeutic drug monitoring contexts, should be extended to all situations where drug levels are assayed. This is particularly so for pharmacokinetic studies and similar research-biased investigations. It is common experience for drug companies to develop an assay for one of its products and to release the assay conditions to external organizations who wish to perform their own experimentation; but all too often communication is lacking between in-house analysts and external investigators concerning the accuracy of the assay procedure. This problem could be resolved by external quality assessment. Evidently it would be desirable for drug-company analysts to produce batches of suitable material for external evaluation of assay accuracy. Such measures would go a long way towards ensuring consistency in the quality of results produced by multi-centered investigations.

References

1. Richens, A. (1975) in *Clinical Pharmacology of Antiepileptic Drugs* (Schneider, H., Janz, D., Gardner-Thorpe, C., Meinard, D. & Sherwin, A.L., eds.), Springer-Verlag, Heidelberg, pp. 293-303.
2. Pipppenger, C.E., Paris-Kutt, H., Penry, J.K. & Daley, D. (1976) *Arch. Neurol. 33*, 351-355.
3. Healy, M.L.R. (1979) *Clin. Chem. 25*, 675-677.
4. Griffiths, A., Hebdige, E., Perucca, E. & Richens, A. (1980) *Ther. Drug Monit. 2*, 51-59.

#G-2

VALIDATION OF BIOANALYTICAL PROCEDURES: AN EXAMPLE

J.A.F. de Silva

Hoffmann-La Roche Inc.
Nutley, NJ 07110, U.S.A.

The present example concerns GC and RIA determination of midazolam and a metabolite in plasma. The validation description complements that given for a midazolam analogue (assayed by HPLC in this laboratory) in Vol. 10 of this series [1], in an article 'GLP in Bioanalytical Method Validation'. That article, which should be consulted as the foundation for the present one, dealt with analytical policies for implementing FDA stipulations for 'Non-clinical Laboratory Studies'. These bear on impending regulations for 'Good Clinical Practices' (GCP) focused on pharmacokinetic studies that need analytical back-up.

Specificity should be demonstrated by selective extraction and/or column chromatography with a specific detection system. For 'clean-up'/selective exclusion, back-extraction into aqueous acid or base can be helpful [cf. #A-1 by R.P. Maickel, this vol.- Ed.]. Accuracy has to be established; factors that govern it include the purity and setting-up of standards, the 'spiking' of drug-free plasma, and analyte stability during storage (as now checked for midazolam) and sample handling. Radiolabelled compounds can help in establishing accuracy, besides being the basis of analytical approaches with particular 'GLP' [1].

Precision evaluation entails replicate analysis of calibration samples, both 'intra-day/assay' and 'inter-day/assay', and analysis of data from quality control samples. Recovery of the analyte from the biological matrix should be reproducible and preferably >85 ±5% for the sake of sensitivity, precision and accuracy; good recovery is aided by having few steps in the assay, by efficient extractions, and by guarding against adsorptive losses. Sensitivity, the limit below which no data should be reported, must be stated along with the sample volume used.

In the midazolam example, ways to settle the calibration regression line are compared for 'fit'.

The choice of an analytical method must be consistent with the foregoing considerations, as expounded earlier in this series [1,2], and with the goals of the specific study in which it is to be used. A pharmacokineticist desires as many data points as possible, for full data-evaluation, whereas the analyst should hesitate to release data that may be questionable due to poor precision, and also should guard against extrapolating data beyond the validated limits of quantitation. Significant pharmacokinetic parameters are risky if based on the least precise and accurate analytical data obtained at the limit of sensitivity. Pharmacokinetic factors that bear on sensitivity include the dose and route of administration, the body-tissue distribution (circulating drug level lowered if extensive), 'first-pass' biotransformation converting parent drug to metabolite(s), and pro-drugs/active metabolites situations.

Midazolam, for which three assay procedures are now considered, is an imidazo-1,4-benzodiazepine maleate (pKa 6.15) under clinical development as a parenterally administered short-acting anaesthetic and orally as a short-acting hypnotic [3]. It undergoes extensive first-pass biotransformation with <0.5% of the dose being excreted in the urine as intact midazolam (Fig. 1). The major metabolite is its 1-hydroxymethyl analogue [II], which is excreted in the urine as a glucuronide conjugate and accounts for 40-50% of the dose following i.v. administration [4] and >60% following oral administration [5]. If values for [II] as well as the parent drug [I] were wanted, assay was by GC with detection by electron-capture (GC-ECD) [4, 5] or by MS with negative-ion chemical ionization, GC-MS(CI) [6]. A radioimmunoassay (RIA) [7] was used to determine [I] *per se* in bioavailability studies utilizing the high sample throughput of the assay to advantage.

A. GC-ECD ASSAY

The automated assay requires extraction of [I] and [II] into 1.0 ml of benzene : acetone : methanol (17:2:1 by vol.) from plasma buffered to pH 9.0. A 0.75 ml aliquot of the extract is transferred into a 2 ml autoinjector vial, and 20 μl are analyzed using a 4-ft. column (2 mm i.d.) containing 3% Poly S-176 (on 80/100 mesh Chromosorb HP) at 265°. The retention times of [I] and flunitrazepam (the internal standard, i.s.) were 7.5 and 13.2 min respectively. The overall recovery of [I] from plasma is quantitative (100%), and the limit of sensitivity is 5.0 ng of [I]/ml plasma. Quantitation of [II] needs prior sample concentration and derivatization with BSTFA [bis-(trimethylsilyl)trifluoroacetamide] in pyridine.

The precision and accuracy of the assay for [I] was validated over a thousand-fold range with 0.1-0.5 ml plasma aliquots, using weighted linear regression analysis. Substitution of the experimentally determined responses of the recovered standards into the theoretical equation for peak-height ratio of analyte to i.s. (Y) *vs.* con-

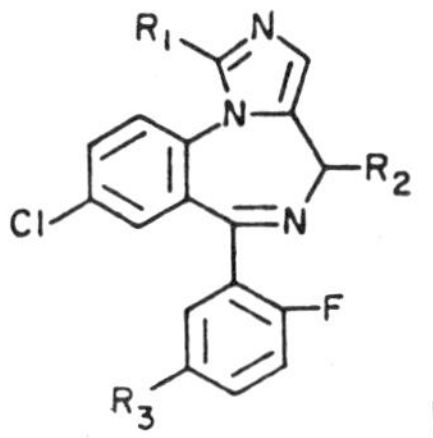

Compound	R_1	R_2	R_3	GC-NCI-MS ANALYSIS $M^{\cdot -}$ Ion Monitored	RIA FOR MIDAZOLAM % Cross Reaction
Midazolam [I]	$-CH_3$	$-H$	$-H$	M/z 325	100
1-Hydroxymethylmidazolam [II]	$-CH_2OH$	$-H$	$-H$	M/z 413 (-OTMS)	7
[I]D_5	$-CD_3$	$-D_2$	$-H$	M/z 330	
[II]D_2	$-CD_2OH$	$-H$	$-H$	M/z 415 (-OTMS)	
4-Hydroxymidazolam [III]	$-CH_3$	$-OH$	$-H$		16
1-Hydroxymethyl-4-Hydroxymidazolam [IV]	$-CH_2OH$	$-OH$	$-H$		3
1-Desmethylmidazolam [V]	$-H$	$-H$	$-H$		6
RIA Hapten	$-CH_3$	$-H$	$-NH_2$		0

Fig. 1. Chemical structures and analytical characteristics of midazolam, its metabolites and analogues as referred to in connection with GC-MS(CI) and RIA (see text).

centration of [I]/ml plasma (X) gives the amount of [I] found. The following values in ng/ml, for mean ±S.D. based on 3-10 determinations (C.V. in parentheses), were obtained for the amounts added:
2.0: 2.0 ±0.3 (12.3%); 5.0: 5.1 ±0.5 (8.9%); 10: 10 ±1.0 (9.7%); 20: 20 ±0.8 (4.1%); 50: 50 ± 2.5 (5.0%); 100: 100 ±2.8 (2.8%); 200: 200 ±7.1 (3.5%); 400: 400 ±3.2 (0.8%); 600: 594 ±7.1 (1.2%); 1000: 1015 ±13.3 (1.3%); 1500: 1511 ±14.5 (0.9%); 2000: 1989 ±10.6 (0.5%). Each mean ±S.D. shows the accuracy, and the average C.V., viz. 3.5% over the whole range, shows the inter-assay precision.

The mode of regression analysis used can influence the precision and accuracy of the data. Calibration data taken over a 40-fold concentration range from 5.0 to 200 ng of [I]/ml were processed by linear and weighted linear regression and by a power (exponential) function (Table 1). Although an r value >0.999 was obtained for each analysis, the mean % deviation from the theoretical concentration obtained upon re-substitution of the experimentally obtained data

Table 1. Regression mode evaluation for GC-ECD assay of midazolam.[*]
'% Dev.' is the deviation between added and observed values (ng/ml).
For Y and X, see text. The lower part of the Table exemplifies
a pharmacokinetic study with different post-administration times.
The limit of sensitivity was 5 ng/ml of blood or plasma.

Added	Linear regression		Weighted linear regres$^{n.}$		Power (exponential)	
	Found	% Dev.	Found	% Dev.	Found	% Dev.
5.0	7.3	+47.0	5.5	+10.8	5.2	+5.2
10.0	12.3	+23.5	10.7	+7.1	10.4	+4.7
20.0	19.7	−1.4	18.3	−8.4	18.1	−9.1
50.0	45.9	−8.0	45.5	−8.9	46.0	−7.9
100.0	97.5	−2.4	98.8	−1.1	101.2	+1.2
200.0	202.1	+1.0	207.0	+3.5	214.3	+7.1
	Mean % Dev. = ±13.9		Mean % Dev. = ±6.6		Mean % Dev. = ±5.9	
	$Y = 0.00896 \cdot X - 0.0198$		$Y = 0.00866 \cdot X - 0.0019$		$Y = 0.00895 \cdot X^{0.9872}$	
	$r = 0.9993$		$r = 0.9991$		$r = 0.9987$	
SAMPLE	Found, ng/ml		Found, ng/ml		Found, ng/ml	
2.5 min	65.7		65.9		67.0	
30.0 min	25.8		24.6		24.6	
2.0 h	19.6		18.2		18.0	
4.0 h	8.3		6.5		6.2	
6.0 h	7.6		5.8		5.5	

[*]#F-8 (J.P. Leppard) in Vol. 10 is relevant, + Corrigenda, Vol. 12.–*Ed.*

points into the computer-generated regression equation was higher
for linear regression than for weighted linear regression or for
the power function. The poor fit using linear regression is due to
a constant absolute deviation being applied at each point, which induces a wide bias at the lower concentration. With weighting, each
data point is treated more equitably to give a better distribution
of the error of the data points (normalization) about the theoretical line for the calibration curve. Thus, with a wide range of
concentrations the weighted linear or the power mode is preferable
to get the best overall 'fit' of the data about the regression line.

B. GC-MS(CI) [NEGATIVE ION] ASSAY, OF METABOLITE AS WELL AS DRUG

The GC-MS assay developed for determining [I] and [II] in human
plasma needs i.s. spiking with a deuterated analogue of each compound, [I]-D$_5$ and [II]-D$_2$ (Fig. 1). The plasma, taken to pH 10,
is extracted with benzene containing 20% (v/v) 1,2-dichloroethane.
The residue from drying down is dissolved in a 2% solution of bis-

[trimethylsilyl]-acetamide (BSA) in acetonitrile. An aliquot is analyzed by GC-MS(CI; methane), monitoring the ion ratios m/z 325/330
and 413/415 for [I] and derivatized [II] respectively. The sensitivity limits were 0.25 and 2.0 ng/ml plasma respectively [6]. Quantitation is based on weighted linear regression analysis for [I] and
a non-linear regression analysis (NONLIN computer program) for [II].

Two validations were performed, first upwards from 2.0 ng [I] in
the 1.0 ml plasma aliquot, then a lower range; for each point there
were respectively 16 (only 4 for highest point) and 4 observations
to get a mean ± S.D. (C.V. also given below, in parentheses). The
zero-spike in each case gave a value of 0.01 ±0.01 ng/ml, attributable to the 'D$_0$' content in the stable isotope analogue used as i.s.

Set at **2**: 2.1 ±0.2 (9.5%); **5**: 5.1 ±0.4 (7.8%); **15**: 15 ± 1 (6.7%);
    30: 30 ±2 (6.7%);   **60**: 60 ±2 (3.3%);   **120**: 120 ±1 (0.8%).

0.25: 0.26 ±0.01 (3.8%); **0.50**: 0.53 ±0.04 (7.6);
1.00: 1.04 ±0.03 (2.9%); **2.50**: 2.43 ±0.05 (2.1).

The mean C.V. for [I] was 5.8% in the first validation and 4.1% in
the second, whereas values of 5.2% and 13.0% were obtained over the
concentration range 2.0-60 ng/ml for [II].

The assay precision was investigated by 16 duplicate determinations, over 2 months, on a 'quality-assurance' plasma sample (prepared by pooling 0.1 ml of plasma from all the subjects who had
received [I] orally, as analyzed for [I] and [II]), in conjunction
with assays on test samples. Thereby inter-day reproducibility was
verified, besides the stability of the experimental plasma samples.
Thus, the observed values (ng/ml) were 27 & 27 for [I] and 13 & 15
for [II] in the first assay, 26 & 27 and 12 & 11 respectively in the
second, 23 & 25 and 12 & 12 in the fifteenth, and 26 & 20 and 12 &
14 in the final assay. For the whole set the respective means were
26 ±2 (C.V. 8%) and 12 ±1 (8%). Reassurance was thus gained for the ·
assay performance over the period of analysis.

C. RADIOIMMUNOASSAY (RIA)

The specific RIA developed [7] for [I] on 20 µl of plasma direct
entails adding [^{3}H]midazolam, pH 7.4 Tris-saline, and the antiserum
(rabbit antiserum to a diazo conjugate of 5'-aminomidazolam and albumin). After 30 min at 4°, polymer-bound goat anti-rabbit IgC in the
assay buffer is added. After 1 h at 40° the sample is centrifuged,
the supernatant aspirated and the pellet suspended in an acetic acid
/Aquasol mixture for scintillometry. The data were analyzed using a
4-parameter logistic curve-fitting program, and unknowns computed
from the fitted line.

As for assays A. and B., calibration curves (14 in all) were
determined on spiked plasma, establishing inter-assay precision: the

overall C.V. was 8.0%. For ng/ml, ±S.D. (and C.V.), the values were:

   2: 2.1 ±0.3 (13.5%); 4: 4.2 ±0.4 (8.3%);   10: 9.5 ±0.8 (8.7%);
   20: 20.0 ± 1.5 (7.3%); 40: 40.7 ±2.0 (4.9%); 80: 80.8 ±5.0 (6.0%);
   160: 157.0 ±12.0 (7.6%).

The sensitivity limit was 2 ng/ml, which produced >10% inhibition of
binding of the radioligand to the antiserum.

   As in **B.**, samples pooled from previously assayed *in vivo* plasmas
were repeatedly assayed along with test samples, with 10-14 observations on each of 6 pools having assayed concentrations (ng/ml) of
21 ±1 (inter-assay C.V. 3.3%), 30 ±3 (11.0%), 80 ±11 (13.0%), 81 ±10
(12.2%), 129 ±20 (15.5%) and 134 ±22 (16.2%). The average of these
C.V. values, 10.2%, served to further validate the inter-day reproducibility (ruggedness) of the assay.

   RIA specificity.- Fig. 1 gives cross-reactivities (3-16%) for
the known metabolites of midazolam, [II]-[IV]. However, [II] is the
predominant metabolite, [III] and [IV] being absent in human plasma
[3]. The attachment point to bovine serum albumin in the hapten
synthesis helps ensure high specificity for the parent drug.

   A more rigorous evaluation of the specificity and accuracy of
the RIA was made by comparison with the GC-ECD and GC-MS procedures
using split plasma samples. The joint determinations were correlated
by linear regression analysis over the three ranges of concentration
(Table 2), and demonstrated equivalence of the RIA to the other procedures as evidenced by each regression line having a slope within
10% of unity. This comparison validates the specificity and accuracy
of the RIA for routinely determining intact [I] in clinical samples.

GC-ECD *vs.* GC-MS(CI) *vs.* RIA: ACCURACY CHECK WITH SPIKED PLASMA

   A three-way cross-over study was performed with fresh control
plasma spiked with midazolam to give 2.1-1000 ng/ml and analyzed by
the three procedures in three separate laboratories. The samples
were 'blinded' as to concentration, hence were assayed as 'unknowns'.

Table 2. Correlation of RIA with GC-ECD and GC-MS assays. From
the collated results (n = no. of sets), linear regression equations
were established; r = correlation coefficient.

Range, ng/ml	n	Equation	r
2-20	45	RIA = 1.06(GC-ECD) + 0.7	0.89
10-50	40	RIA = 0.95(GC-ECD) + 3.5	0.92
50-575	51	RIA = 0.99(GC-ECD) + 5.4	0.96
2-25	10	RIA = 1.10(GC-MS) + 0.5	0.96

For the complete concentration range, 2.1-1000 ng/ml, RIA (n= 3) gave an average C.V. of 9.2% and ±12.9% deviation; GC-ECD (n = 3) gave 10.0% C.V. and ±5.5% deviation; and GC-MS (n =4) 4.6% C.V., ±7.6% deviation. This indicates that GC-MS is ~twice as precise as RIA or GC-ECD. However, at the lowest concentrations, viz. 2.1 and 3.6 ng/ ml, RIA and GC-ECD have a C.V. exceeding 20%; all three precisions are equivalent for the concentration range 8.4-1000 ng:- RIA, 5.0% C.V., ±12.4% deviation; GC-ECD, 4.1% C.V., ±4.2% deviation; GC-MS, 4.2% C.V., ±7.1% deviation.

The above values for % deviation (from theoretical concentration) indicate that equivalent concentrations for the 'unknowns' were obtained by the three methods. This equivalence in accuracy is reassuring.

STORAGE STABILITY IN FROZEN BLOOD OR PLASMA

Stability of the parent drug in the biological matrix to be assayed must be demonstrated over the time elapsed between sampling in the clinic and eventual analysis. Stability can be documented during method development using drug-free matrix suitably spiked, and eventually with authentic *in vivo* clinical samples.

Three sets of patient samples were later re-assayed to assess the stability of midazolam during storage at $-17°$ to $-20°$. The con-conclusions hinged on linear regression equations where x refers to the first assay and y to the second.-

    PLASMA, 2 months' storage (GC-ECD for x, GC-MS for y):
       y = 0.92x + 4.66; r = 0.988 (n = 20).
    PLASMA, 7 months' storage (GC-ECD for x, GC-MS for y):
       y = 1.01x + 8.14; r = 0.990 (n = 14).
    BLOOD, 4 years' storage (manual GC-ECD for x, automated GC-ECD for y):
       y = 0.82x − 1.25; r = 0.986 (n = 12).

The drug is evidently stable during 2 or 7 months' storage, as shown by the slope values for the regression line (which differ from unity by 10%). Storage for 4 years entailed ~20% degradation based on the slope value of 0.82.

GENERAL OUTCOME

The various criteria summarized at the start of this article have been documented using GC-ECD, GC-MS(CI) and RIA procedures, as applied in the clinical pharmacokinetic evaluation of midazolam.

Acknowledgements

The author is indebted to Dr. W.A. Garland for the GC-MS(CI) data, Dr. R. Dixon for the RIA data, Mr. R.E. Weinfeld for the GC-ECD data, Dr. M.A. Brooks for his valuable advice and discussions

on the statistical analysis of the data, and Ms. S. Christopher for
the preparation of the manuscript.

References

1. de Silva, J.A.F. (1981) in *Trace-Organic Sample Handling* [Vol.
    10, this series] (Reid, E., ed.) Horwood, Chichester, pp. 298-310.
2. de Silva, J.A.F. (1978) in *Blood Drugs and Other Analytical
    Challenges* [Vol. 7, this series] *as for* 1., pp. 7-28.
3. Various contributors (1981) [Proc. Symp. on Midazolam, held
    June 1980] *Arzneim. Forsch. 31 (Special Issue 12a)*, 2177-2288.
4. Puglisi, C.V., Meyer, J.C., D'Arconte, L., Brooks, M.A. & de
    Silva, J.A.F. (1978) *J. Chromatog. 145*, 81-96.
5. Heizmann, P. & von Alten, R. (1981) *J. High Resol. Chromatog.
    & Chromatog. Comm. 4*, 266-269.
6. Rubio, F., Miwa, B.J. & Garland, W.A. (1982) *J. Chromatog. 233*,
    157-165.
7. Dixon, W.R., Lucek, R., Todd, D. & Walser, A. (1982) *Res. Comm.
    Chem. Path. Pharmacol. 37*, 11-20.

#NC(G)

NOTES and COMMENTS relating to

Analytical quality assurance

Comments related to particular contributions:

#G-1, #G-2, #NC(G)-1 & #NC(G)-2, p. 405

#NC(G)-1

A Note on

GLP IN A CHEMICAL-PHARMACEUTICAL COMPANY: CURRENT STATUS AND EXPERIENCES IN ANALYTICAL AND DRUG DISPOSITION DEPARTMENT

H. de Bree, H. Keuker and A. Peters

Duphar Research Laboratories
Weesp P.O. Box 2,   The Netherlands

In the early 1970s there were serious troubles with a registered drug on the U.S. market. The compound appeared to have certain toxic properties which were not reported by the manufacturer. An inspection by the FDA revealed that pivotal data in safety studies were incomplete or falsified. This experience was one of the main reasons that led the FDA to set up the so-called Good Laboratory Practice (GLP) regulations [1]. These rules have to warrant the quality and integrity of safety studies. (For expositions of GLP see J.A.F. de Silva, preceding art. and #F-6 in Vol. 10.- *Ed.*)

The heart of GLP is the requirement to describe procedures and responsibilities. Since June 1979 the GLP rules have had the status of law, and pharmaceutical companies submitting non-clinical safety data to the U.S. Food & Drug Administration, to get a new drug authorized, have to comply with Good Laboratory Practices. In principle, the rules cope only with toxicological studies; but departments which carry out supportive work, e.g. Analytical and Drug Disposition Departments, have to comply as well. Our management has chosen to adopt an integral implementation of GLP, also embracing activities for which GLP would not be a strict requirement in U.S. law. The rules were extended to the following activities:
- CLINICAL STUDIES
- EFFICACY STUDIES
- STUDIES AIMED AT SCIENTIFIC PUBLICATION
- MICROBIOLOGY
- CROP PROTECTION DIVISION

In 1978, anticipating the introduction of GLP, a functionary was charged with implementing GLP rules and a Quality Assurance Unit

(QAU) was formed for inspection and checking compliance within the
R & D organization. Departmental master-files were compiled in
which the facilities and responsibilities are described, under the
following headings:-

A. General Provisions
B. Organization/Personnel – Organizational Chart
                                – Responsibilities
                                – *Curricula vitae*
                                – On-the-job Training forms
C. Buildings/Facilities – Design
                              – Air control
                              – Sanitation
D. Apparatus/Equipment – Inventory list
                             – Calibration/Maintenance

E. SOP's (Standard Operating Procedures) – Inventory list

All existing procedures were recorded in SOP's. We distinguish
three levels of SOP's: general R & D SOP's, applying to more than
one department; departmental SOP's, applying to groups within a
department; and group SOP's. A general example of SOP format is
given in Fig. 1.

Some of the pivotal SOP's are about the Study Director, protocols, sample identification, laboratory notebooks, instrument calibration, raw data handling and archiving. Eventually, on completing
the study and reporting the results, the raw data have to be stored
for 15 years, in such a way that studies are reconstructable. We
have adopted an archiving system as follows:-

I.   WORKING (GROUP) ARCHIVE
II.   DEPARTMENTAL ARCHIVE
III. QAU ARCHIVE

The working archive (I) contains the original raw data or micro-
film copies thereof. The archive is chronological per study. The
departmental archive (II) is updated once every 6 months with copies
of the microfiches for each study. The QAU archive (III) contains
microfiche copies of the data related to finished studies, arranged
logically for each study.

In the first years, GLP activities have led to a cost increase
of ~30%; since 1981 this has gradually decreased to a now constant
amount of 15%. As well as satisfying the FDA requirements we have
noted clear benefits to the company, including easier project control, quick and clear reporting, and better data retrieval.

Date	Title	Number

Date of former copy		Page

---

1. Summary (in English)

2. Summary (in native language)

3. List of copy-holders

4. Contents

Signature (dated)   QUA

Duphar B.V.   Type of SOP

Fig. 1. Layout of SOP's.

Reference

1. Federal Register (1978) *GLP-Regulations,* #43, No. 247, Washington, D.C., pp. 59986-60025.

#NC(G)-2

A Note on

QUALITY CONTROL SYSTEMS FOR ROUTINE DRUG ANALYSIS

[1]P. Hajdu and [2]J. Chamberlain

[1]Hoechst A.G.
 Pharma Forschung
 Postfach 80 03 20
 6230 Frankfurt (Main) 80
 W. Germany

[2]Hoechst Pharmaceutical
 Research Laboratories
 Walton Manor
 Walton, Milton Keynes
 Bucks. MK7 7AJ, U.K.

Hoechst has a number of analytical laboratories, throughout the world, engaged in assaying drugs in biological materials. This type of work involves both developing methods for such analyses, and applying these methods to the analysis of samples from research projects, clinical trials, volunteer trials and toxicological studies. When only a limited number of samples are involved, the scientist who developed the method may continue to carry out the analyses; however in the later stages of drug development when many hundreds, if not thousands, of samples must be assayed, the task needs to be treated differently. For this reason the Hoechst analytical laboratories at Frankfurt and Milton Keynes have set up small groups charged with carrying out routine assays, using previously established methods, as quickly, efficiently and economically as possible. These groups are relatively small (6 technicians each) but have a high level of automation and computerization.

Table 1 shows the work executed in these laboratories over several years. The two laboratories do similar work, except that the Milton Keynes laboratory includes radioimmunoassay (RIA) work for local development activities (13,000 samples/year) whereas RIA in Frankfurt is done in other laboratories. The increase in the number of samples analyzed over the years implies increasing automation rather than staffing. Most analyses other than RIA are now chromatographic, with emphasis on autosampling devices and on-line computing integrators. Wherever the latter are used, the chromatogram and the calculated concentrations appear together on the computer print-out. Thus, this output can be treated as raw data and archived for GLP purposes. In this article we describe the quality-control aspects of the running of such a laboratory.

400   P. Hajdu & J. Chamberlain   [NC(G)-2]

Table 1. Drug analyses during 1973–1981 in Hoechst bioanalytical laboratories, in Frankfurt and *(italicized) Milton Keynes*. 'Total' refers to the number of non-RIA analyte values reported. *Right: M.K. RIA total.* Other values are % assayed by each technique (Fluor = fluorimetric, Polar = polarimetric). For RIA in Frankfurt, see text.

Year	Fluor	GC	HPLC	TLC	Polar	GC/MS	TOTAL	RIA
1973	99	0.5	0.5	–	–	–	6550	
1974	86.5	6	7.5	–	–	–	7480	
1975	85	6	7.5	–	1.5	–	8200	
	15	*66*	*11*	*–*	*8*	*–*	*1725*	
1976	68	6	6.5	11.5	8.5	–	9600	
	30	*43.5*	*23.5*	*–*	*2.5*	*0.5*	*5156*	
1977	70	7.5	10.5	10	2	–	13340	
	19	*43.5*	*27.5*	*–*	*–*	*–*	*7069*	
1978	42	24	20.5	11.5	2	–	11890	
	16	*48*	*30*	*–*	*–*	*6*	*10559*	
1979	18	18	51	12	–	1	11530	
	18	*42*	*40*	*–*	*–*	*–*	*8428*	*351*
1980	18	8	55.5	18	–	0.5	18550	
	0.2	*38*	*44*	*18*	*–*	*–*	*17683*	*9947*
1981	8	15	67	10	–	–	15370	
	0.2	*53*	*30*	*17*	*–*	*–*	*14410*	*12778*

RIA listed (as total) only for Milton Keynes

REFERENCE MATERIAL

By definition, reference material has one or more properties sufficiently well established to serve for calibrating an apparatus or verifying a measurement method. There are two types in routine analysis: calibration standards and control specimens. Normally both types can be regarded as primary standard solutions, i.e. solutions produced by weighing the primary standard and dissolving it in a suitable solvent. Thus for the purpose of drug analysis the batch of drug used must be of primary-standard quality. Usually this means that it has an analytical certificate with stated levels of impurities. Yet such a certificate will relate to impurities normally arising from by-products or precursors of the synthesis, or from chemical degradation, and not to impurities which may be comparable to metabolites in analytical behaviour; the analyst should therefore check this aspect in the analytical systems chosen.

When standards have to be prepared using a new batch of drug, it is advisable to run both old and new standards in parallel to ensure the validity of the new standards.

In principle the reference material can be prepared by dissolving weighed drug in any matrix, from ultra-pure solvents to complex
biological matrices. We usually employ, for both calibration standard and quality-control specimen, a matrix corresponding as closely
as possible to the test specimens. Often the drug is dissolved in a
small amount of organic solvent prior to dispersal in the desired
matrix; the analyst must then be alert to possible influences of
the solvent and check the standard accordingly. Similarly the counter-ion of the salt form of the pure drug may introduce an unexpected
difference between prepared standard and drug-containing biological
specimen.

The importance of accuracy in preparing the reference materials
cannot be over-emphasized, as judgement of the ultimate standard of
analysis rests on this. The skill of a particular analyst could conceivably surpass the conscientiousness of the person making up the
reference material. Thus all weighings performed in making it up
must be recorded, and we also recommend that a UV-absorption or fluorescence spectrum be run on an appropriate dilution of the dissolved
standard to guard against gross errors at this stage.

INTERNAL QUALITY CONTROL

To prepare quality-control (QC) specimens, the weighed and pre-
dissolved standard is diluted with the appropriate matrix to give 3
different concentrations covering the expected range. Portions each
sufficient for one analysis are dispensed into appropriate vessels
and immediately deep-frozen. This phase of the operation should be
entrusted to the most skilled personnel (cf. above). As the storage
period may be long, stability must be established by analyzing such
QC specimens and freshly prepared specimens in parallel; this will
also help establish storage conditions for the particular drug. For
routine use, the selected QC specimens are thawed and analyzed along
with the unknowns and appropriate calibration samples. In a large
batch every 8th sample is a control; small batches have 2 controls.

We favour disclosure of the QC values to the analyst, who can
thus take remedial steps if necessary. Certain criteria must be laid
down for acceptance or rejection of results of an analytical batch,
based on the found values for the QC samples. Warning limits and
(wider) control limits are set; if only one QC result is outside the
former, this is not yet of great importance. If the next control
result is again outside in the same direction, the usual investigations and preventive measures must be taken immediately. The assay
also has to be investigated and corrective measures taken if –
two successive results for the same sample are outside the warning
limit in the same direction;
three successive QC results are outside it in either direction;
seven successive QC results are above, below or exactly on the target
(nominal) value, or show a steady rise or fall.

In principle the warning limits are set at 2 S.D.'s about the mean (or target) value and the control limits at 3 S.D.'s about this value. However, this assumes a constant level of precision for the analyst, irrespective of the analyst, and an alternative is to set the warning limit at ±10% of the target value and the control at ±15%.

CONTROL CHARTS

In both laboratories, individual analysts maintain their own control charts as part of documentation for demonstrating analytical quality. These charts include *all* QC values including rejected ones. Retrospective values for accuracy and precision can be calculated at the completion of a study. The precision thus calculated is rarely coincident with that obtained in method development, but is the more realistic, being obtained under genuine routine conditions. The archiving of such control charts, which may be produced by modern programmable integrators from all analyses associated with the study, is a matter of genuine good laboratory practice.

EXTERNAL QUALITY CONTROL

In clinical chemistry laboratories it is common practice to use QC specimens provided by an outside laboratory. [See #G-1.-*Ed*.] For drug analyses in companies this option is limited and mostly amounts to comparison of results from two different laboratories. Yet the analyst should always attempt some sort of external QC exercise as this can reveal unexpected shortcomings when the analysis and control are all contained in the same laboratory. External QC is particularly important when taking over a method from a different laboratory.

Two kinds of material are normally used for external QC purposes: duplicate samples from the same study or samples spiked with the drug. Each type has its advantages and disadvantages: specimens from actual study will resemble those that will be met with subsequently but the true concentrations are unknown; spiked samples facilitate accuracy comparisons but are in all respects identical with the internal QC samples, hence matrix effects that may occur in clinical samples will not be apparent.

PRE–LABORATORY ERROR

Many gross errors may occur before the sample reaches the analytical laboratory. [Especial attention was given to this aspect in Vol. 10, e.g. #NC(F)-3.- *Ed*.] These may include: sampling at the wrong time, incorrect sample identification, and deterioration of the sample due to inappropriate storage. Such errors are extremely difficult to detect and may only be suspected if the analytical results are inconsistent with expected results, or with previous experience. This can be a very serious problem, especially where the analytical laboratory is remote from the originating laboratory

and hence may be unaware of the latter's facilities and working
practices. The problem is all the more serious because it is often
unrecognized.

Apart from gross errors, other pre-laboratory factors may
influence the analysis. These include the presence of other drugs
or their metabolites, variability in the collection technique, and
the handling of the specimen before storage and shipment. There is
only one way to minimize the pre-laboratory errors, viz. to include
detailed specifications for obtaining and handling of analytical
specimens in the study protocol. Important aspects include:
- the container(s) for sampling and handling of the specimen, including the use of anticoagulants and preservatives (not all types of
vials suitable for 'clinical chemistry' samples are appropriate for
samples destined for drug analysis); if the analyst suspects interferences he should make suitable recommendations;
- pre-shipment sample manipulation, including the permissible time
for the sample to stay at room temperature in daylight or laboratory
light during clotting and centrifugation operations, storage temperature and time limit; the analytical chemist should always provide
appropriate stability data;
- shipment conditions, the use of sufficient coolant being specified
in view of possible shipping delays; for particularly delicate or
critical samples the use of the old-fashioned personal carrier
service is advisable, with individual responsibility for collection,
carriage and final delivery.

RECAPITULATION

The context is the operation of laboratories engaged in the
routine analysis of large numbers of samples from clinical, volunteer
and toxicology studies. Such laboratories have a high degree of
automation and computerized data handling. Quality control (QC)
procedures are described above, and the characteristics and use of
QC samples are discussed, including use for independent checking in
a second laboratory.

Comments on material in #G

Comments on #G-1, J. Williams – EXTERNAL QUALITY ASSESSMENT SCHEME

In reply to J. Ramsey, R.M. Lee: concerning expression of concentrations, a majority of participating laboratories are reluctant to change from mass units (mg/ml) to molar units (mmol/L) such as we employ. *Note by Senior Editor:* this book series is deliberately old-fashioned in this respect, even to the extent of stipulating to authors that 'µg/ml' be used rather than 'µg.cm^{-3}' (typographical burden!). J. Williams, *answering* J. Chamberlain.- It is indeed a recognized problem that if the mean for all laboratories were low compared with the spiked-in value, a laboratory reporting exactly the 'right' value could receive a low rating. However, all quality-control schemes accept the consensus value (i.e. the mean of all returns) as the yardstick, rather than the spiked-in value.

Comments on #G-2, J.A.F. de Silva – VALIDATION OF PROCEDURES

Replies to G.E. von Unruh, H. de Bree.- With a calibration curve spanning, say, 3 orders of magnitude, we establish a C.V. for each of 5 concentrations, as we pool the results; but 2 concentrations can suffice if they are representative of the concentration range measured. In deciding whether or not the chosen calibration points should be equidistant, we take the nature of the samples etc. into account.

Comments on #NC(G)-1, H. de Bree, & #NC(G)-2, J. Chamberlain – GLP

H. de Bree, *replying to* J. Chamberlain.- It is only microfilmed raw data that the Q.A. Archives hold; the original hard copy is kept in Group Archives, for 15 years. P. Hajdu, *replying to* H. de Bree. - If computer malfunction blocks the automated analytical output, all further samples are saved, and a technician starts looking for the cause of the failure.

Analyte Index

Below: **Inorganics, including complexes** *(no such list in past vols.)*

Overleaf: **Organics** *(as in past vols.; Cumulative Index in Vol. 12)*
— together with **INDEXING PRINCIPLES** *and, to collate analytes according to analytically relevant common features (see Preface), a 10-category* **CHEMICAL CLASSIFICATION**. *Organometals =* 'Inorganic'.

====================

Accompaniments to some page entries
besides superscripts (explained overleaf):-

'-' *to indicate a* **major entry**, *e.g.* 17-;

parenthetical indication of **sample type**, *where applicable to the text citation, e.g.* (urn), *with abbreviations as follows:-*

aq = *aqueous samples, e.g. effluents*
bld = *blood, usually plasma or serum*
tss = *tissues, including incubates of liver etc.*
urn = *urine.*

Inorganics, incl. complexes
For MIP-detectable elements see p. 134

Anions, as in drug salts: 350, 355
Boron compounds, incl. steroidal
  carboranes: 132, 135
Chromium (III) β-diketonates: 136
Cobalt(III) complexes (*for Vit.* B_{12}
  see Organics, #IIIy): 136

Gold complexes, e.g. auranofin
  (Ridaura): 137, 156 (bld, etc.),
  158 (urn, synovial fluid), 197
  [*Organic moieties include*
  Thiomalate, #Ia, & Thioglucose, #Iy]
Halocarbons: see Fluoro...., #Iz

Lead 'species': 125, 135
- trialkyl: 129 (aq), 197
Nitrite: 350 (saliva), 365-
  [*cf.* Nitro...., #Ia & #Iz: detection may entail generating
  nitrite or nitrate]

Platinum complexes, e.g. cisplatin
  (*syn.:* CDDP, *cis*-platin): 139-,
  142 (bld), 146- (bld, urn),
  161- (bld, urn), 197 (tss, urn)
  [CBDCA, TNO-6, *etc.: names, p.* 145]
Thiocyanate: 350 (saliva)

ASSIGNMENT 'CATECHISM', especially for organics

\# **Metabolites** are signified by a superscript in entry for the parent molecule: *Phase I,*[1] or, if including *N*-de(s)alkyl, [1]; *conjugates,* [c].

\# **Parent molecules as indexed** are, where applicable, generic names as listed (with formulae) in the *Merck Index*. But some are comprehended in a class title, e.g. Steroid hormones, Peptides (incl. proteins). **Biogenic amines**, e.g. dopamine, are mostly listed under Phenylethylamine (**IIb'** or **IIb"**; superscript system inapplicable); but a deaminated metabolite would appear under Carboxylic acids – phenolic (**Ia**; e.g. VMA) or Glycols (**Iy**; e.g. HMPG).

\# **Assignment as 'acidic'** (to **Ia**, **IIa** or **IIIa**) applies where:
– pK_a <6 (excludes phenols); conjugates excluded *(see above)*;
– analyte is an **ester** yielding an acidic group in the main moiety if hydrolyzed (as would happen *in vivo*).

\# **Cyclic** *N* (never treated as 'amino') is 'imide' (put in **Ic**?) if –CO–N–CO–.

\# **Consult Cumulative Index** (in Vol. 12) to find past mentions; same categorization, except that **'Inorganics (incl. complexes)'** is an innovation – see preceding page (which also lists **Sample types**, e.g.'bld').

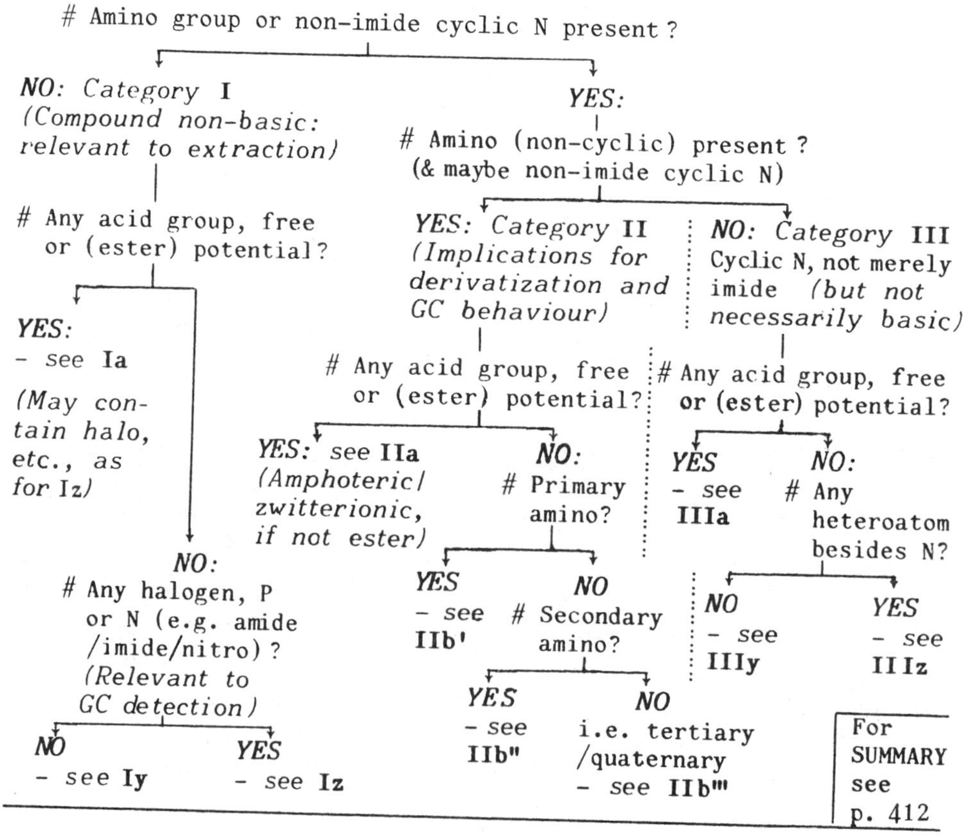

\# Amino group or non-imide cyclic N present ?

NO: *Category* **I**
(Compound non-basic: relevant to extraction)

\# Any acid group, free or (ester) potential?

YES:
– see **Ia**

(May contain halo, etc., as for Iz)

NO:
\# Any halogen, P or N (e.g. amide /imide/nitro) ? *(Relevant to GC detection)*

NO – see **Iy**   *YES* – see **Iz**

YES:
\# Amino (non-cyclic) present ?
(& maybe non-imide cyclic N)

YES: Category **II**
(Implications for derivatization and GC behaviour)

\# Any acid group, free or (ester) potential?

YES: see **IIa**
(Amphoteric/ zwitterionic, if not ester)

NO:
\# Primary amino?

YES – see **IIb'**   *NO* \# Secondary amino?

YES – see **IIb"**   *NO* i.e. tertiary /quaternary – see **IIb'''**

NO: Category **III**
Cyclic N, not merely imide *(but not necessarily basic)*

\# Any acid group, free or (ester) potential?

YES – see **IIIa**   *NO:* \# Any heteroatom besides N?

NO – see **IIIy**   *YES* – see **IIIz**

For SUMMARY see p. 412

CATEGORY II (amino, not in a ring)

#IIa: acid *other than conjugate,*
  or ester *(main moiety = acid)*

#IIb': primary amino; *no* acid
  (unless conjugate) or ester

#IIb": secondary amino, not in a
   ring; otherwise as for #IIb'

#IIb''': tert. amino or quaternary
   ammonium, not in a ring; otherwise as for #IIb'

CATEGORY III (cyclic N, not merely imide; *no* amino)

#IIIa: acid *other than conjugate, or ester (main moiety = acid)*

Captopril: 117, 196 (bld)
Cocaine: 264[1] (bld, urn), 268, 281[1], 373 (bld) [incl. refs. to ecognine]
Fenbendazole: 197[1]
5-Hydroxyindoleacetic acid (5HIAA): 321- (tss), 371 (tss)
Indomethacin: 62 (bld, urn), 94 (urn)
D-Penicillamine, *N*-acetyl: 118

#IIIy: only N-hetero (not merely imide); *no* amino, & main moiety *not* acid or potential acid

Antipyrine: 96
Benzodiazepines (*see also* Chlordiazepoxide in #IIb", & Flurazepam in #IIb'''): 23 (bld), 263 (bld, urn), 297- (bld), 316, 352; cf. 295
- Carbamazepine & its epoxide: 35- (bld), 75- (bld; incl. epox.), 378 & 383 (bld; incl. epox.)
- Clobazam: 184 (bld, urn), 197=[1], 299-[1] (bld)
- Clonazepam: 376 (bld)
- Demoxipam: 184 (bld, urn)
- Diazepam: 25 (tss), 62, 263 (bld, urn), 299-[1] (bld), 347 (bld) [Desmethyld'm *included here*]
- Flunitrazepam: 264 (bld), 299- (bld)
- Midazolam: 386-[1] (bld, urn)
- Oxazepam: 298
- Pinazepam: 373[1] (bld, urn)
- Temazepam (3-Hydroxydiazepam): 263 (bld, urn), 347 (bld)
Benzoylecognine: Cocaine *entry,* #IIIa
Bisantrene (ADCA): 374 (bld)
Caffeine: 21 (bld)
Celiptium: 258[c] (urn)
Haloperidol: 316[1] (bld)
Isoniazid: 374

....
Abbreviations, etc: pp. 407 & 408

LSD: 20 (bld, urn), 264 (bld, urn), 318 (urn), 360
Mercaptopurine: 117, 258
Paraquat: 266 (bld, urn), 347 (bld)
Physostigmine: 189 & 190 (bld)
Primidone (Primaclone): 21 (bld), 75- (bld), 378 & 383 (bld)

Quinine: 349
Riboflavin: 178
TCNP (→ TCN, a tricyclic nucleoside): 258 (bld, urn, bile)
Theophylline: 21, 54- (bld, urn), 62 (bld)
Vitamin B$_{12}$: 136 (bld), 197

#IIIz: heteroatom besides N; otherwise as for #IIIy

Chlormethiazole: 25 (tss)
Codeine: 265, 281
Morphine: 23 (bld), 104, 265 (bld, urn), 271 (bld), 280- (bld), 313-[c] (bld), 349 (urn), 350 (bld), 371 (tss), 372, 374
Opiates, various (& see Codeine, Morphine): 265 (bld, urn)
- Diamorphine: 265, 313
Pemoline: 373 (urn)
Phenothiazines (*see also* #IIb''' *entry*): 184, 298 (bld)
- Fluphenazine: 43, 316 (bld)
- Thioridazine: 25 (tss)
- Trifluoperazine: 40- (bld)

SUMMARY OF CATEGORIES for organics			
	I	II	III
Amino? Non-imide hetero-N?	no no	✓ maybe	no ✓
Acid/Ester?[*]	✓ = Ia	✓ = IIa	✓ = IIIa
- no!	Halo, P or N?	Primary amino?	Hetero atom besides N?
.... *criterion amplified on p. 408*	- no: Iy - ✓ = Iz	✓ = IIb' If no: 2^y= IIb" 3^y or 4^y =IIb'''	- no = IIIy, ✓ = IIIz

General Index

This Index deals especially with phenomena and points of technique, indexed similarly to previous volumes (5, 7, 10 & 12 in the 'Analysis' subseries) so as to facilitate retrieval. The application of techniques such as HPLC is not comprehensively indexed. There are a few entries for therapeutic classes of drugs, e.g. Anticonvulsants; but classes such as Benzodiazepines are to be found along with individual analytes in the preceding Analyte Index.
Page entries such as 25- signify that the ensuing pages are also relevant, i.e. the - denotes a major entry.

Corrections to Vol. 12, *Drug Metabolite Isolation & Determination*

p. 90, line 3 from foot: letter missing, **should be** precision

p. 172, line 4 from foot: multipet **should read** multiplet

p. 206, ref. 2: (Vol. 11.... **should read** (Vol. 10....

p. 275 (Analyte Index, **#Iz**): DOPA **to be shifted to** **#IIa**

Corrections to Vol. 13, *Investigation of Membrane-located Receptors*

p. vi (Preface), line 8: p. 540 **should read** p. 546

p. 308, line 6: [23, 23] **should read** [22, 23]

p. 425: **insert missing Art. no., #NC(D)-3**